PROCESSES IN MARINE REMOTE SENSING

The Belle W. Baruch Library in Marine Science

THE BELLE W. BARUCH LIBRARY IN MARINE SCIENCE NUMBER 12

Processes in Marine Remote Sensing

Edited by F. John Vernberg and Ferdinand P. Diemer

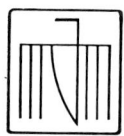

Published for the Belle W. Baruch Institute for Marine Biology and

Coastal Research by the

UNIVERSITY OF SOUTH CAROLINA PRESS

Copyright © University of South Carolina 1982

First Edition

Published in Columbia, South Carolina, 1982

Manufactured in the United States of America

Library of Congress Cataloging in Publication Data
Main entry under title:

Processes in marine remote sensing.

   (The Belle W. Baruch library in marine science;
no. 12)
   Includes index.
   1. Oceanography—Remote sensing—Congresses.
I. Vernberg, F. John, 1925-   . II. Diemer,
Ferdinand P.   III. Belle W. Baruch Institute for
Marine Biology and Coastal Research.   IV. Series.
GC10.4.R4P76   551.46′0028   81-16214
ISBN 0-87249-411-X   AACR2

PARTICIPANTS

CONTRIBUTORS

BRUNO APOLLONI, Universita della Calabria, Dipartimento de Sistemi, Cosenza, Italy 80736.

PETER C. BADGLEY, Arctic and Earth Sciences, Office of Naval Research, Department of the Navy, Division Code 460, Arlington, VA 22217.

PETER E. BAYLIS, The University of Dundee, Department of Electrical Engineering and Electronics, Dundee DD1 4HN, Scotland, U.K.

F. BECKER, Department of Physics, Universite Louis Pasteur, Strasbourg, 7 Rue De L'Universite, Strasbourg 67000, France.

D.G. BOWERS, Department of Physical Oceanography, University College of North Wales, Marine Science Laboratories, Menai Bridge, Anglesey LL59 5EY, U.K.

D.E. CARTWRIGHT, Natural Environment Research Council, Institute of Oceanographic Sciences, Bidston Observatory, Merseyside L43 7RA, Bidston, U.K.

A.P. CRACKNELL, Carnegie Laboratory of Physics, University of Dundee, Dundee DD1 4HN, Scotland, U.K.

I. DALTON, Carter Observatory, Wellington 1, New Zealand.

DOUGLAS J. DE PRIEST, Statistics and Probability Program, Office of Naval Research, Department of the Navy, Arlington, VA 22217.

ARDASH DEEPAK, Institute for Atmospheric Optics and Remote Sensing, PO Box P, Hampton, VA 23666.

R.J. DODD, Carter Observatory, Wellington 1, New Zealand.

P.B. FELLGETT, The University of Reading, Department of Cybernetics, 3 Early Gate, Whiteknights, Reading RG6 2AL, U.K.

C.R. FRANCIS, British Aerospace Dynamics, Space System Group, PO Box 5, Filton House, Bristol, England BS12 7QW, U.K.

P. GABORSKI, U. S. Naval Oceanographic Office, NSTL Station, Bay St. Louis, MS 39522.

JAMES J. GALLAGHER, U. S. Navy Underwater Systems Center, New London Laboratory, New London, CT 06320.

DONALD P. GAVER, Naval Postgraduate School, Department of Operations Research, Monterey, CA 93940.

D. GERSON, Code TOT, Defense Mapping Agency Hydrographic Topographic Center, 6500 Brookes Lane, Washington, DC 20315.

DAG T. GJESSING, Remote Sensing Technology Programme, Royal Norwegian Council for Scientific and Industrial Research, PO Box 25 N2007, Kjeller, Norway.

J.F. GRAINGER, Institute of Science and Technology, The University of Manchester, Department of Pure & Applied Physics, PO Box 88, Manchester, England M60 1QD, U.K.

E. KHEDOURI, U. S. Naval Oceanographic Office, NSTL Station, Bay St. Louis, MS 39522.

PAUL M. MATHER, The University of Nottingham, Department of Geography, University Park, Nottingham, England NG7 2RD, U.K.

EDWARD C. MONAHAN, Physical Oceanography, Department of Oceanography, University College, Galway, Ireland.

N.G. PACE, University of Bath, England, U.K.

LAWRENCE J. ROUSE, JR., Lousiana State University, Coastal Studies Institute, Center for Wetland Resources, Baton Rouge, LA 70803.

ARCHIE E. ROY, The University, Department of Astronomy, Glasgow G12 8QQ, Scotland, U.K.

RONALD L. SCHWIESOW, Environmental Research Laboratories, National Oceanic and Atmospheric Administration, Boulder, CO 80303.

J.H. SIMPSON, Department of Physical Oceanography, University College of North Wales, Marine Science Laboratories, Menai Bridge, Anglesey LL59 5EY, U.K.

D.P. THOMAS, British Aerospace Dynamics, Space System Group, PO Box 5, Filton House, Bristol, England BS12 7QW, U.K.

C. TOMASSINI, European Space Organisation Centre, Darmstadt, Federal Republic of Germany.

DENNIS B. TRIZNA, Naval Research Laboratory, Washington, DC 20390.

SUEO UENO, Kanazawa-South, Nonoichimachi, Ogigaoka, Ishikawa 921, Japan.

KLAUS A. ULBRICHT, DFVLR-Institut fur Nachrichtentechnik, German Aerospace Research Establishment, 8031 Oberpfaffenhofen, Germany.

F. JOHN VERNBERG, Belle W. Baruch Institute for Marine Biology and Coastal Research, University of South Carolina, Columbia, SC 29208.

EDWARD J. WEGMAN, Statistics and Probability Program, Office of Naval Research, Department of the Navy, Arlington, VA 22217.

PREFACE

A continuing need exists to understand the complex interactions of the various processes that govern the behavior of the
world's ocean. Because of the enormity of the seas, it is impera-
tive to utilize the best available technology to obtain data necessary to interpret oceanic activities. Remote sensing is one
promising research tool. To explore the use of remote sensing to
determine the features and properties of the sea surface and
related coastal regions, a workshop was held at the University of
Manchester, United Kingdom from June 26 to June 30, 1979. A multi-
disciplinary group of European and United States scientists and
engineers met to review recent advances in the application of
remote sensing to the marine environment, to identify problems, and
to recommend future research and development direction. The major
topics included: Physical Targets, Transmission Medium, Signals
and Information, Sensors and Instrument Systems Applications,

Environmental Studies, and Observational Studies and Processing.

To maintain the high quality of the Belle W. Baruch Library in Marine Science series, all papers have been peer-reviewed. Special thanks are extended to the scientists who served as reviewers.

We gratefully acknowledge the support of the United States Department of Navy, Office of Naval Research. Through the Navy Ocean Science Office, F. Diemer initiated the idea for this workshop and with the valuable assistance of colleagues developed the program. The on-site travel logistics and arrangements were handled most excellently by Professor Z. Kopal and his staff in the Department of Astronomy, University of Manchester. Doctors E. Wegman and H. Soloman, ONR, assisted in financial arrangements and coordination.

F. J. Vernberg

F. P. Diemer

CONTENTS

DATA LINKS RECEPTION AND PROCESSING

SENSORS AND INSTRUMENT SYSTEMS APPLICATIONS

OVERVIEW AND PERSPECTIVE

WORKING GROUP REPORTS

Multiple-Mirror Telescope Systems in Astronomy

J.F. Grainger

ABSTRACT

The function of a telescope is to optimise the transfer of the information of interest from a source of finite angular size to the analytical device which is going to extract it. It is far from obvious that the conventional large telescope is the optimum fore-optical system for looking through a turbulent atmosphere. There are considerable scientific and financial advantages in going to multiple imaging-elements and several groups in different parts of the world are working in this direction.

INTRODUCTION

The role played by the atmosphere in ground-based astronomical observations is very different from that which it plays in space-

1

borne observations of the earth. In the latter, the turbulent medium is a relatively thin layer in close contact with the object - the two being observed from a distance. In many cases the distortions introduced by the turbulence are barely resolvable by the space-borne optics and the view obtained is hardly degraded.

The astronomer however is constrained to observe distant objects with the turbulent medium in close proximity to the image-forming optics. Moreover, in order to obtain the large collecting area needed for the weak signals he is dealing with, the astronomer uses imaging optics which are large compared with the scale of the distortions introduced into the incoming wavefront. The surface of constant phase, instead of being plane, is crumpled. The effect of this on the 'image' of a star depends on the time for which the light is integrated and on the character of the crumpling. Very short exposure photographs show the diffraction pattern of the telescope aperture convoluted with the source geometry, repeated at numerous points in the focal plane spread over an angular size many times that of the Rayleigh limit. The particular positions of these points, or "speckles", are determined by the instantaneous two-dimensional spatial frequency spectrum of the wavefront crumpl-ing, and therefore in subsequent instantaneous photographs are quite different. In the technique known as Speckle Interferometry, the constant factor within these photographs - namely the diffrac-tion pattern of the telescope aperture convoluted with the source geometry - may be recovered, and in this sense, the effects of the turbulent medium may be overcome. However, for the majority of astronomical measurements, the observation times are necessarily long and the integrated effect of the randomly-dancing speckles must be regarded as the telescope's response function. The form of this function clearly depends upon the exact nature of the turbu-lence at the site of the Observatory and upon the aperture of the telescope concerned. Much work has been done on the degradation of resolving power caused by observing through a turbulent medium, e.g. Tatarski (1961), Hufnagel and Stanley (1964), Fried (1966), Fried and Mevers (1974), and Young (1974). Fried (1966) has shown that the effects of the turbulence can be characterized by a para-

meter having the dimensions of a length. This length, which Fried
called "r_o", has the following significance:

    (i) The scale of the wavefront crumpling is determined by it.
Over a circle of diameter r_o, the wavefront distortion
from plane has an r.m.s. value of almost exactly $\frac{\lambda}{2\pi}$.

    (ii) The resolution of an image averaged over a time long
compared with the time-scale of the wavefront fluctuations, cannot exceed the free-space resolution of a
diffraction-limited system with aperture r_o, no matter
how large the actual aperture.

It is interesting to examine the way in which the attainable resolution approaches this limiting value as telescope size is increased (see Figure 1). It can be seen that the attainable resolution is 90% of the limiting value for an aperture of $6r_o$ and 95% of
the limiting value for an aperture of $12r_o$. There is little point
therefore, on the grounds of resolving power, in having apertures
more than an order of magnitude greater than r_o.

The full impact of this conclusion is only felt when the value
of r_o is determined at what are generally regarded as good observatory sites. Fried and Mevers (1974) using data for Flagstaff and
Kitt Peak, have shown that r_o at these sites is ~7cms! Hence, from
the point of view of resolution, 70cms aperture would be the biggest worth having at either of these sites. At any site, there
will be a Largest Worthwhile Aperture (LWA) and it is clear from
the foregoing that the LWA is going to be totally inadequate as far
as light-gathering power is concerned. Now it might be argued that
a large conventional telescope, although far bigger than the LWA,
does at least provide the light-gathering power. Surely this is
sufficient justification for building telescopes this way? There
are several reasons why the answer to this question must be 'no'.
Consider for example that the total cost of an observatory goes
approximately as the 2.7th power of the length of the telescope
involved. For meaningful results, we must compare telescopes of
the same f-number, and so their lengths are proportional to their
apertures. Large aperture telescopes are therefore a very expensive way of achieving greater light-gathering power than the LWA

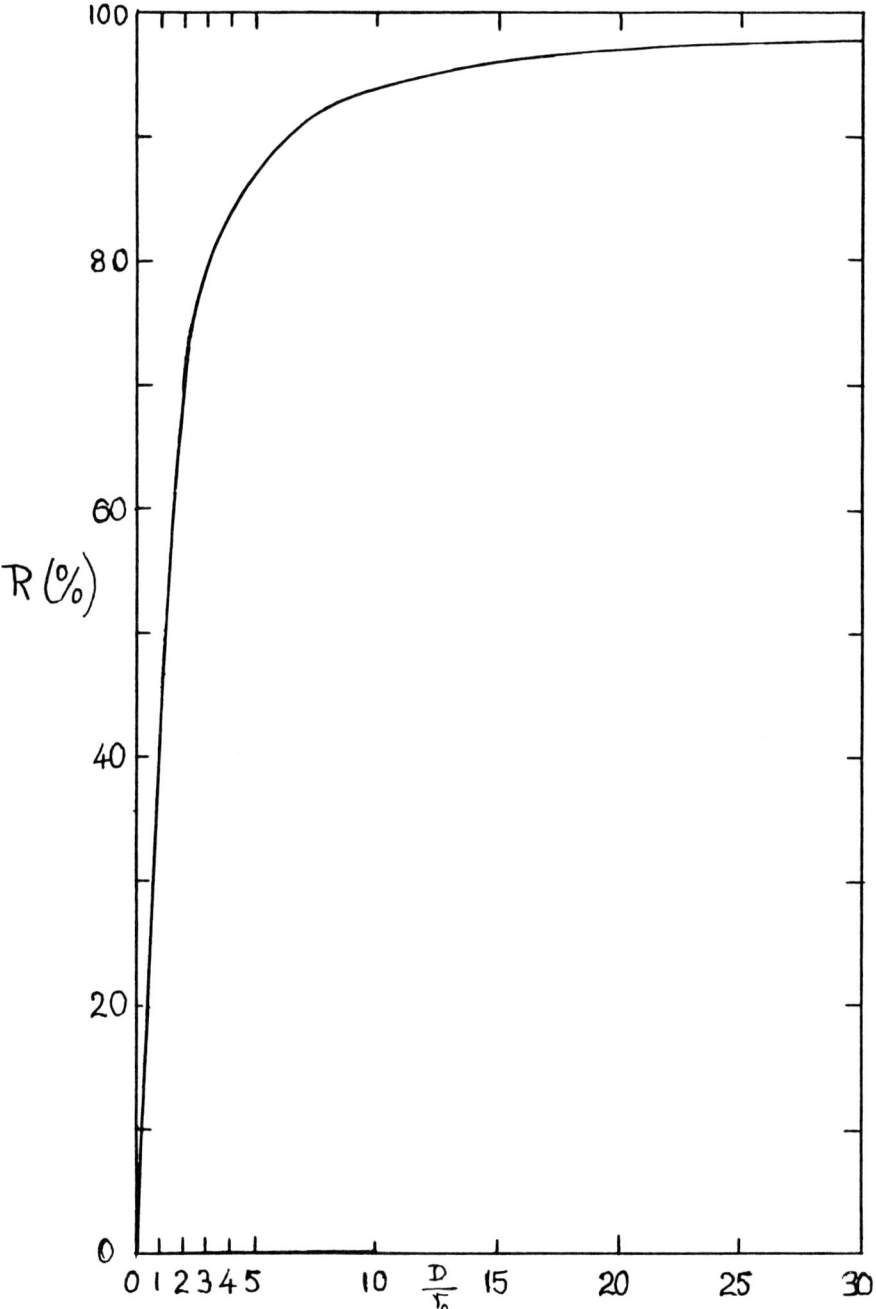

FIGURE 1: Resolution R attainable by a telescope of diameter D observing through a turbulent atmosphere with an integration time long compared with the time-scale of the turbulent fluctuations (after Fried 1966). R is expressed as a percentage of the maximum achievable and D is expressed as a multiple of r_o (see text).

for the site. For example, for the cost of a 150-inch telescope in a site with an LWA of 30 inches, at least 75 30-inch telescopes could be built. Their combined collecting area would be three times that of the 150-inch and, moreover, for objects not needing the large collecting area, 75 sources in different parts of the sky could be simultaneously observed with a variety of measuring techniques. Alternatively, the same collecting area as the 150-inch telescope could be achieved for one-third of the cost by building twenty-five 30-inch telescopes.

Of course, some experiments in astronomy need the large primary image scale of the conventional telescope. However, these form a relatively small percentage of the astronomer's work and there are already many such telescopes in existence. On these grounds alone then, the case for building no more large conventional telescopes for a long time to come is very strong.

MULTIPLE-MIRROR SYSTEMS

The above arguments are based on a straight comparison of large and small conventional observatories, i.e. each of the 30-inch telescopes referred to was assumed to be separately mounted in its own Dome. However, the idea of building up collecting area from a number of imaging elements each no bigger than the LWA at the site leads naturally to the notion of using these elements in an array on a single steerable mounting in a single building. This not only affords further cost advantages, but the separate small images thus formed are ideally suited to feeding a spectrometer slit - thereby avoiding the image-slicer problem associated with large single-imaging-element collectors.

A multiple imaging-element fore-optical system based on these arguments was mooted by the present writer over a decade ago and has led to the production of a Multi-Aperture Telescope, or MAT, consisting of seven 15-inch Cassegrain systems in hexagonal array (see Figure 2 and Grainger 1971 and 1979). The separate images are juxtaposed by a set of radially disposed periscope systems. The intrument has a collecting area equivalent to a 40-inch conven-

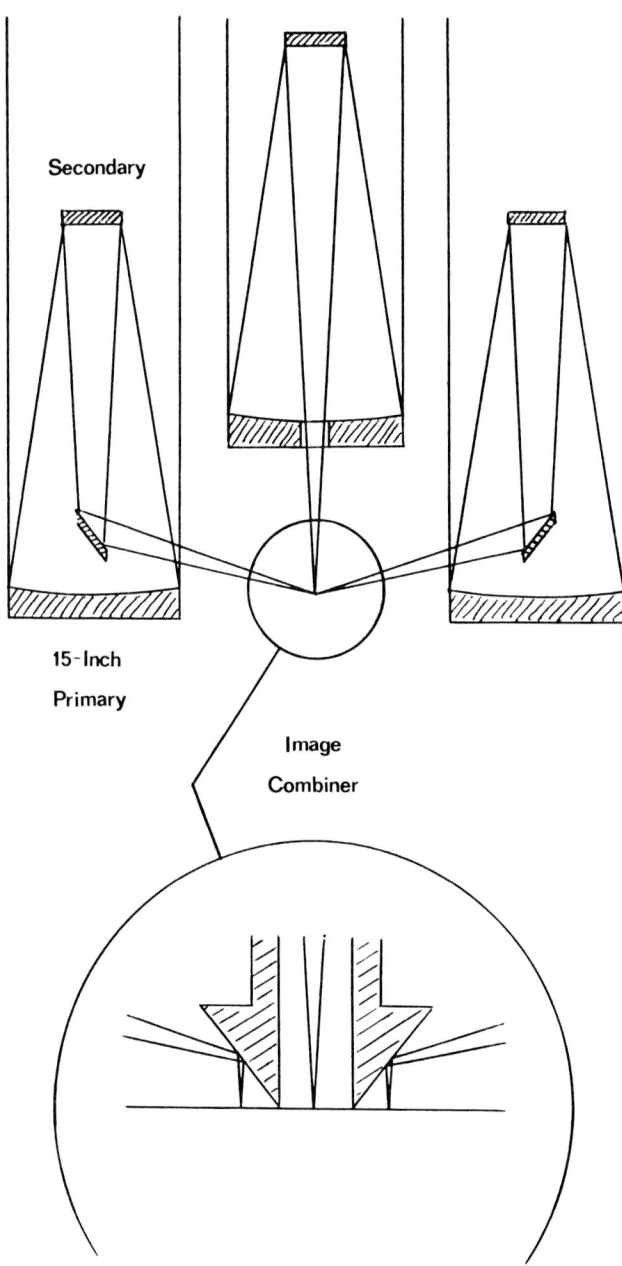

FIGURE 2: Optical configuration of the 7 x 15-inch Multi-Aperture
Telescope referred to in the text.

tional telescope, but because the light is pre-packaged in seven small images it must be thought of as having a built-in seven-bite image-slicer capability. Total cost of the instrument and its building etc. is about £60,000. Its 15-inch apertures are designed to meet the LWA condition at good European sites and at £60K, such an instrument is well within the grasp of individual university departments.

The idea of multiple imaging elements has been proposed by other workers; but based on different lines of argument. The University of Arizona and the Smithsonian Institution have recently opened their Multiple Mirror Telescope (or MMT) - also proposed about a decade ago as a means of cutting the cost of large collecting area telescopes. This instrument consists of six 72inch Cassegrain systems, (see Figure 3) the separate images from which are to be superimposed and held in phase by an active optical system - thus synthesising the effect of a 176-inch single imaging element whilst having the cost savings of a short structure. The instrument is designed to operate in the infrared as well as the visible region of the spectrum. Total cost of the programme is about £3 1/2 million. The size of the individual mirrors was determined by the rather unusual - but very persuasive - criterion that there just happened to be six such blanks available on the surplus market! From our earlier analysis, it is clear that they are far bigger than can be justified on the grounds of resolution, though of course they do provide a large collecting area. Nevertheless, proposals are afoot for an MMT consisting of seven monolithic 10-metre mirrors (the biggest monolithic mirror so-far built is the 5-metre at Mt. Palomar).

Multiple Mirror Systems in the USSR have followed yet another approach. Here, the array of primary imaging elements are arranged in a mosaic to synthesise a single optical surface. Each element operates at very high f-number, and to avoid off-axis problems, all surfaces are spherical - the aberrations being corrected at the secondary mirror and at the final image where necessary. The individual imaging elements are servo-controlled to maintain their images superimposed. A prototype consisting of six 40cm mirrors

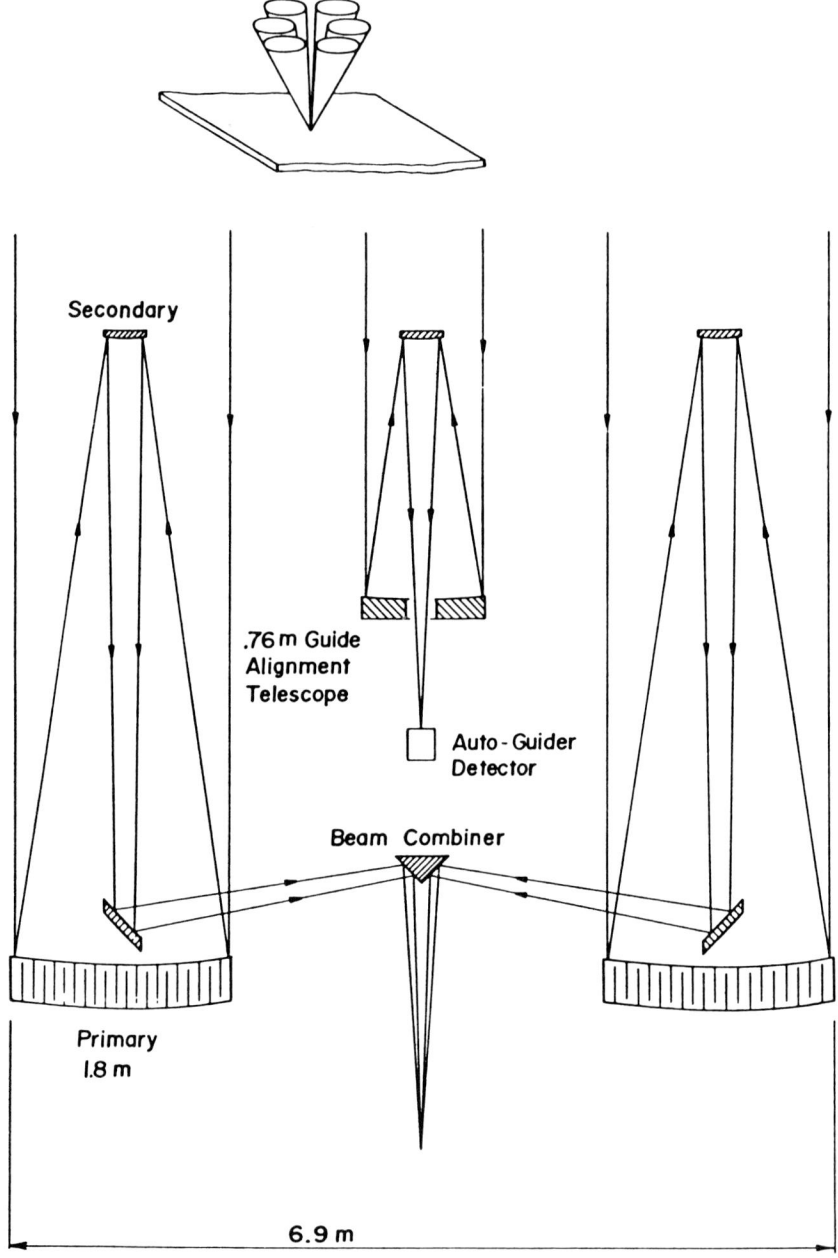

FIGURE 3: Optical configuration of the 6 x 72-inch Multi Mirror
Telescope referred to in the text.

has been built. Each mirror blank is hexagonal and the servo-
control system has worked satisfactorily. Based on these trials
(though using a completely different system of servo-control) a
mosaic-mirrored telescope of 25 meters aperture is projected, each
imaging element being a hexagon 1.2 meters across (Steshenko,
1979).

CONCLUSION

The function of a telescope is to optimise the transfer of the
information of interest, from a source of finite angular size, to
the analytical device which is going to extract it. Telescopes -
or fore-optical systems - are therefore only one link in the chain
connecting the source and the analyser. The whole chain must be
considered in the optimisation process and, for ground-based observatories, the crucial optical link is the atmosphere over which we
have no control. It is far from obvious that the conventional
large telescope is always the optimum fore-optical system for
looking through this turbulent medium. We have seen that there are
considerable scientific and financial advantages in going to multiple mirror systems and several groups in different parts of the
world are working in this direction. This activity, combined with
advances in coherent processing of the incoming wavefront not
mentioned in this paper, show that ground-based astronomy is very
much alive and several revolutionary new observatories are likely
to appear before the end of the century.

REFERENCES

Fried, D. L. 1966. "Optical resolution through a randomly inhomo-
    geneous medium for very long and very short exposures," J.
    Opt. Soc. Amer., 56, 1372-1379.
Fried, D. L. and Mevers, G. E. 1974. "Evaluation of r_0 for propa-
    gation down through the atmosphere," Appl. Opt., 13 2620-2622.
Grainger, J. F. 1971. "The design of fore-optics for analytical
    optical equipment used in astronomy," J. Phys. E.: Scientific
    Instruments, 4, 713-718.
Grainger, J. F. 1979. "The UMIST multiple-telescope programme,"
    Smithsonian Astrophysical Observatory Special Report 385,
    199-207.

Hufnagel, R. E. and Stanley, N. R. 1964. "Modulation transfer function associated with image transmission through turbulent media," J. Opt. Soc. Amer., $\underline{54}$ 52-61.

Steshenko, N. V. 1979. "On the feasibility of the 25-meter optical telescope," Smithsonian Astrophysical Observatory Special Report 385, 191-197.

Tatarski, V. I. 1961. "Wave Propagation in a Turbulent Medium," McGraw-Hill.

Young, A. T. 1974. "Seeing: Its cause and cure," Astrophysical Journal, $\underline{189}$, 587-604.

Astronomical and Non-Astronomical Uses of the Cosmos Machine

R.J. Dodd and I. Dalton

ABSTRACT

Each sky limited photographic plate taken with the U.K. 1.2 meter Schmidt telescope records about a million star and galaxy images. It is not feasible to measure these plates using manual machines and this led to the development of COSMOS; an automatic machine capable of finding and measuring the Co-Ordinates, Sizes, Magnitudes, Orientations, and Shapes of all the images on a single plate in less than a day. Information obtained from these measurements enables computer discrimination between star and galaxy images, and hence permits quantitative studies of the properties and distributions of the two kinds of objects to be carried out. However, COSMOS is not restricted solely to astronomical problems and it has been used in the fields of topography and image restoration and enhancement.

11

INTRODUCTION

A typical sky limited 1.2 meter Schmidt telescope photographic plate contains about a million star and galaxy images. If a suitably dedicated astronomer could be found who would be prepared to measure, with a manual iris photometer, all the star images on such a plate, it would take him or her some six years allowing time off for weekends and vacations! That it is not practical to measure and analyze these photographs using manual machines led the Royal Observatory Edinburgh and Computer Applications Services of Heriot-Watt University to develop an automatic machine, COSMOS (Co-Ordinate, Sizes, Magnitude, Orientations, and Shapes), capable of finding and measuring all the images on a single plate in less than a day.

In addition to astronomical measurements, COSMOS is regularly used to tackle non-astronomical problems in such fields as topography and the restoration and enhancement of images.

THE COSMOS MACHINE

The COSMOS system (Pratt, 1977) consists of a measuring machine, a hard wired control section, a Honeywell 316 computer, and magnetic tape, teletype, VDU, and Versatec printer-plotter peripherals.

The measuring machine has a compound plate carriage driven in the X and Y directions by hydraulic rams with a Moiré fringe grating system which allows the carriage position to be determined to ± 0.5 μm. A linear scanning densitometer consisting of a Ferranti microspot CRT projects a focussed, optically reduced light spot on to the plate below which are mounted diffuse collection optics and a photomultiplier. The system is of the dual beam type allowing the continuous monitoring and correction of the spot brightness.

The control unit houses the H316 computer, present position indicators, monitor oscilloscope, setting up switches, and power supplies. The magnetic tape unit is directly interfaced with the control hardware. Nine track, phase encoded 1600 b.p.i. magnetic tape is the normal data recording medium.

Some off-line data processing is necessary before meaningful data are available. This is currently carried out either on an in-house VAX 11/780 or GEC 4082 at the Royal Observatory or on a pair of IBM 360/195's at the Rutherford Computing Laboratories using a GEC 2050 based remote data entry system at the Royal Observatory.

MEASUREMENT MODES

At present there are two measurement modes in use; the mapping mode (MM) and the threshold mapping mode (TM).

In the mapping mode, which is suitable for the detailed examination of small areas of plate, a TV-type raster is generated by the combination of linear scan CRT and drifting the plate carriage in the Y-direction. A rectangular raster pattern is produced on the photographic plate by having the scan slightly tilted from the direction of the X-axis. Timing is arranged so that the individual scans are separated by intervals equal to the size of the scanning spot. The photomultiplier output, measuring the light transmitted through an element of plate, can be observed on the monitor oscilloscope.

The scan is divided into 128 increments, whose size may be chosen from 8 μm, 16 μm, or 32 μm. The projected spot size matches the chosen interval. The plate is thus scanned in strips 1.024, 2.048, or 4.096 mm wide. The transmission of the plate at each increment is digitized into one of 256 "T" levels. The relation between T and transmission is linear. The digitized transmission values are transferred to magnetic tape in blocks of 1024 bytes. A 625 mm² area measured at a resolution of 8 μm fills one 750 m magnetic tape in eight minutes and contains some 10 million pixels.

For the TM mode, the digitized transmissions are compared with a computer generated threshold transmission. Each lane being pre-scanned with a 32 μm spot step to set up a data matrix for computing, via a digital filtering technique (Martin and Lutz,

1979), a value for the local background transmission. A threshold some way below (usually ~ 10%) the plate background is selected to avoid misidentification of noise as signal. Image elements with transmissions below this threshold are combined using a software alogorithm (Lutz, 1979) to yield, after some off-line reduction, the following parameters per image:

(X,Y)	Centroid and image extents
(A)	image area
(Tm)	minimum transmission in the image
(ΣI)	intensity sum of the pixels in the image
(θ)	image orientation
(e)	image ellipticity.

In this mode, some 2-3 magnetic tapes, each containing around 400,000 images, are required to store the information extracted from one 35 cm square plate, the measurement time being typically around 18 hours.

DATA PROCESSING

In order to provide data in a form which may be readily understood by users, a series of interpretive and display programs has been written. Some of these are outlined below.

Alphanumeric mapping

The region to be displayed is mapped onto a 256 x 256 element array each T value being plotted as a number (0-9) or character (A-Z + symbols). An area 2 mm square measured with an 8 μm spot can thus be displayed without loss of resolution. This kind of display allows the calculation of profiles, or integrated intensities of small features without the use of a large computer. A modification of this technique is to use false color rather than numbers and letters for the coding.

Contour mapping

Contour maps may be produced from mapping data. While visually more acceptable than alphanumeric maps, the contour plot is less useful for data analysis. The contour lines may be simply

black on white or printed in different colors above a black background.

"Three dimensional" mapping

Borrowing data display methods commonly used by radio-
astronomers 3-D plot programs were written, and whilst in general
they show well how the background plate transmission level varies,
this type of display tends to be cosmetic rather than practical.

Cross sections

This software allows the user to specify sections along which
the intensity distribution is to be displayed and numerically
output (Brand and Zealey, 1978).

In the thresholded mapping mode, software exists for four main
applications namely field identification, distribution of objects
in the field, the separation of stellar and non-stellar images and
magnitude calibration.

Field identification

In order to locate reference objects in the TM data, it is
possible to plot out up to 500 images on the Versatec printer.
Each image can be plotted as its escribed rectangle, as a square of
side $\sqrt{(\log A)}$, or as a best fitting ellipse, in each scaled to the
display size and numbered (Stobie et al., 1979). In parallel, a
list of numbered images with their COSMOS output is provided.

Distribution of objects

To study the large scale distribution of images on a plate,
software has been written which plots the 100,000 brightest images,
with equal weights, as points in the (X,Y) plane. The program also
provides a number-density map of the distribution. Subsequent
programs produce isodensity contour maps of the region
(MacGillivray and Dodd, 1979). Among the uses to which this software is being put is the study of clusters of galaxies and areas of
high galactic and extragalactic obscuration.

Star-galaxy separation

Star images may be separated from those of galaxies by plott-
ing $\log A$ against Tm for each image. For a given $\log A$ value
galaxies, until their images become center saturated, have a higher

value of minimum transmission Tm than do star images (MacGillivray
et al., 1976).

Magnitude calibration

In order to tie COSMOS measurements of stellar brightness to
an externally accepted magnitude scale, plots of log A or log ΣI
against photo-electrically determined magnitudes of standard stars
may be made and a least squares best fitting orthogonal polynomial
determined (Dodd et al., 1979).

A block diagram of the astronomical measures made by COSMOS
and their applications is shown in Figure 1.

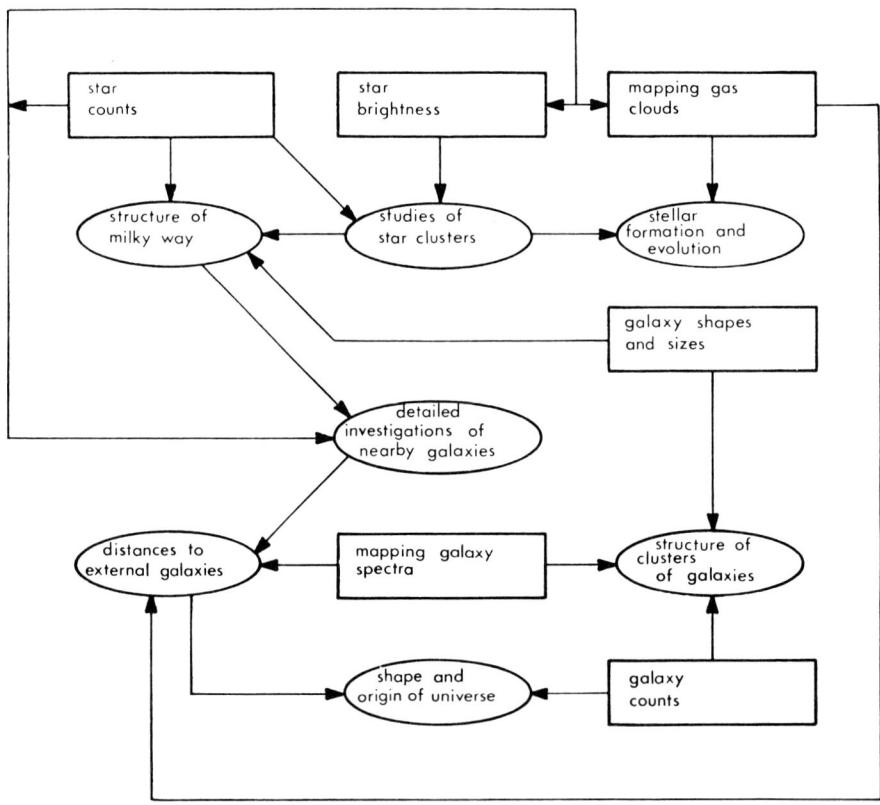

FIGURE 1: COSMOS measurements and their astronomical applications.
The machine measurements are shown in rectangular boxes and their
applications in the elliptical boxes.

NON-ASTRONOMICAL APPLICATIONS

Non-astronomical applications of COSMOS, which are being developed by Computer Applications Services of Heriot-Watt University, presently fall into two main catageories, namely:

1. Topographical (digital data bases, terrain classifications, and feature detection)

2. Image restoration and/or enhancement (signal to noise enhancement, feature enhancement, and feature restoration).

Regarding topographical applications, the concern is essentially with the mapping of data to a given resolution from any form of translucent material (at present we are even experimenting with geological slices prepared for microscopic examination with a view to calculating porosity characteristics, etc., for oil assessments).

Normally the data bases are constructed by scanning high resolution air photographs (e.g., urban mapping to a resolution of 1 meter on the ground from aerial surveys at 20,000 meters with an f6 camera) or side scan radar or sonar traces. The data base can then be analyzed or manipulated to provide a variety of presentations and information such as maps, feature classification (woods, pastures, buildings, cars, etc.) calculations of areas, number counts, etc.

In general, whilst this approach involves the scanning of larger volumes of material or large areas, it does offer a high speed, rigorous, and high resolution scanning, detection and measurement capability which provides a quality of survey not previously available. Examples include urban mapping, traffic surveys, feature detection, radar reflectivity, land use, etc.

In the case of image restoration and enhancement, the scanning phase is used as the basis for either improving the visual quality of the image by raising the contrasts, removing motion blur and distortions due to perspective or a partially obscuring medium, or editing the picture by the selective removal of noise (that is

unwanted features in any particular image) in order to highlight
the significant objects. Examples of this category of work are
photographs taken under difficult conditions of poor lighting
involving moving objects (e.g., the Loch Ness 'monster' pictures).

CONCLUSIONS

As may be seen the COSMOS machine is a fast, yet versatile
measuring machine capable of tackling both astronomical and non-
astronomical problems. Currently the former occupy some 80% of
machine time and the latter the remainder.

REFERENCES

Brand, P. W. J. L. and W. J. Zealey 1978. Dust clouds in HII
     regions. The dragon in M8. Astron. Astrophys., $\underline{63}$: 345-352.
Dodd, R. J., H. T. MacGillivray, G. M. Smith and L. A. Ellery 1979.
     Photographic photometry of stars and galaxies from COSMOS
     machine measures. International Workshops on Image Processing
     in Astronomy. Trieste, Italy. Ed G. Sedmak, M. Capaccioli
     and R. J. Allen pp 360-370.
Lutz, R. K. 1979. On the real-time analysis of astronomical
     images. International Workshop on Image Processing in
     Astronomy. Trieste, Italy. Ed G. Sedmak, M. Capaccioli and
     R. J. Allen pp 218-232.
MacGillivray, H. T. and R. J. Dodd, 1979. Properties of faint
     galaxies in two fields near the South Galactic Pole. Mon.
     Not. R. astr. Soc., $\underline{186}$: 69-84.
MacGillivray, H. T., R. Martin, N. M. Pratt, V. C. Reddish, H.
     Seddon, L. W. G. Alexander, G. S. Walker and P. R. Williams
     1976. A method for the automatic separation of the images of
     galaxies and stars from measurements made with the COSMOS
     machine. Mon. Not. R. astr. Soc., $\underline{176}$: 265-274.
Martin R., and R. K. Lutz 1979. A digital filtering technique for
     background following on Schmidt plates. International
     Workshop on Image Processing in Astronomy. Trieste, Italy.
     Ed G. Sedmak, M. Capaccioli and R. J. Allen pp 211-217.
Pratt, N. M. 1977. The COSMOS measuring machine. Vistas Astr.,
     $\underline{21}$: 1-42.
Stobie, R. S., G. M. Smith, R. K. Lutz and R. Martin 1979. The
     performance of the COSMOS machine. International Workshop on
     Image Processing in Astronomy. Trieste, Italy. Ed G.
     Sedmark, M. Capaccioli and R. J. Allen pp 48-64.

Detecting the Gulf Stream from Digital Infrared Data Pattern Recognition

D. Gerson, E. Khedouri and P. Gaborski

ABSTRACT

Detection of oceanic fronts from satellite sea surface temperature (SST) data is presently a manual, subjective procedure.
Automated SST products are oriented towards large area temperature
analyses which tend to smooth out fronts rather than detect them.
A numerical method for automated detection of oceanic fronts was
developed using digital infrared imagery from the GOES satellite.
The method consists of a sequential decision approach with the
decisions involving geographic location, a priori frontal location,
and two stages of imagery pattern recognition. Selection of statistical features for pattern recognition stages occupied the major
investigative effort. Features tested included first order statistics, Fourier analysis, co-occurence matrix analysis, difference
histograms, and gradient analysis. The resultant techniques were

successfully applied to the western North Atlantic to map the Gulf
Stream. Results were in excellent agreement with frontal locations
obtained by manual methods, thereby establishing the feasibility of
using automated methods for frontal detection.

INTRODUCTION

Detailed charts of major oceanic fronts, such as the Gulf
Stream, are currently prepared on a regular basis by professional
oceanographers using satellite imagery as a prime source of data.
Use of automation is primarily for enhancement of imagery to
visually aid the analyst in his interpretation effort. Until now,
fully automated products have been oriented toward large area
temperature analyses which tend to smooth out fronts rather than
detect them.

The purpose of this paper is to describe an objective, fully
automated procedure, for locating ocean front and other major ocean
features from digital satellite data. The proposed procedure
consists mainly of imagery pattern recognition methods, such as
those described by Duda and Hart, (1973), Haralick et al., (1973)
and Gerson (1975). The pattern recognition features used for
frontal identification are automatically evaluated for their effectiveness, and the best feature set is then selected for a particular case. Because an infinite number of potential pattern recognition features exist, an exhaustive study is impossible. The
features tested, however, are relatively complete and their evaluation clearly indicates the most useful ones.

DATA

Digital infrared data from the GOES II (Geostationary
Operational Environmental Satellite) were used for this study. A
sample of the full disc image is shown in Figure 1. Five such
pictures are available daily in digital form on magnetic tape. The
satellite is centered at the equator and 75°W and covers an area of
approximately 89° of latitude and 99° of longitude. The spatial

resolution is approximately 8 km; that is, each pixel (picture element) represents the average temperature over an 8 x 8 km square on the earth's surface. There are 256 gray scales for each pixel; when converted to temperatures, they cover a range from -110.2°C to +56.8°C. It is, therefore, possible to truncate both the high and low ends without affecting the temperature resolution and to obtain a useful temperature range of approximately 32°C.

A package of computer programs known as XAP was used for manipulating the satellite pictures. Details are described in

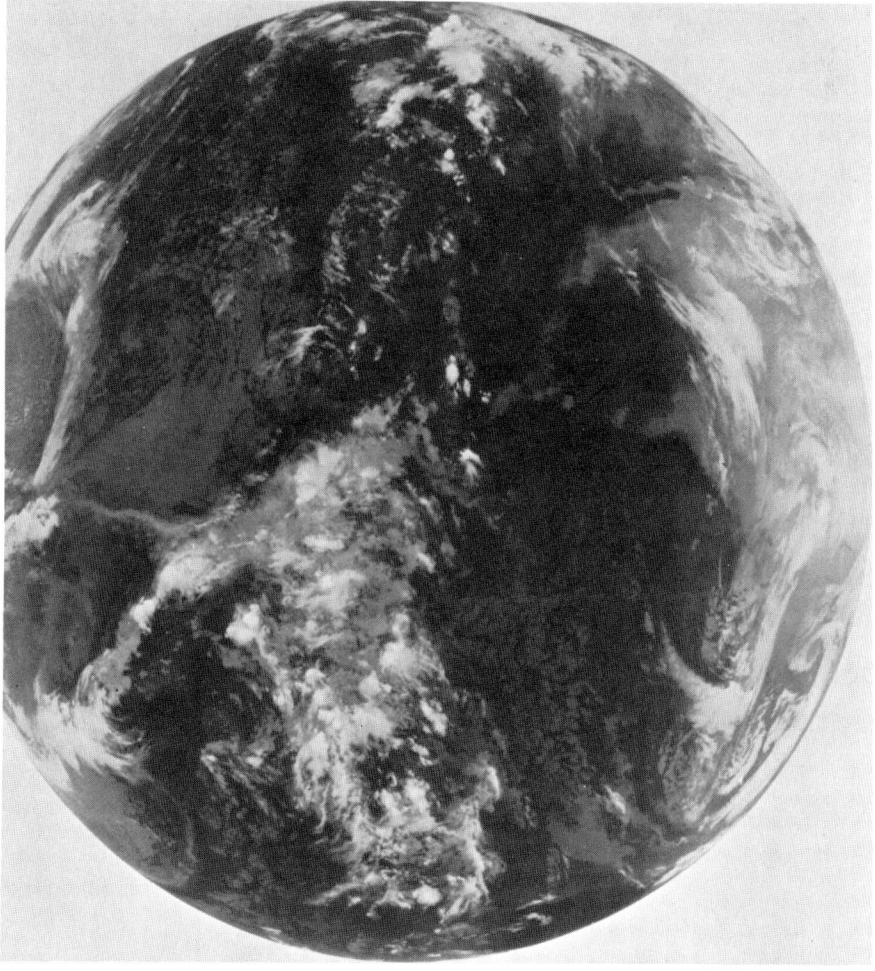

FIGURE 1: GOES SMS-2 infrared image for 26 February 1977.

Hayes (1975). These programs facilitate the following operations: selecting an area of specified size from a picture, adding, subtracting or algebraically combining pictures, etc. These programs were very valuable in developing the techniques for this study.

CLOUD REDUCTION

Clouds present a problem in automated frontal detection as they do in the manual operation. A technique has been developed, however, to reduce their effect. In the Gulf Stream region, clouds are almost always colder than water. Therefore, a temperature threshold can be selected to identify clouds, but this gives no information about the sea surface temperatures below the clouds. Fortunately, clouds move much more rapidly than sea surface features, therefore, clouds can be reduced by overlaying several daily pictures and compositing them by selecting the warmest temperatures at each point. Using this method, clouds have been eliminated entirely in many pictures or reduced substantially. When the clouds are extremely persistent, the overlay process can be extended to several days. This latter process may degrade the results because corresponding points can represent temperatures from different days but it has produced useable images from very poor data.

A sample picture from February 26, 1977, is shown in Figure 2 and a composite picture, produced from five images for that day, is shown in Figure 3. It can be seen that a substantial amount of clouds were removed by compositing. In this case, a sizable segment of the Gulf Stream, seen on the composite picture, was not evident in any of the individual images. Note that these pictures are actually computer printouts, and the shading is produced by over-printing letters. The darker shades represent colder temperatures. The quality of the pictures is immaterial because the final results are not dependent on visual effects. These type pictures are used only to examine some interim results.

The disadvantage of the above method is that sometimes cold features embedded in warm water, such as cold eddies, can become

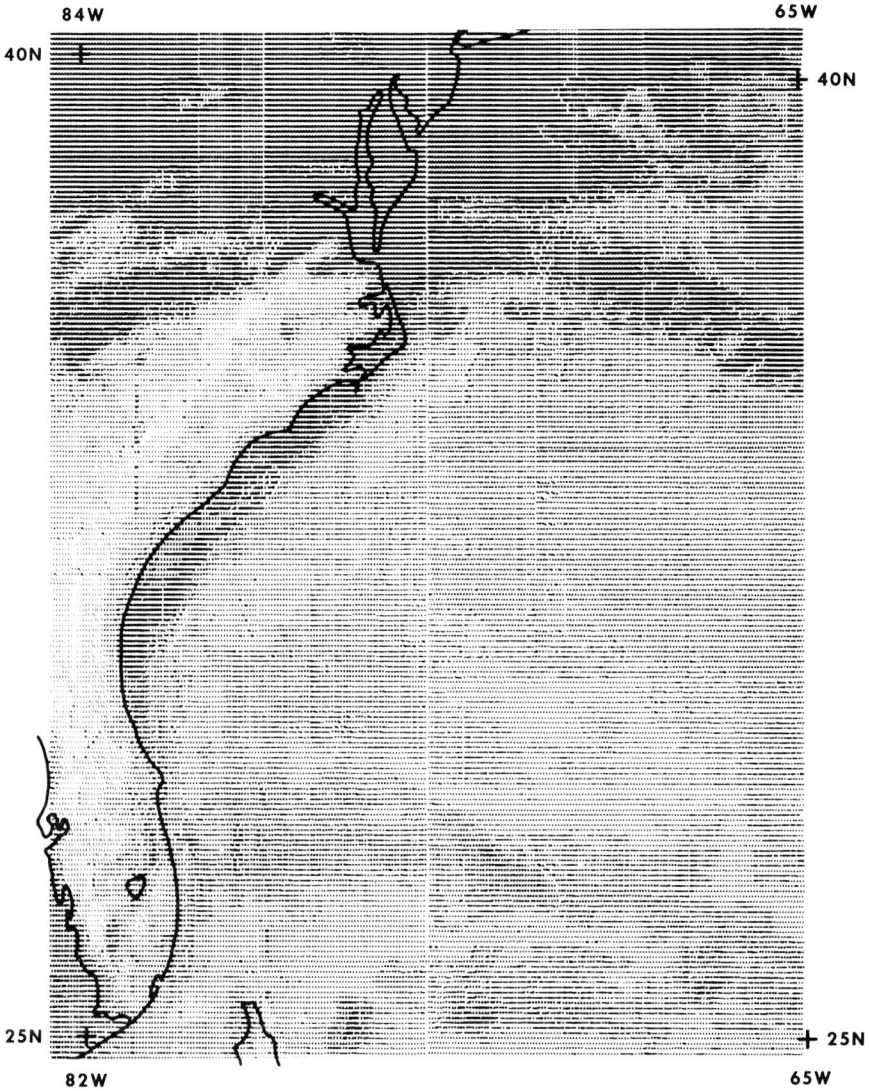

FIGURE 2: Sample picture produced from GOES digital data for 26 February 1977.

suppressed because compositing consists of selecting the warmest temperature from each picture and it is, therefore, possible for a cold eddy to be masked because of afternoon heating. To avoid this problem, each point identified as a cloud, can be set to maximum temperature at each point. In this method the cold features are accentuated rather than suppressed.

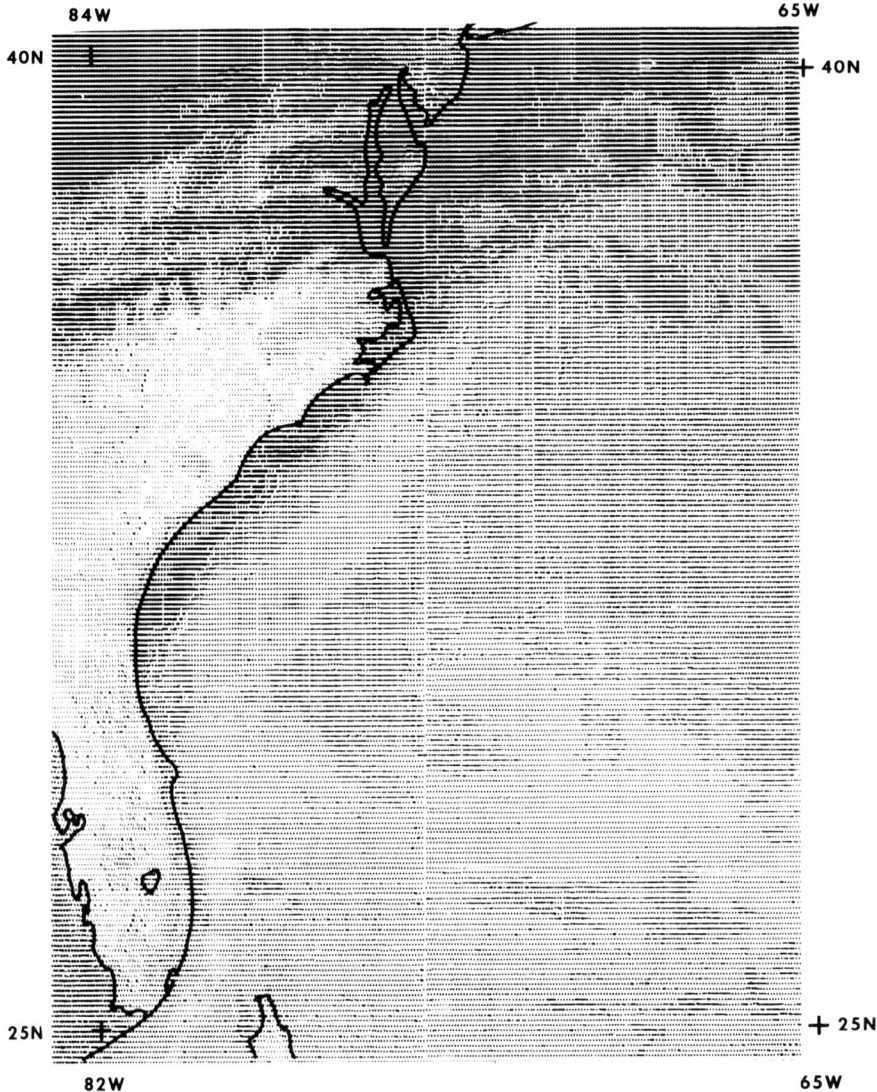

FIGURE 3: Sample of cloud reduced picture produced by compositing
five pictures for 26 February 1977.

PATTERN RECOGNITION

Once a relatively cloud free picture is produced, it is sub-
divided into overlapping 16 x 16 pixel (approximately 128 x 128 km)
frames and each frame is evaluated in sequences shown in the hier-

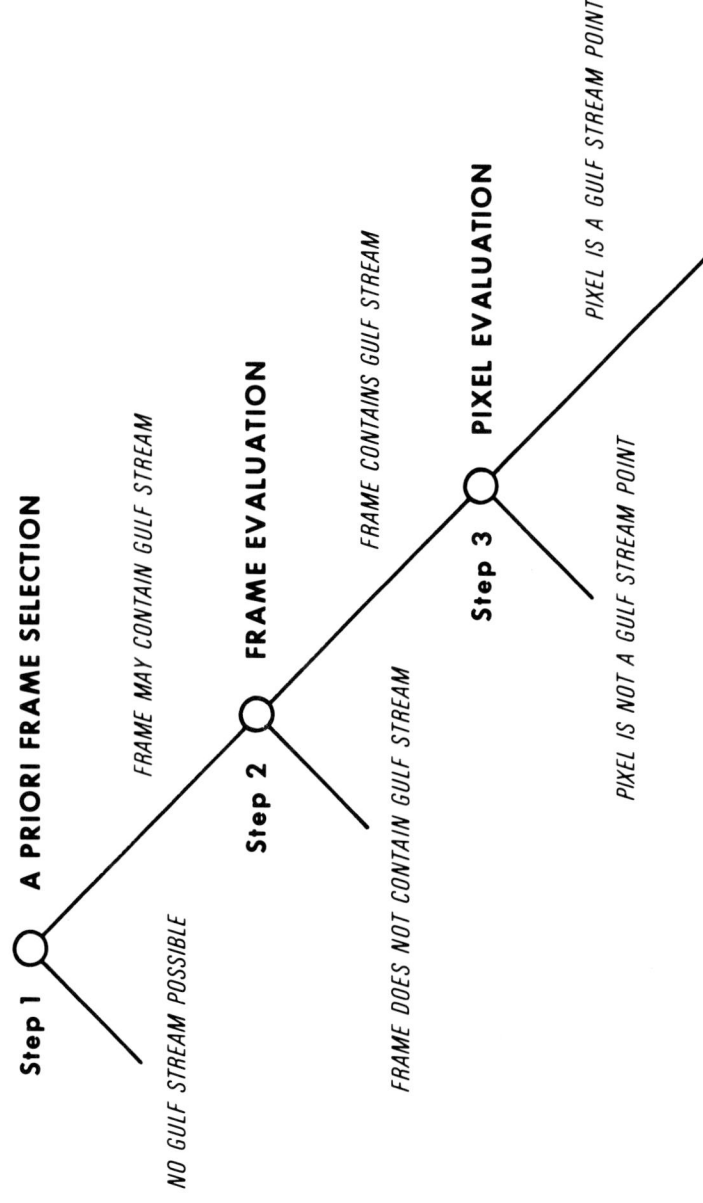

FIGURE 4: Hierarchical decision tree.

archical decision tree depicted in Figure 4. The first step is to
decide whether a frame, because of its geographical position, could
possibly contain the Gulf Stream; this is done by using a priori
knowledge about Gulf Stream location and movement. The next step
is to examine the promising frames and decide whether they do, in
fact, contain the Stream; this is done by water mass classification. Development of classification scheme for this step was the
major experimental effort and involved the use of pattern recognition techniques including decision theory necessary to evaluate and
implement the classification scheme. The third step in the decision tree is pixel evaluation. In this step, each pixel within the
frames which are known to contain the Stream, is evaluated to
determine the exact frontal location.

Step 1 - A Priori Frame Selection

Because the historical geographical boundaries of the Gulf
Stream are known, all the frames which are outside those boundaries
can be eliminated. Similarly, by knowing the previous position of
the front, and knowing the maximum possible fluctuation within the
time period involved, frames which could not possibly contain the
Gulf Stream are eliminated. Figure 5 shows the frames which are
selected after eliminating all the frames which could not possibly
contain the front.

Step 2 - Frame Evaluation

Several statistical features were evaluated to decide which of
the frames selected in Step 1, do in fact, contain the Stream. The
statistical features were evaluated for their ability to classify
imagery into three classes: (1) Gulf Stream Water, (2) Slope Water
(water north and west of the Gulf Stream), and (3) Sargasso Water
(water south and east of the Stream). To evaluate the ability of a
particular statistical feature to accomplish the above, Fisher's
linear discriminant was used. The choice was based on its computational simplicity and availability of computer programs.

(a) Fisher's Linear Discriminant

A complete discussion of the discriminant is given in Duda and
Hart (1973). The method can briefly be described as follows.
Suppose we have two classes (crosses and circles) plotted as a

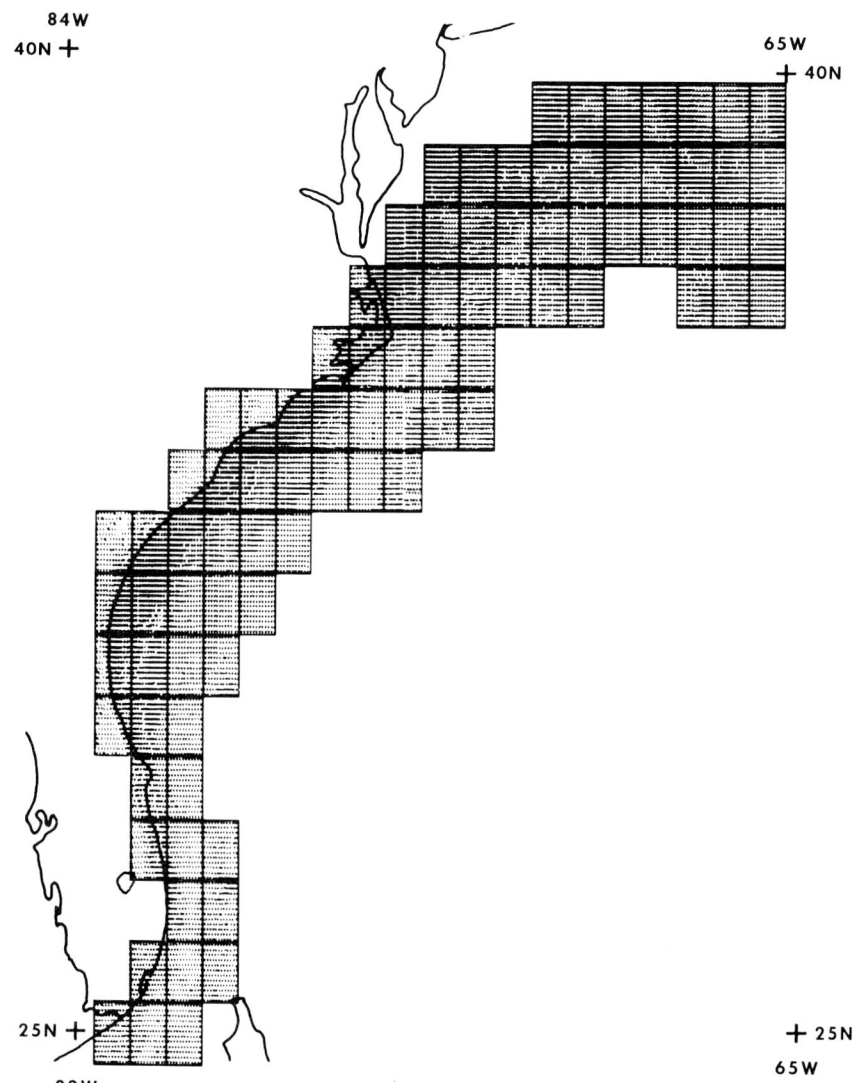

FIGURE 5: Frames which could possibly contain the Gulf Stream based on a priori information.

function of two statistical features, X_1 and X_2. If the samples are projected onto a line of correct orientation (W), as in Figure 6a, there will be a good separation between the clusters, but if they are projected onto an arbitrary line as in Figure 6b, the separation will not be evident. To evaluate a pattern being tested, it is necessary to know how well the clusters of data

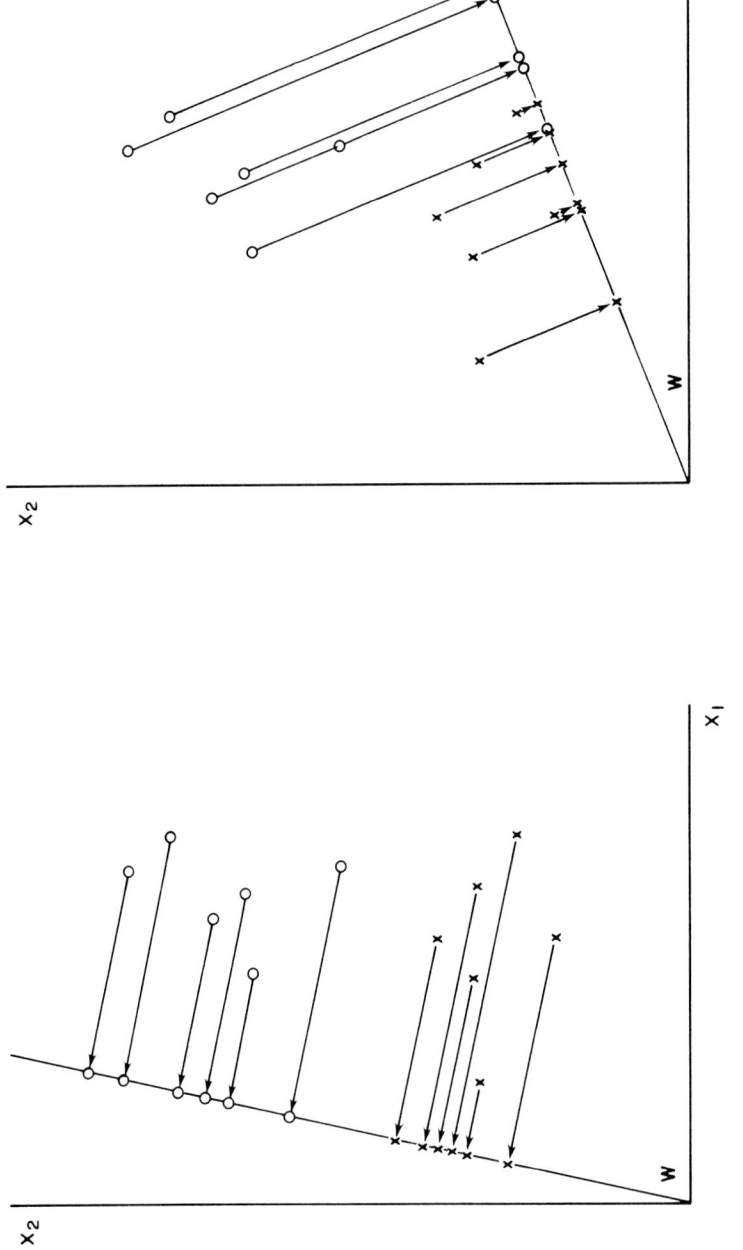

FIGURE 6: Projection of samples onto a line.

projected into a correctly oriented line are separated. The best
statistical features are those which result in greatest separation
between classes and smallest dispersion within a class. A convenient measure for the above two criteria is the Fisher distance
(FD) defined as follows:

$$FD_{1,2} = \frac{|M_1 - M_2|}{S_1 + S_2}$$

where M_1 and M_2 are the mean values and S_1 and S_2 are the standard
deviations. The statistical features resulting in the largest
Fisher distances for any two classes, are considered to be the best
statistics for discriminating between those two classes. The
minimum error classification into two classes is obtained using the
threshold value (T) defined as follows:

$$T_{1,2} = \frac{M_1 S_2 + M_2 S_1}{S_1 + S_2}$$

All points falling on one side of T are identified with one class,
while those on the other side are identified with the second class.

(b) Statistical Features

Statistical features evaluated by Fisher's discriminant for
this study summarized in Table 1. Complete description of each
statistic is given in Gerson and Gaborski (1977), a brief description is as follows:

Mean and Standard Deviation - Two very simple, though sometimes very effective, statistics are mean and standard deviation of the temperature values. These statistics should give
good results in separating Slope Water from Sargasso Water
because there is a consistent difference in temperatures
between the two water masses and because temperature variation
is small within Sargasso and large within Slope Water.

Histogram Analysis - Examination of histograms of frames
containing the Gulf shows that skewness changes from positive

to negative as the Gulf Stream is crossed from west to east. Skewness measures the asymmetrical distortion about the mean.

TABLE 1: Statistics used for water mass classification.

Mean
Standard Deviation
Histogram Analysis
Power Spectrum Analysis
Co-occurence Matrix Analysis
        (1)   Angular Second Moment
        (2)   Contrast
        (3)   Correlation
        (4)   Entrony
Gradient Statistics
        Mean and Standard Deviation of:
        (1)   Horizontal differences
        (2)   Vertical differences
        (3)   Maximum differences
Difference Histograms
        (1)   Mean
        (2)   Contrast
        (3)   Angular Second Moment
        (4)   Entropy
        (5)   Maximum
        (6)   Mode
        (7)   Inverse Difference Moment

Power Spectrum - If a picture is considered to be a function of two variables, its Fourier transform can be obtained and thus its power spectrum. The power spectrum plot of a picture shows the low frequencies, such as flat unchanging areas, near the center of the plot. Any high frequencies, such as sharp edges or rapidly changing temperatures, would appear further from the center.

Co-occurence Matrix Analysis - A useful method for examining the texture of a picture is to study the co-occurence of temperature values. A co-occurence matrix is an ordered arrangement of the frequencies with which the various possible temperatures occur adjacent to or at some fixed distance from

one another. Four types of matrices are formed depending on
the direction of the adjacency: horizontal, vertical, left
diagonal, and right diagonal. Four features selected for
analyzing these matrices were: angular second moment, contrast, correlation, and entropy. For a complete discussion
see Haralick et al. (1973).

Gradient Analysis - An analysis of gradients in a picture can
provide information which may be useful in classification.
The following gradient statistics computed were: mean and
standard deviation of horizontal, vertical, and maximum
differences.

Difference Histograms - Another set of features selected were
difference histogram statistics. Difference histograms,
similar to co-occurence matrices, measure the frequency with
which temperature differences occur adjacent to or at a fixed
distance between pixels. These histograms can be computed
over four directions: horizontal, vertical, right diagonal,
and left diagonal. Seven statistics were generated from these
histograms: mean, contrast, angular second moment, entropy,
maximum, mode and inverse difference moment.

(c) Results

The above statistics were applied to experimental data consisting of 85 samples which were classified by an experienced
analyst into three classes: Gulf Stream, Sargasso Water, and Slope
Water. Fisher distances and classification error rates were calculated for all the statistical features and are given in Gerson and
Gaborski (1977). Statistics giving the best results (greatest
Fisher distances and smallest error rates) are summarized in Table
2. The statistic "mean" gave the best results for classifying
Stream and Slope Water. The implication is that the two water
masses have consistently different temperatures. Several statistics gave good results in separating Stream and Sargasso waters.
The best was 'difference histogram maximum entropy" implying that
more energy is present within the Stream than within Sargasso

Water. "Standard deviation" also gave good results implying that
more variability is present in the Stream than Sargasso Water.
Finally for separation of Sargasso and Slope Water, statistic
"mean" gave the best results implying that the mean temperature of
the two water masses is consistently different. While it should be
technically correct to test all possible pairs and triplets of
features, computational limits preclude such extensive tests.
Several of the most promising pairs and triplets were tested,
however, and the best results were obtained with triplet "mean",
"standard deviation", and "difference histogram maximum entropy".
The results gave 100 percent correct classification for all three
water masses.

TABLE 2:   Sample of Best Classification Statistics.

WATER MASSES	STATISTIC	FISHER DISTANCE	PERCENT ERROR
Gulf Stream vs. Slope	Mean	2.08	2
Gulf Stream vs Sargasso	Difference Histogram Maximum		
	Entropy	1.78	2
	Standard Deviation	1.76	0
Sargasso vs Slope	Mean	2.84	0

Once the best feature set is determined, it is used to
identify the water masses within each frame tagged in Step 1 as
"possibly containing Gulf Stream". Results depicted in Figure 7
show only the frames identified as containing Gulf Stream water.
In this case, a discriminant function consisting of the three
statistics discussed above was used for classification. This
constitutes the second step in hierarchical decision approach. To
accomplish the third step, it is necessary to locate the precise
position of the Gulf Stream edge within each frame by evaluating
each pixel.

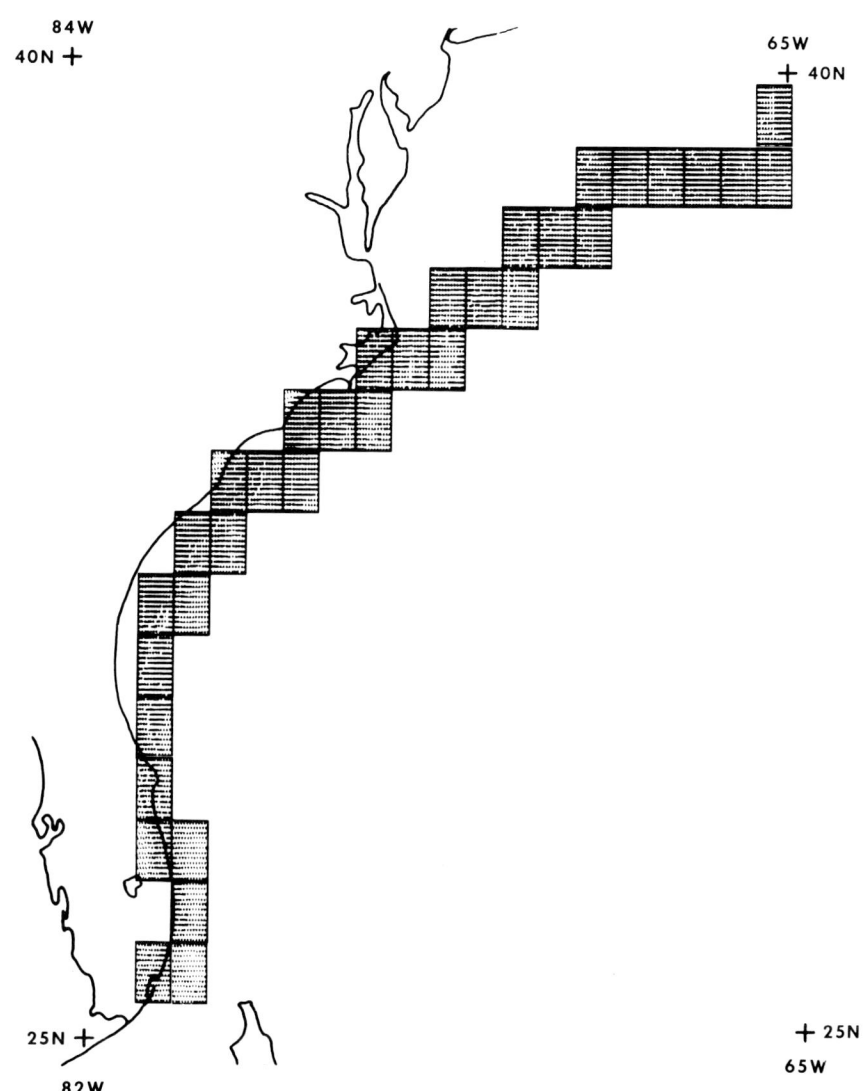

84W
40N +

65W
+ 40N

25N +
82W

+ 25N
65W

FIGURE 7: Frames determined by pattern analysis to contain the Gulf Stream.

Step 3 - Pixel Evaluation

(a) Procedure

This step was successfully accomplished by scanning each frame with a 5 x 5 pixel square and applying some of the statistical features discussed previously. The scanning process is illustrated

in Figure 8. Of the statistics tested, it was found that skewness
of the histogram which changes from positive to negative as the
Gulf Stream is crossed from west to east gave the best results.
The procedure developed and illustrated in Figures 8 and 9 is as
follows:

> First, scan each 16 x 16 pixel frame using a 5 x 5 pixel
> square and calculate the histogram and skewness of the 25
> points.

> Next, set all positive values of skewness to '1' and all
> negative values to '0'. The result is the binary picture in
> Figure 9b obtained from 9a.

> Finally, apply an edge detector to the binary picture. The
> edge detector inserts a frame of 1's around the binary picture
> and then tests all internal 1's; if the 1's are situated next
> to 0's, they are edge points. Isolated or pairs of 0's in the
> binary picture are ignored since they would be surrounded by
> an edge. This procedure produced the edges shown in Figure 9c
> from 9b.

> (b) Results

> The edge picture in Figure 9c shows the northern and southern
> edges of the Gulf Stream. This completes all the steps in the
> hierarchical decision approach. The only remaining require-
> ment is to connect the points from all the frames for the
> desired front. For example, if the location of the Gulf
> Stream northern edge is desired, all the edge points from the
> left side of every frame are connected producing the results
> shown in Figure 10.

DISCUSSION AND CONCLUSIONS

The results based on 85 samples show that it is possible to
use an automated system to obtain frontal locations similar to
those depicted by a trained analyst. Whether it is possible to
improve on the manual method is a subject for future investigation.
It is entirely possible, for example, that some surface manifestation of a front may not be readily apparent to the eye, but may

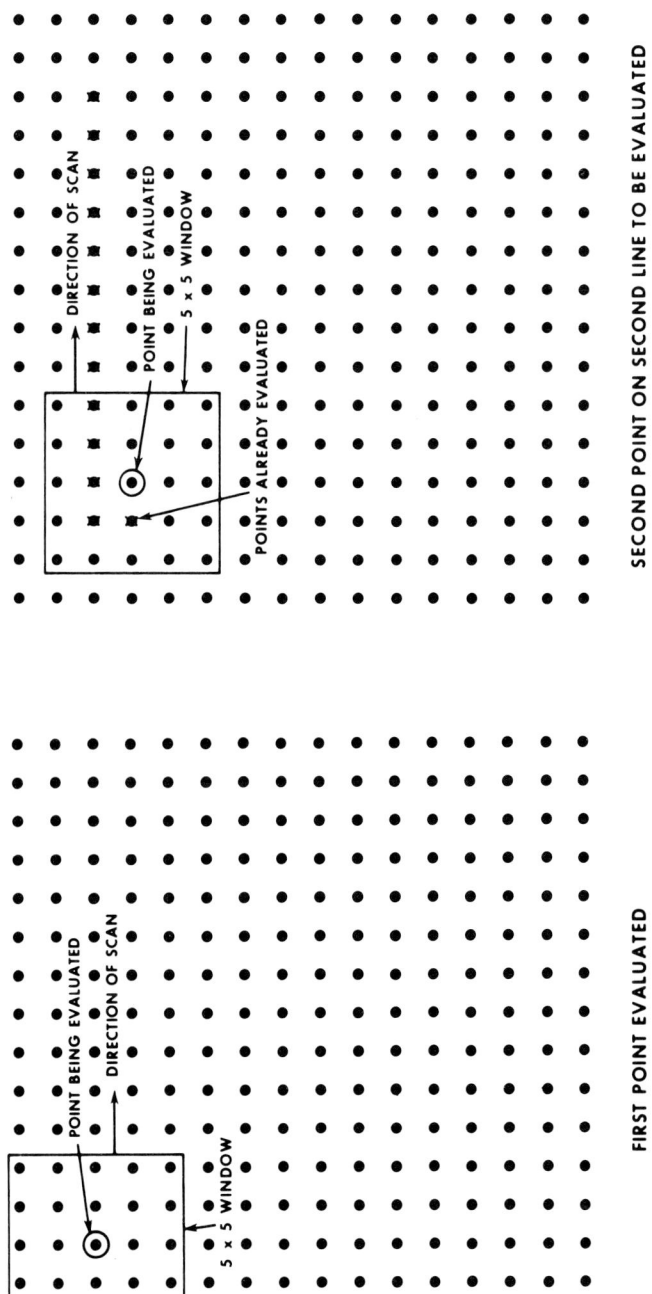

FIGURE 8: Scanning procedure for pixel evaluation.

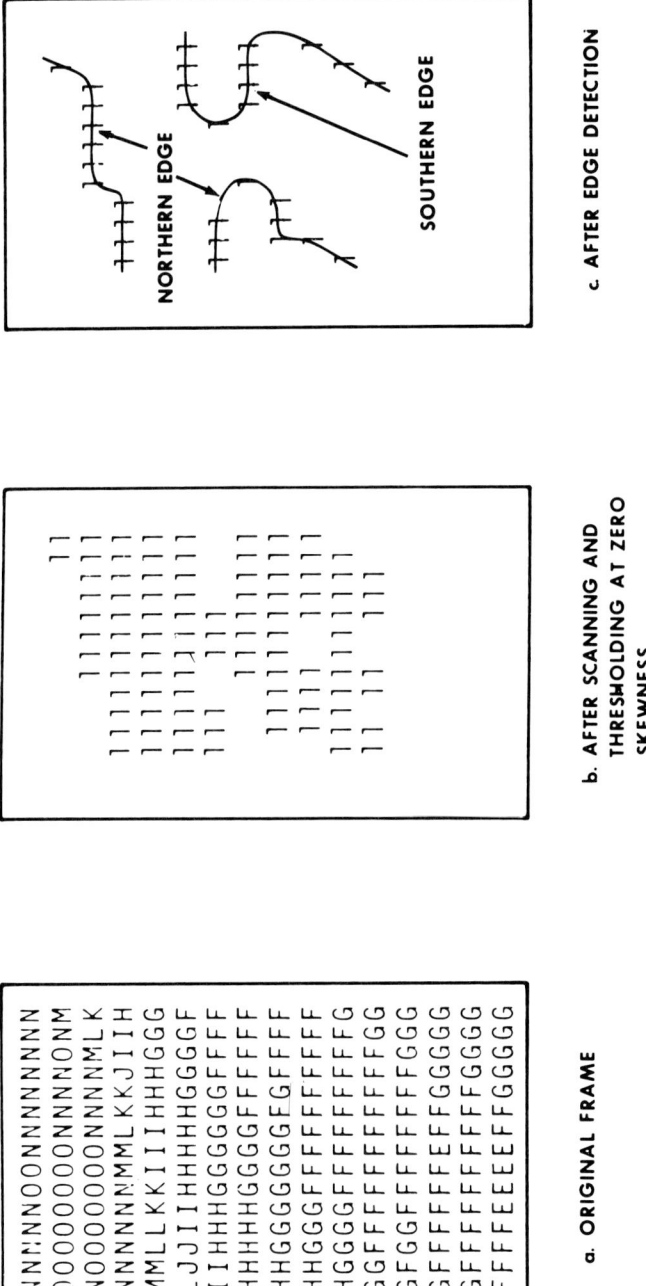

FIGURE 9: Pixel evaluation and edge detection.

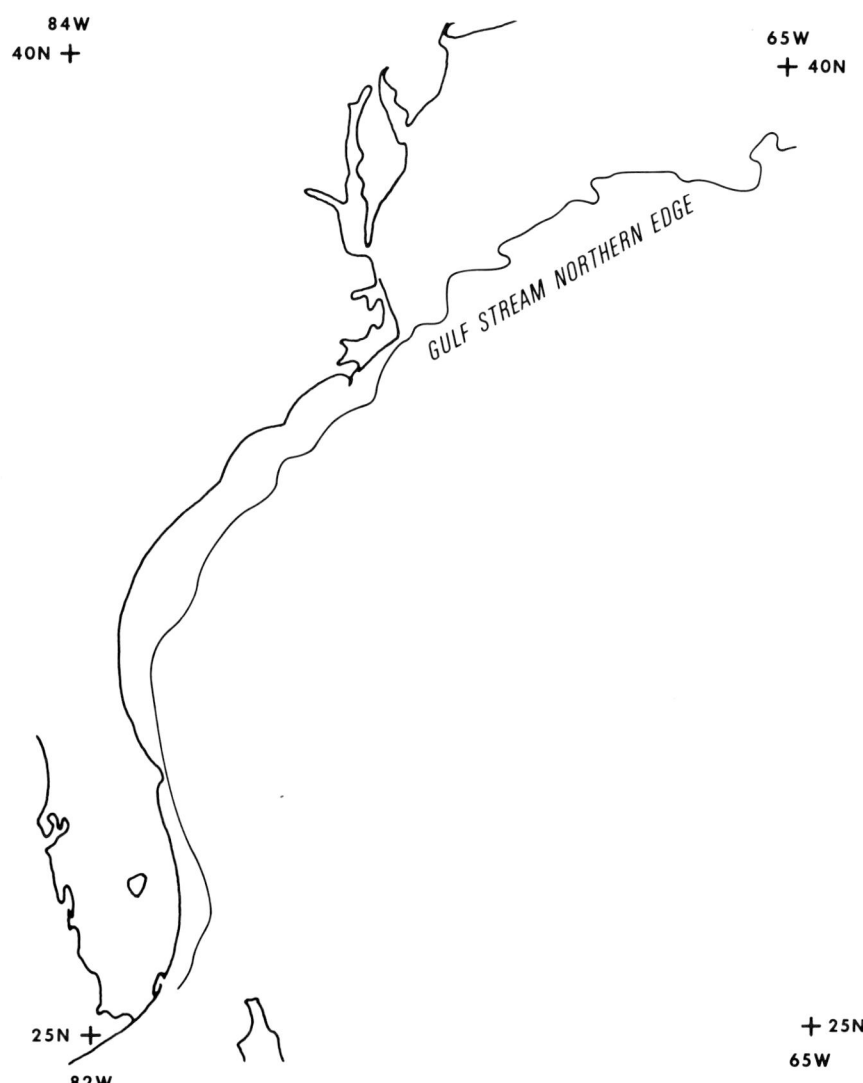

FIGURE 10: Gulf Stream location for 26 February 1977.

respond to some statistics, such as co-occurence matrix analysis, which detect fine variations in the texture.

Although the statistical features derived here are applicable only to the Gulf Stream region, the proposed method has the inherent flexibility to be applied to any major front. Also, the method is applicable if different criteria are used to determine what constitutes a front. In the case of the Gulf Stream, for example, the deep front is better correlated to maximum surface temperature than it is to the surface gradient as described by Khedouri et al. (1976). This could be built into the system by determining which statistical features will give best results for finding the front based on these criteria and revising the required threhold values for Fisher's discriminant. This will be the subject of future work.

The following conclusions can be reached from work completed up to this point:

1) In the Gulf Stream region, clouds can be reduced or removed entirely by compositing several satellite pictures.

2) Fisher's linear discriminant is a simple and effective way to evaluate the statistical features for their ability to classify water masses.

3) The triplet composed of "mean", "Standard deviation" and "difference histogram maximum entropy" resulted in 100 percent correct water mass classification on the 85 test samples.

4) Change in skewness of the temperature histogram, from positive to negative as the Gulf Stream is crossed from west to east, is an effective statistical feature for determining the exact frontal location.

ACKNOWLEDGEMENTS

The authors acknowledge the efforts of Mr. M. K. Shank, Jr. and Dr. R. W. James for their administrative support and technical guidance; Mr. W. George for his assistance in computer programming; Mr. G. Voorheis for the final illustrations; and Mrs. L. Allen for preparation of the manuscript.

REFERENCES

Duda, R. O., and P. E. Hart 1973. Pattern Classification and Scene Analysis. 482 pp. New York: J. Wiley & Sons.
Gerson, D. J. 1975. Computer Estimation of the Presence of Sea Ice in Satellite Pictures. 85 pp. College Park, Maryland.

University of Maryland Computer Science Center, Technical Report 366.

Gerson, D. J., and P. Gaborski 1977. Pattern Analysis for Automatic Location of Ocean Fronts in Digital Satellite Imagery. 67 pp. Bay St. Louis, Mississippi. U.S. Naval Oceanographic Office Technical Note 3700-65-77.

Haralick, R. M., K. Shanmugam, and I. Dinstein 1973. Textural Features for Image Classification, IEEE Transactions on Systems, Man, and Cybernetics, Vol. SMC-3, No. 6: 610-621.

Hayes, K. C. XAP Users' Manual 1975. 95 pp. College Park, Maryland. Computer Science Center Technical Report 348.

Khedouri, E., W. Gemmill, and M. Shank, Jr. 1976. Statistical Summary of Oceanic Fronts and Water masses in the Western North Atlantic 24 pp. Bay St. Louis, Mississippi. U.S. Naval Oceanographic Office, Reference Publication No. 9.

In-Shore Ecosystem and Biochemical Studies and Data Collection

F. John Vernberg

ABSTRACT

An ever increasing interest in the coastal regions of the world's ocean is being shown by international, national, and local government agencies, industry, and scientists. The need exists for an understanding of the complex dynamic processes controlling this coastal ecosystem. Although remote sensing techniques have been used to help provide some of the data base necessary for analysis of ecosystems and individual organisms, a need exists for intensive development and application of a more sophisticated technology.

INTRODUCTION

In recent years, remote sensing techniques used to scan the entire earth's surface have developed rapidly. The focus on the present symposium is on developments as they pertain to the ocean, particularly the coastal regions, rather than on a review of advances relating to the terrestrial habitat. The scientists in attendance are a highly capable group who have responded to the request of government, industry, and other scientists to provide the technology to meet the need for specific information that might best be gathered by remote sensing techniques. The purpose of my paper is to emphasize some informational needs perceived by a marine scientist oriented toward coastal problems, with a view toward obtaining information about the feasibility of using remote sensing techniques for the acquisition of data. It is perhaps a truism to note that the interaction between users and developers is vital for scientific and technological advancement.

The coastal regions of the world are extremely important to our understanding the ecological dynamics of the world's ocean, for it is here that most of our living marine resources are found. Further, since most of the large population centers of the world are located in the coastal areas, it is here that human society exerts its greatest observable impact on oceanic ecosystems. The importance of the coastal zone has been recognized throughout the world, and in the United States special attention by the Congress has resulted in major legislative acts dealing with this ecological region. Two examples are the Coastal Zone Management Act of 1972 (Public Law 92-583) and the Fisheries Conservation and Management Act of 1976 (Public Law 94-265). Both legislative mandates require the development of resource management plans, however, it has become abundantly clear that the necessary scientific data bases are lacking to enable the formulation of comprehensive and meaning-ful management plans. One of the principal objectives of the Belle W. Baruch Institute for Marine Biology and Coastal Research, University of South Carolina is to gain an understanding of the ecological dynamics of the coastal zone which would help develop the required management plans. To achieve this objective our Institute uses the expertise of approximately 50 senior scientists and scholars representing various scientific disciplines (biology, chemistry, computer science, engineering, environmental health,

geography, geology, physics and statistics) and other subject areas
concerned with the coastal areas (economics, law, political science
and social sciences). In the present paper I will emphasize the
potential role of remote sensing in solving the scientific problems
of the coastal zone, which I define as the environmental continuum
from where rivers become estuaries to a point approximately 200
miles offshore, and ranging from adjacent terrestrial habitat to
benthic regions of the continental shelf.

COASTAL ECOSYSTEMS

     For purposes of study the coastal ecosystem can be subdivided
into two major subsystems: the estuarine-marsh subsystem and the
near-shore oceanic subsystem. Figure 1 represents a simplistic
model of the North Inlet estuarine-marsh subsystem. This subsystem
is located in South Carolina, but it is characteristic of most of
the southeastern region of the United States as well as other
regions of the world. The functioning of this subsystem is the
result of the dynamic interaction of various biotic and abiotic
factors (Summers and McKellar, 1979).
     One of the problems in understanding ecosystem dynamics
revolves around the ability to monitor various environmental parameters, for the relative importance of the numerous parameters,
acting either independently or in concert, are at best poorly
understood. Decisions about what to measure and how, often are
critical, for some significant events in the estuarine-marsh ecosystem occur within a matter of minutes, while others may require
years. To further complicate the process of ecosystem analysis, we
also need to know the specific site where measurements should be
made. For example, the North Inlet Estuary system is relatively
small (a rectangular area approximately 2 miles wide and 6 miles
long) compared either to other estuaries or to near-shore systems
which may be an area 200 miles wide by 600 miles long. Thus, it is
critical to know where representative measurements within these
systems can be made. Remote sensing, whether from satellite,
aircraft, platforms, or instruments in the environment, offers the

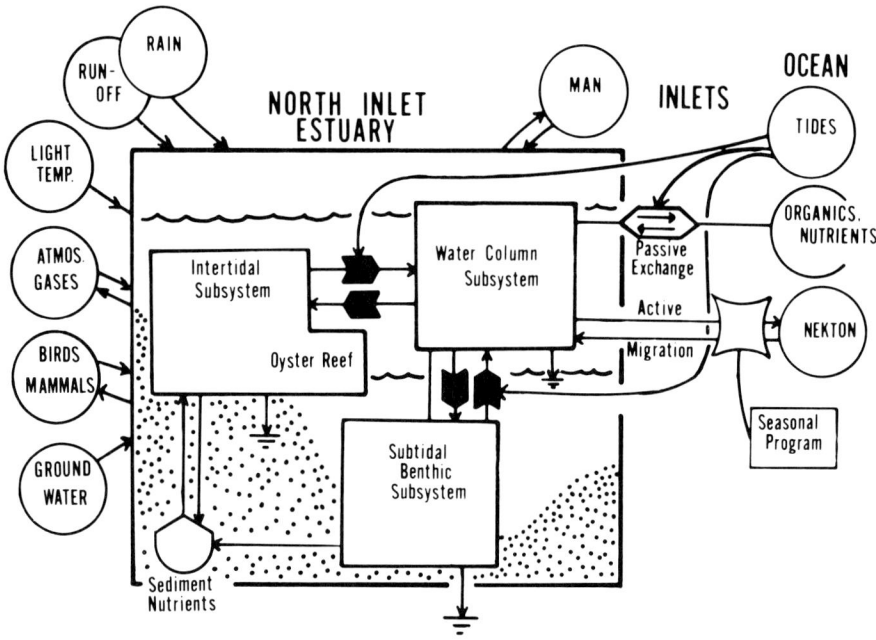

FIGURE 1: Interactions Between Subsystems of the North Inlet
Marsh-Estuarine Ecosystem.

possibility of generating quantities of essential data with minimal
manpower. But perhaps of even greater significance, aerial remote
sensing can monitor large areas instantaneously, so that the
behavior of all parts of a system can be examined at time zero.

Enough data are available to demonstrate that remote sensing
techniques can be used successfully in coastal zone research.
Examples of such research follow, as well as a description of
future needs.

COASTAL ZONE MANAGEMENT

One important use of remote sensing to the various coastal
zone management programs has been in mapping the coastal wetlands
in terms of vegetational types and the amount of each in acres
(e.g., wildlife habitat, mosquito breeding habitats) and the impact
of man-made structure on wetlands. Such studies can be done from

aircraft (Egan and Hair, 1973; Reimold et al., 1973) and skylab imagery and LANDSAT multispectral scanner (Erb, 1974; Anderson et al., 1975), although a number of problems remain to be resolved (Klemas et al., 1980).

PRIMARY PRODUCTIVITY IN COASTAL-ESTUARINE WATERS

One important biochemical function underlying the energetics of the coastal ecosystem is the primary production of organic matter by photosynthetic activity, principally by phytoplankton, vascular plants (such as Spartina), and benthic algae. Traditional methods of measuring primary production require labor intensive field studies which generally sample a relatively small segment of the environment at a specific time. Hence little information on spatial heterogeniety is obtained and estimates of production for a large water mass may have limited validity. Some examples of the application of remote sensing technique to determine primary production follow.

Phytoplankton

In aquatic systems, the density and diversity of phytoplankton is an indicator of the productivity of a given water mass. To help obtain a synoptic measurement of phytoplankton productivity, a remote sensing device, an airborne fluorosensor, was developed by NASA, Langley Research Center and tested in Narragansett Bay, with ground truth provided by personnel from the University of Rhode Island who manned 11 field stations (Farmer et al., 1980). Fluorosensor data were obtained from a helicopter either when hovering over a boat station or when flying between stations. Remote data obtained from the hover position showed excellent correlation with in situ measured in vivo fluorescence, chlorophyll a, and total cell count; data obtained in flight correlated well with in situ fluorescence. The distribution of phytoplankton was not uniform with more patchiness in certain regions than in others. In addition, different species appeared to occupy various regions of the bay, i.e., golden-brown population was found throughout the western half with green species found elsewhere. This technique holds

promise in answering the need of the ecosystem scientist for estimating phytoplankton productivity.

Salt Marsh Vascular Plants

Not only have remote sensing techniques been used to map salt
marsh vegetation types, but they are used to detect seasonal
changes in canopy cover and in biomass determinations (Bartlett and
Klemas, 1977; Klemas et al., 1978). Using tonal variations in
color-infrared aerial photography, Reimold et al. (1973) reported
that general biomass classes of Spartina could be established.
Klemas et al. (1980) felt that remote sensing techniques holds
promise for estimating emergent green biomass and would permit more
cost-effective and extensive sampling than by current methods of
harvesting quadrants.

Benthic Algae

The contribution of benthic algae to the total primary production of estuaries is relatively unknown compared to studies on
vascular plants and phytoplankton. One of the principal reasons
for this paucity of data is: because microalgae are not distributed homogeneously, a large number of samples would have to be
taken and analyzed to provide a suitable estimate of the production
of the total area. This would be labor intensive and costly. To
test the feasibility of using remote sensing techniques to estimate
benthic algal production, Jobson et al. (1980) used a multispectral
scanner mounted on a tower at the edge of the North Inlet Estuary
near Georgetown, South Carolina. Briefly the experimental procedure consisted of running the scanner across the surface of the
intertidal mud flat during low tide on either a uniformly overcast
or a clear day. Following this initial run, a series of cores were
removed from the area which had been scanned and these cores were
analyzed for chlorophyll (chlorophyll concentrations is an
indicator of algal biomass). At each site that a core was removed,
a white styrofoam cup was inserted in the resultant depression and
then the scanner was run through a second cycle. The chlorophyll
analysis provided the necessary ground truth to calibrate the
scanner which could then estimate the distribution of chlorophyll

over the entire mud flat transect. Although a number of problems were encountered, Jobson et al. (1980) found a statistically significant linear inverse relationship between chlorophyll concentration and radiance levels in a blue spectral band. Their results suggested that it might be possible to use aircraft remote sensing to estimate benthic microalgae production in a wider sector of the estuary than possible with a scanner mounted on a fixed tower.

Other Ecosystem Components

In addition to estimating the level of primary productivity on a seasonal, yearly, and long-term basis, other segments of the ecosystem need to be studied and remote sensing techniques could be useful. Estimates of the zooplankton and carnivore standing crop and diversity are needed to determine grazing levels, secondary production, and predator-prey relations. The movements of nektonic species between estuaries and the open ocean represent a dynamic flux of biomass which is extremely difficult to estimate by current methods. Acoustical techniques are being developed but we need new levels of sophistication (Harden Jones, 1980; Holliday, 1980). Time-series analysis of estimates of various physical and chemical parameters are required to understand the dynamic interaction of various ecosystem components.

Organismic Studies

The functional responses of marine organisms monitored in the field under "normal" conditions have not become common in the physiological ecology literature, although the movements of some animals and their responses to environmental changes have been studied (various marine animals, Myrberg, 1980; whales, Watkins et al., 1980). At the physiological-biochemical level, the in situ determination of heart rate, blood flow, body temperature, and osmoregulation have been determined by various remote sensing instruments. However, a definite need exists to employ existing technology and to develop more sensitive and "field hardy" instruments for determining the functional responses of organisms to changing environments.

ACKNOWLEDGMENTS

A portion of this work was supported by a grant (DEB 7683010) from the National Science Foundation.

REFERENCES

Anderson, R. R., L. Alsid, and V. Carter, 1975. Comparative utility of LANDSAT-1 and Skylab data for coastal wetland mapping and ecological studies. In: Proc. of NASA Earth Resources Survey Symp., Vol. III, Summary Reports, pp. 469-478. NASA, Houston, Texas.

Bartlett, D. and V. Klemas, 1977. Variability of wetland reflectance and its effect on automatic categorization of satellite imagery. Proc. Am. Soc. Photogram. 43rd Annual Meeting, pp. 70-89. Washington, D. C.

Egan, W. G. and M. E. Hair, 1973. Automated delineation of wetlands in photographical remote sensing. In: Pro. of the Seventh International Symposium on Remote Sensing of the Environment, pp. 2231-2251. University of Michigan, Ann Arbor, Michigan.

Erb, R. B., 1974. The ERTS-1 investigation (ER-600): ERTS-1 coastal/estuarine analysis. NASA Report TMX-58118. Houston, Texas. 258 pp.

Farmer F. H., C. A. Brown, Jr., O. Jarrett, Jr., J. W. Campbell, and W. L. Staton, 1980. Remote Sensing of Phytoplankton Density and Diversity Using an Airborne Fluorosensor In: Advanced Concepts in Ocean Measurements for Marine Biology. F. P. Diemer, F. J. Vernberg, and D. L. Mirkes, eds. pp 151-168. University of South Carolina Press, Columbia, SC.

Harden Jones, F. R., 1980. Acoustics and the Fisheries: Recent Work with Sector-Scanning Sonar at the Lowestoft Laboratory. In: Advanced Concepts in Ocean Measurements for Marine Biology. F. P. Diemer, F. J. Vernberg, and D. L. Mirkes, eds. pp 409-422. University of South Carolina Press, Columbia, SC.

Holliday, D. V., 1980. Use of Acoustic Frequency Diversity for Marine Biological Measurements. In: Advanced Concepts in Ocean Measurements for Marine Biology. F. P. Diemer, F. J. Vernberg, and D. L. Mirkes, eds. pp 423-460. University of South Carolina Press, Columbia, SC.

Jobson, D. J., R. G. Zingmark, and S. J. Katzberg, 1980. Remote Sensing of Benthic Microalgal Biomass with a Tower-Mounted Multispectral Scanner. Remote Sensing of Environment 9: 351-362.

Klemas, V., D. S. Bartlett, and W. Philpott, 1978. Remote sensing of coastal environment and resources. Proc. Coastal Mapping Symp., Am. Soc. Photogrammetry.

Klemas, V., D. S. Bartlett, W. D. Philpot, 1980. Remote Sensing of Marine Fisheries Resources. In: Advanced Concepts in Ocean Measurements for Marine Biology. F. P. Diemer, F. J.

Vernberg, and D. L. Mirkes, eds. pp 205-226. University of South Carolina Press, Columbia, SC.

Myrberg, A. A., 1980. Ocean Noise and the Behavior of Marine Animals: Relationships and Implications In: Advanced Concepts in Ocean Measurements for Marine Biology. F. P. Diemer, F. J. Vernberg, and D. L. Mirkes, eds. pp 461-492. University of South Carolina Press, Columbia, SC.

Reimold, R. J., J. L. Gallagher, and D. E. Thompson, 1973. Remote sensing of tidal marsh. Photogrammetric Engineering 39: 477-489.

Summers, J. K. and M. McKellar, Jr., 1979. A simulation model of estuarine subsystem coupling and carbon exchange with the sea. I. Model structure. In: State-of-the-art in ecological modelling. S. E. Jorgensen (ed.) pp 323-366. Pergamon Press, Ltd. Oxford.

Watkins, W. A., D. Wartzok, H. B. Martin, III, and R. R. Maiefski, 1980. A Radio Whale Tag In: Advanced Concepts in Ocean Measurements for Marine Biology. F. P. Diemer, F. J. Vernberg, and D. L. Mirkes, eds. pp 227-241. University of South Carolina Press, Columbia, SC.

Considerations in Using the Truncated Normal Distribution for Remote Sensing Data

Douglas J. de Priest

ABSTRACT

The presence of clouds has been a formidable problem when estimating ocean parameters from satellite derived data. This paper discusses the truncated normal model which has an intuitive appeal for estimating under certain restrictive conditions. In particular, a procedure is described for estimating clear radiances in the presence of clouds when IR sensors are utilized and goodness-of-fit statistics are used for testing that the satellite data are distributed as a truncated normal distribution. Some results of a Monte Carlo power study for some goodness-of-fit statistics are presented for specified alternative distributions.

51

INTRODUCTION

Satellites with infrared (IR) radiometers have been used to provide both day and night coverage of the earth. For example, IR sensor systems have been used for several years to assist in the determination of sea surface temperatures in cloud free areas. On earlier systems, IR images of various ocean parameters were of questionable quality. There were several factors which contributed to the problem. These limiting factors included the following: (1) limited spatial resolution, (2) electronic instrument noise which introduced significant spatial and temporal errors, (3) atmospheric absorption of the IR signals, and (4) cloud cover of the sea surface (Legeckis, 1978).

This paper describes a procedure for estimating clear radiances in the presence of clouds when IR radiometers are utilized as sensors and goodness-of-fit statistics are used to test that the data are distributed as a truncated normal distribution. One of the advantages of using the truncated normal distribution is that it is a relatively simple model to use to estimate clear radiances when the fields of view are partially covered by clouds. The key idea is to use measurements from the clear areas to estimate clear radiance associated with certain ocean parameters. A motivation for considering the truncated normal model can be understood by studying Figures 1, 2, and 3. Figure 1 (Bower et al., 1971) shows a comparison of histograms or sample frequency distribution of IR temperature values with and without contamination. Particular attention should be given to part (b) of Figure 1 because it characterizes the situation this paper attempts to address. Figure 2 is a histogram of counts of satellite data (NOAA Satellite) which appear to be approximately uniformly distributed. Here the data are radiance measurements from counts which are easily converted to correspond to parameters of interest. Figure 3 consists of a histogram of counts of satellite data which appear to be normally distributed in the tail (truncated normal).

Parameter Estimation

The following discussion considers a procedure for estimating

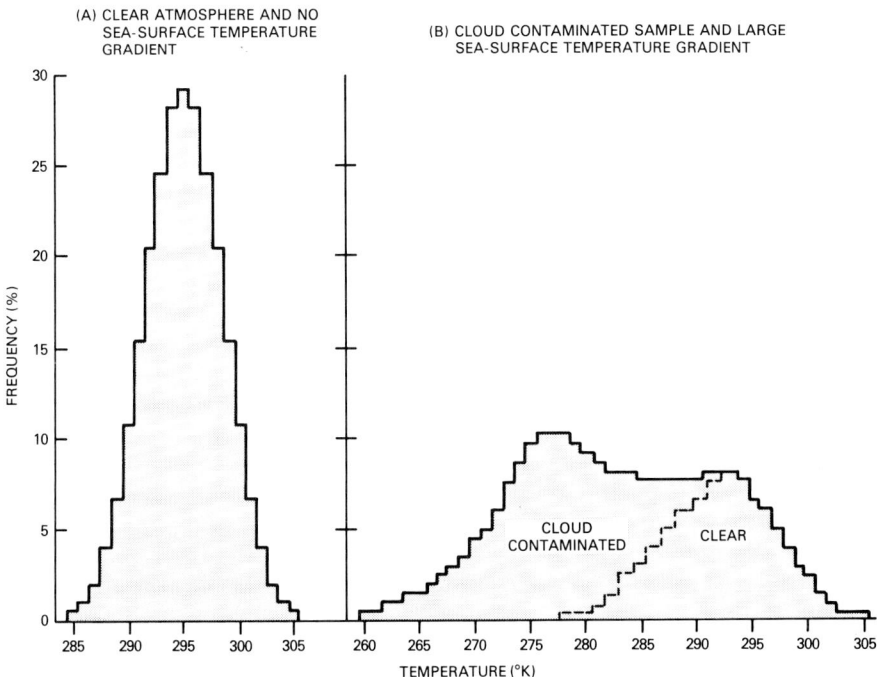

FIGURE 1: Sample frequency distributions of IR temperature values with and without cloud contamination.

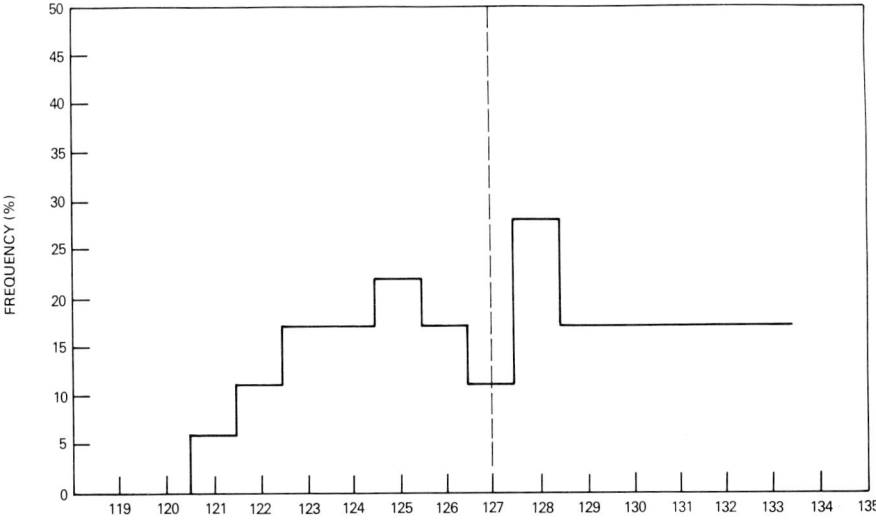

FIGURE 2: Histogram of counts of satellite data which appears to be approximately uniformly distributed.

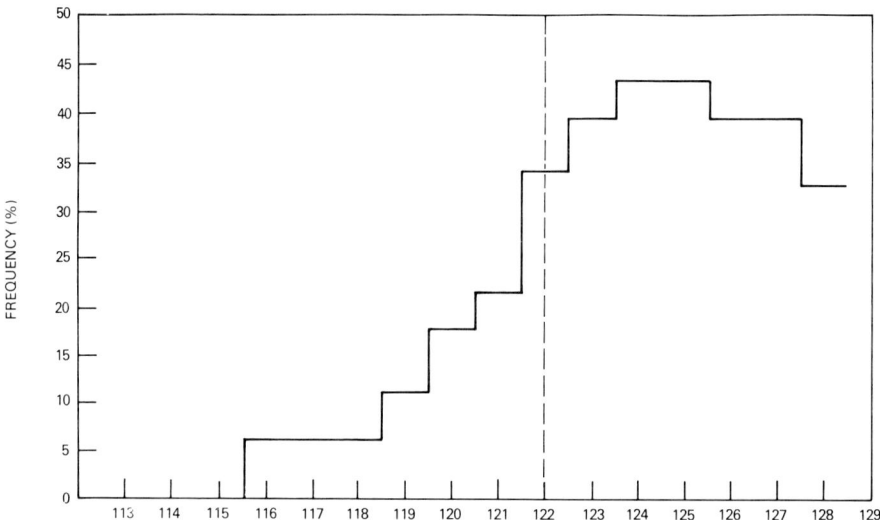

FIGURE 3: Histogram of counts of satellite data which appears to be normally distributed in the tail.

the clear radiance when the normal model is contaminated by data from cloudy areas. The truncated normal model is appropriately used when the following conditions are satisfied: (1) there is a sufficient number of neighboring measurements where the radiance is the same in the absence of clouds, (2) the presence of clouds lowers the radiance, (3) some fields-of-view are cloud free, and (4) the difference between radiances from two cloudfree fields-of-view is due only to instrument noise whose distributions are assumed normal with known standard deviation. (Crosby, et al., 1976).

The density of radiance Y from the clear areas is assumed to be normal

$$f(y) = (2\pi\sigma^2)^{-\frac{1}{2}} \exp \{-(y-\mu)^2/2\sigma^2\} \quad -\infty < y < \infty \qquad (1)$$

the standard deviation σ is assumed to be known and is a function only of the instrument noise. When clouds are present the truncation point T is approximated or can be computed. The problem then is to obtain a good estimate of μ. The parameter μ in equation 1 represents an estimate of the radiance if there were no noise and the fields-of-view were cloud free. More specifically, let Y be the radiance above the truncation point T. Then Y has density

$$p(y) = \exp\{-(y-\mu)^2/2\sigma^2\}/\int_T^\infty \exp\{-(t-\mu)^2/2\sigma^2\}dt, \quad y > T \qquad (2)$$

$$= 0, \text{ otherwise.}$$

The cumulative distribution of Y is given by

$$F(y) = \{\phi((y-\mu)/\sigma) - \phi((T-\mu)/\sigma)\}/\{1-\phi((T-\mu)/\sigma)\} \quad y > T \qquad (3)$$

The function ϕ is the standard cumulative normal given by

$$\phi(x) = (2\pi)^{-\frac{1}{2}} \int_{-\infty}^x \exp(-t^2/2) \, dt \qquad (4)$$

In equation (2), $p(y)$ is called a truncated normal density. For a comparison of the standard normal and the standard truncated normal distributions see Figure 4.

To estimate the mean μ of the uncontaminated distribution, maximum likelihood methods are used on the truncated normal distribution. In order to simplify the notation in equation (2), we make a change of variables by letting $Z = (Y-T)/\sigma$ and $-V = (\mu-T)/\sigma$. Then the density of Z is given by

$$q(z) = \exp\{-(Z+v)^2/2\}/\int_0^\infty \exp\{-(s + v)^2/2\}ds, \quad z > 0 \qquad (5)$$

Given n measurements of $Z(Z_i = (Y_i-T)/\sigma, \ i = 1,2,\ldots,n)$, the problem now is to estimate v. It is not difficult to show that the maximum likelihood estimate of v, denoted $\hat{v}$, is given by the solution to the equation

$$-\hat{v} + \exp(-\hat{v}^2/2)/\int_{\hat{v}}^\infty \exp(-s^2/2)ds = \bar{Z} \qquad (6)$$

where
$$\bar{Z} = \sum_{i=1}^n \bar{Z}_i/n.$$

The functional relationship between $\hat{v}$ and $\bar{Z}$ is plotted for a range of $\bar{Z}$ values in Figure 5. The estimate $\hat{\mu}$ of the true clear column radiance is found by

$$\hat{\mu} = -\sigma\hat{v} + T. \qquad (7)$$

Equation (7) is a consequence of the invariance property of maximum likelihood estimators. A more general discussion of estimating parameters of the truncated normal distribution I presented in Kotz and Johnson (1971).

DENSITY FUNCTIONS

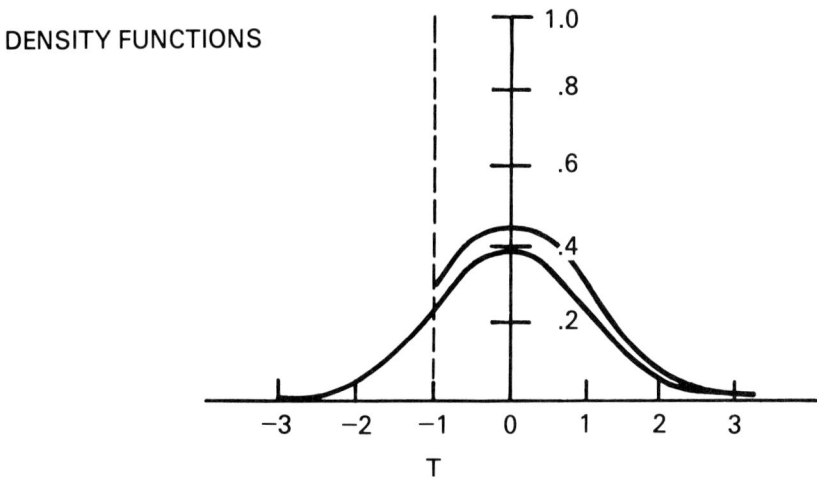

CUMULATIVE DISTRIBUTION
FUNCTIONS

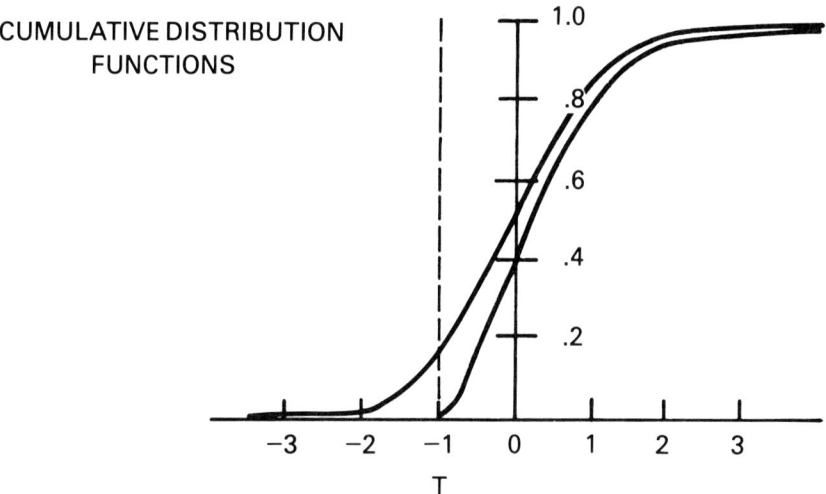

FIGURE 4: Comparison of Standard Normal and Standard Truncated
Normal Distributions.

Testing Goodness of Fit

In the discussion of the model, it was assumed that the data
are distributed as the truncated normal distribution. In this
section, goodness-of-fit tests are used to validate this assumption
(goodness-of-fit tests can be used to assess which sets of satel-

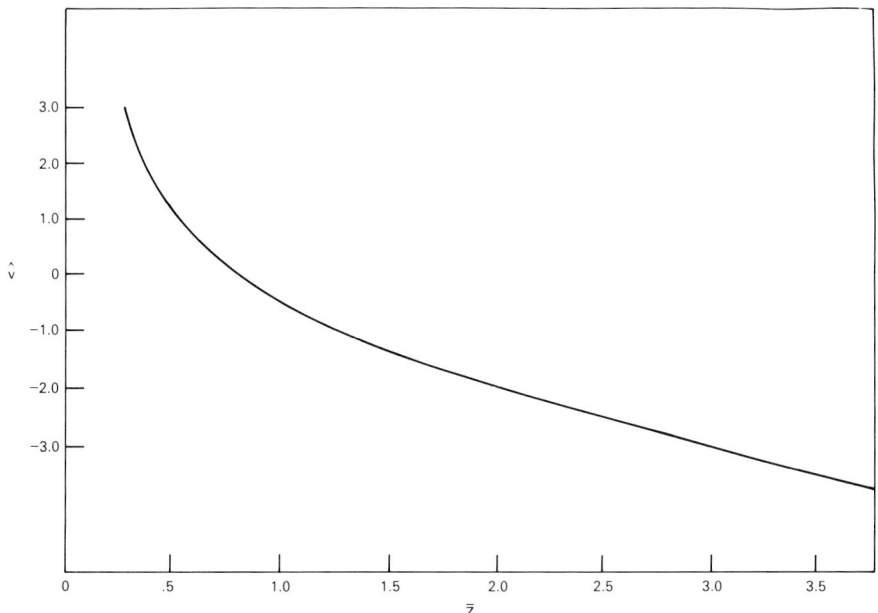

FIGURE 5:   Graphical illustration of the Functional Relationship between $\bar{Z}$ and $\hat{v}$.

lite data are too contaminated for operational use). Statistically, the problem can be formulated in the following way: let x_1, $x_2, \ldots, x_n$ be a random sample of size n from some unknown distribution $F(x)$ where x_i are greater than some known truncation point T. It is desirable to test the null hypothesis H_o that $F(x)$ is a truncated normal distribution with σ known, that is

$$Ho: \quad F(x) = \bar{F}(x)$$

Now let $F(x)$ be estimated by

$$F(x) = \{\Phi((x-\hat{\mu})/\sigma) - \Phi((T-\hat{\mu})/\sigma\} / \{1-\Phi((T-\hat{\mu})/\sigma)\} \quad x > T \qquad (8)$$

Where $\hat{\mu} = -\sigma \hat{v} + T$, $\hat{v}$ is the maximum likelihood estimate of v given by the solution to equation (7) and Φ is the standard cumulative normal.

The goodness-of-fit statistics used to test the null hypothesis, Ho are the following:

(a)   Chi-square $$X^2 = \sum_{i=1}^{k} (x_{oi}-m_{oi})^2/m_{oi}$$

(b) Kolmogorov-Smirnov

$$D = \sup_{x} \left| \bar{F}(x) - S_n(x) \right|$$

$$= \max \{\max(\bar{F}(x_{(i)}) - (i-1)/n), \max(i/n - \bar{F}(x_{(i)}))\}$$

(c) Cramer-Smirnov-Von Mises

$$NW2 = n \int_{-\infty}^{\infty} \{\bar{F}(x) - S_n(x)\}^2 d\,\bar{F}(x)$$

$$= 1/12n + \sum_{i=1}^{n} \{\bar{F}(x_{(i)}) - (2i-1)/2n\}^2$$

(d) Anderson-Darling Weighted NW2

$$WNW2 = \int_{-\infty}^{\infty} \{[\bar{F}(x) - S_n(x)]^2 / [\bar{F}(x)\,(1-\bar{F}(x)(1-\bar{F}(x))]\}d\,\bar{F}(x)$$

$$= -n - \sum_{i=1}^{n} \{(2i-1)\log_e \bar{F}(x_{(i)}) + (2n-2i+1)\log_e(1-\bar{F}(x_{(i)}))\}/n$$

(e) Kuiper

$$VN = \sup_{x}(F(x) - S_n(x)) + \sup_{x}(S_n(x) - F(x))$$

$$= \max(F(x_{(i)}) - (i-1)/n) + \max(i/n - F(x_{(i)}))$$

In (b) through (e) above, $S_n(x)$ is the empirical distribution function (EDF) defined as

$$S_n(x) = \begin{array}{ll} 0 & x < X_{(1)} \\ r/n & X_{(r)} \leq x < X_{(r+1)} \\ 1 & X_{(n)} \leq x \end{array}$$

where $X_{(r)}$ are the order statistics.

It is known in statistics that when parameters of the hypothetical distribution are estimated, the established tables of critical values for all the goodness-of-fit statistics considered in this paper are not appropriate except the chi-square statistic. In the chi-square test, by simply reducing the number of degrees of freedom by the number of parameters estimated and replacing the unknown parameters with consistent estimators, the test can be performed in the usual way. The chi-square statistics, X^2, is computed using equal class intervals. For a specific level of significance α, the chi-square test rejects Ho if the computed X^2 is larger than the critical value found in an appropriate table.

When the parameters of the hypothetical distribution are

estimated, the distribution functions of these statistics are very
difficult to derive. To circumvent these difficulties, Monte Carlo
experiments were conducted to approximate the distributions of
these statistics. From these approximate distributions, critical
values in terms of quantiles were obtained to test goodness-of-fit.

In all, 5,000 samples of size n were generated from a standardized normal distribution and truncated at several specified
points T. The sample values above the truncated point were used to
estimate the mean of the original normal distribution using maximum
likelihood procedures. The values of the statistics D, NW2, WNW2
and VN were obtained and ordered from maximum to minimum to get the
approximate critical values corresponding to .01, .05, .10, .15,
and .20 significance levels. These critical values are presented
in tables in terms of quantiles p where p=1 - α. Tables 1 through
4 represent approximate critical values in terms of quantiles p of
the D, NW2, WNW2 and VN statistics respectively when sampling from
the standard truncated normal where the point of truncation is -1
and the sample size is n. In using, say the D statistic to test
the null hypothesis Ho (i.e., the random sample of size n has the
truncated normal distribution with unspecified mean when T = -1),
the procedure is to reject Ho at the approximate level of significance α if the computed D is greater than the pth quantile as
given in Table 1. The same procedure is used to test goodness-of-
fit under the hypothesis Ho with the other EDF statistics. For
further discussion on EDF statistics for testing goodness-of-fit,
see the reference by Stephens (1974). The chi-square test is given
a thorough development in a paper by Cochran (1952).

Results from the Power Study of Some EDF Statistics

In considering the various statistics available for testing
goodness-of-fit, it is reasonable to ask which one is preferred
over the others. Of particular interest is to know how well a test
discriminates or performs under the assumption of some specified
alternative hypothesis. This brings us to the area of investigating the power. The power is a measure of how good the test, based
on a certain statistic, rejects Ho when the true distribution is
assumed to be some specified alternative under Ha.

Monte Carlo experiments were also conducted to study the power of each test statistic based on the empirical distribution function (EDF). Additional 1,000 samples of sizes 20, 25, 30, 40, and 50 were generated from the uniform, exponential, and truncated compound normal distributions. Tables 5 and 6 show the powers at the 5 percent and 10 percent significance levels respectively when the truncation point T = -1. In simulating the power for a test statistic when a specified alternative, Ha, is true, the major change in the computer program occurs in the sampling algorithm. That is, the algorithm which generates truncated normal values is replaced

TABLE 1: QUANTILES OF THE D STATISTIC WHEN T = -1.

p n	.99	.95	.90	.85	.80
5	.4590	.3975	.3649	.3451	.3291
10	.3248	.2852	.2618	.2475	.2363
11	.3123	.2733	.2524	.2372	.2268
12	.3015	.2614	.2415	.2281	.2179
13	.2920	.2527	.2326	.2199	.2086
14	.2819	.2446	.2228	.2105	.2015
15	.2699	.2368	.2169	.2048	.1963
16	.2650	.2269	.2096	.1975	.1895
17	.2619	.2236	.2046	.1920	.1841
18	.2525	.2152	.1979	.1869	.1787
19	.2497	.2115	.1935	.1813	.1737
20	.2402	.2022	.1869	.1756	.1690
25	.2189	.1855	.1686	.1599	.1529
30	.1995	.1702	.1553	.1470	.1402
35	.1866	.1571	.1448	.1363	.1302
40	.1778	.1469	.1355	.1270	.1210
50	.1474	.1316	.1202	.1126	.1069
75	.1237	.1073	.0972	.0904	.0858
100	.1045	.0921	.0827	.0774	.0741
over 100 n	$1.045/n^{\frac{1}{2}}$	$.921/n^{\frac{1}{2}}$	$.827/n^{\frac{1}{2}}$	$.774/n^{\frac{1}{2}}$	$.741/n^{\frac{1}{2}}$

TABLE 2: QUANTILES OF THE NW2 STATISTIC WHEN T = -1.

p n	.99	.95	.90	.85	.80
5	.2491	.1758	.1433	.1242	.1097
10	.2277	.1642	.1353	.1178	.1050
11	.2342	.1652	.1332	.1158	.1052
12	.2300	.1630	.1331	.1162	.1033
13	.2282	.1663	.1324	.1153	.1029
14	.2289	.1654	.1331	.1166	.1030
15	.2276	.1648	.1351	.1174	.1042
16	.2382	.1624	.1333	.1159	.1021
17	.2362	.1668	.1348	.1165	.1050
18	.2302	.1666	.1347	.1169	.1040
19	.2358	.1651	.1364	.1178	.1041
20	.2315	.1611	.1346	.1168	.1021
25	.2318	.1627	.1336	.1171	.1046
30	.2381	.1660	.1378	.1191	.1066
35	.2437	.1654	.1366	.1183	.1050
40	.2580	.1678	.1351	.1175	.1042
50	.2316	.1692	.1339	.1110	.0977
75	.2277	.1592	.1233	.1049	.0937
100	.2136	.1594	.1361	.1155	.1009

with one which generates random values for the specified alternative distribution. Since the power tables are the percentage of significances out of one-thousand samples of size n, an algorithm for counting significant values was also included as part of the computer program.

TABLE 3: QUANTILES OF THE WNW2 STATISTIC WHEN T = -1.

p n	.99	.95	.90	.85	.80
5	1.6277	1.1402	.8926	.7823	.7017
10	1.5125	1.0578	.8638	.7600	.6833
11	1.5001	1.0628	.8608	.7548	.6866
12	1.5144	1.0460	.8390	.7387	.6741
13	1.5284	1.0686	.8602	.7467	.6733
14	1.4828	1.0543	.8578	.7493	,6791
15	1.5449	1.0585	.8685	.7605	.6867
16	1.4900	1.0357	.8588	.7504	.6746
17	1.4901	1.0882	.8772	.7629	.6887
18	1.5163	1.0642	.8745	.7711	.6848
19	1.5124	1.0779	.8844	.7716	.6879
20	1.4490	1.0441	.8844	.7664	.6808
25	1.4960	1.0782	.8780	.7782	.6898
30	1.5487	1.1042	.9107	.7887	.7009
35	1.6052	1.1007	.9162	.7935	.7002
40	1.6832	1.1196	.9061	.7845	.7045
50	1.6475	1.1245	.9187	.7795	.6840
75	1.5643	1.0428	.8253	.7358	.6588
100	1.5356	1.0878	.7909	.6899	.6293

TABLE 4:   QUANTILES OF THE VN STATISTIC WHEN T = -1

p n	.99	.95	.90	.85	.80
5	.7379	.6753	.6292	.5948	.5654
10	.5490	.4877	.4559	.4320	.4134
11	.5328	.4692	.4339	.4116	.3946
12	.5132	.4459	.4150	.3949	.3773
13	.4890	.4321	.3992	.3794	.3645
14	.4780	.4166	.3868	.3620	.3516
15	.4560	.4050	.3774	.3567	.3430
16	.4339	.3914	.3636	.3439	.3297
17	.4329	.3840	.3547	.3365	.3227
18	.4201	.3701	.3441	.3273	.3132
19	.4163	.3637	.3375	.3189	.3060
20	.3972	.3500	.3246	.3088	.2966
25	.3622	.3167	.2936	.2799	.2680
30	.3315	.2925	.2709	.2574	.2463
35	.3056	.2715	.2514	.2380	.2290
40	.2937	.2536	.2339	.2214	.2130
50	.2486	.2265	.2065	.1954	.1860
75	.2032	.1831	.1695	.1582	.1503
100	.1750	.1629	.1467	.1377	.1309
over 100 n	$1.75/n^{\frac{1}{2}}$	$1.63/n^{\frac{1}{2}}$	$1.47/n^{\frac{1}{2}}$	$1.38/n^{\frac{1}{2}}$	$1.31/n^{\frac{1}{2}}$

TABLE 5: Percentage Power for D, NW2, WNW2, and VN as Tests for
Truncated Normality When the Mean is Estimated, p = .95 and T = -1.

		D	NW2	WNW2	VN
Alternative	n				
	20	.31	.39	.52	.37
	30	.42	.53	.66	.51
Uniform(-1,3)	40	.57	.69	.80	.67
	50	.70	.80	.89	.79
	75	.87	.95	.98	.94
	20	.45	.46	.53	.30
	30	.54	.58	.63	.40
Exponential (-1,6)	40	.64	.68	.72	.49
	50	.74	.78	.81	.59
	75	.87	.90	.93	.77
	20	.19	.22	.33	.20
	30	.24	.30	.44	.27
Truncated Compound	40	.32	.38	.52	.35
Normal (-1,5)					
	50	.42	.50	.62	.46
	75	.60	.69	.80	.64

TABLE 6: Percentage Power for D, NW2, WNW2, and VN as Tests for Truncated Normality When the Mean is Estimated, p = .90 and T = -1.

Alternative	n	D	NW2	WNW2	VN
	20	.42	.48	.60	.47
	30	.55	.63	.75	.61
Uniform(-1,3)	40	.68	.79	.88	.77
	50	.81	.88	.93	.87
	75	.94	.98	.99	.97
	20	.53	.54	.60	.40
	30	.64	.65	.70	.49
Exponential (-1,6)	40	.73	.76	.80	.60
	50	.81	.84	.86	.71
	75	.92	.95	.96	.87
	20	.28	.30	.41	.28
	30	.36	.41	.52	.38
Truncated Compound	40	.45	.52	.64	.48
Normal (-1,5)					
	50	.57	.63	.71	.61
	75	.72	.80	.86	.74

SUMMARY

In summary, it should be mentioned that various constraints and other considerations have been described in utilizing the truncated normal model for remotely sensed data. This paper considered only the singly truncated normal distribution. However, it should be mentioned that procedures exist for estimating parameters of a normal population from doubly truncated samples (Cohen, 1950). The physical problem of estimating ocean parameters in the presence of clouds led to considering the singly truncated normal model

which gives reasonably good estimates. Specifically, algorithms
have been developed which estimate the parameter μ (mean clear
radiance) and test goodness-of-fit of a random sample of size n
assumed distributed as a singly truncated normal distribution.
Monte Carlo experiments were conducted to generate appropriate
tables of critical values for the given EDF statistics. In comparing the EDF statistics, it is apparent that the Anderson-Darling
Weighted Statistic generally gives better power results for the
alternative distributions considered in this paper.

REFERENCES

Bower, R., B. Goddard B. J. Lesse, and W. Pichel 1971. An experi-
    mental model for automated detection, measurement and quality
    control of seasurface temperatures from ITOS-IR data, Proceed-
    ings of the Seventh International Symposium on Remote Sensing
    of Environment: 625-647.
Cochran, W. G. 1952. The X^2 test of goodness-of-fit, Annals of
    Mathematical Statistics, 23: 315-345.
Cohen, A. C. 1950. Estimating the mean and variance of normal
    population from singly truncated and doubly truncated samples,
    Annals of Mathematical Statistics, 21: 557-69.
Crosby, D. S., D. J. de Priest 1976. Estimating clear radiances, a
    report on a new decision rule, Seventh Conference on Aerospace
    and Aeronautical Meteorology and Symposium on Remote Sensing
    from Satellites: 101-102.
Kotz, S., N. Johnson 1971. Continuous Univariate Distributions,
    Vol. 1. Houghton Mifflin Company.
Legeckis, R. 1978. A survey of worldwide sea surface temperature
    fronts detected by environmental satellites, Journal of
    Geophysical Research, 83: 4501-4522.
Stephens, M. A. 1974. EDF statistics for goodness-of-fit and some
    comparisons, Journal of the American Statistical Association,
    69: 730-737.

Infrared Transmittance of Marine Atmospheres

Lawrence J. Rouse, Jr.

ABSTRACT

The accuracy with which single-channel satellite infrared radiometers can detect sea-surface temperature differences is discussed. A relationship among sea-surface temperature differences, satellite measures of these differences, atmospheric transmittance in the infrared, and atmospheric water vapor content is presented. A comparison of model calculations with experimental data reveals differences that require further investigation. The dangers associated with infrared remote sensing of sea-surface temperatures in the temperate and tropic regions are also discussed.

67

INTRODUCTION

That remotely sensed thermal infrared data do not always give true values of sea-surface temperature (SST) has been known for many years (Saunders, 1967). The first implication that something other than a simple additive correction was necessary appeared in a paper by Maul and Sidran (1973), who presented the results of theoretical calculations of the effects of atmospheric state, look angle, and random noise on temperature data from the NOAA series satellites. These calculations showed that temperature differences at the sea surface could be sharply reduced in magnitude when determined by satellite-borne radiometers. A most compelling satellite image acquired by the NOAA-3 satellite on 4 April 1975 (Fig. 1) underscored the seriousness of the problem. In this image a rare cloud-free frontal passage over the Florida peninsula reveals the clear-sky obliteration of the temperature differences associated with the Gulf Stream by humid, tropical maritime air.

That SST differences should not always be accurately reproduced by satellite infrared radiometers is evident when the radiative transfer equation is examined. The integral form of this equation in an absorbing and emitting but non-scattering atmosphere is

$$N_o = {}_o\!\int^\infty \phi(\nu)\ N(T_s,\nu)\ \tau_s(\nu)\ d\nu + {}_o\!\int^\infty {}_P\!\int^o_s \phi(\nu)\ N(T,\ P,\nu)$$

$$\frac{\partial \tau_p(\nu)}{\partial p}\ dpd\nu \qquad\qquad (1)$$

where N_o is the radiance received by the sensor, N is the radiance emitted by a blackbody and given by Planck's law, ϕ is the spectral response of the sensor, and τ_p is the transmittance of the atmosphere from the pressure level P to the satellite. The first term on the right side of the equation represents the contribution to the total due to surface emission, while the second term represents the contribution of the atmosphere. Integrating this equation, applying the mean value theorem, and assuming that radiance is a

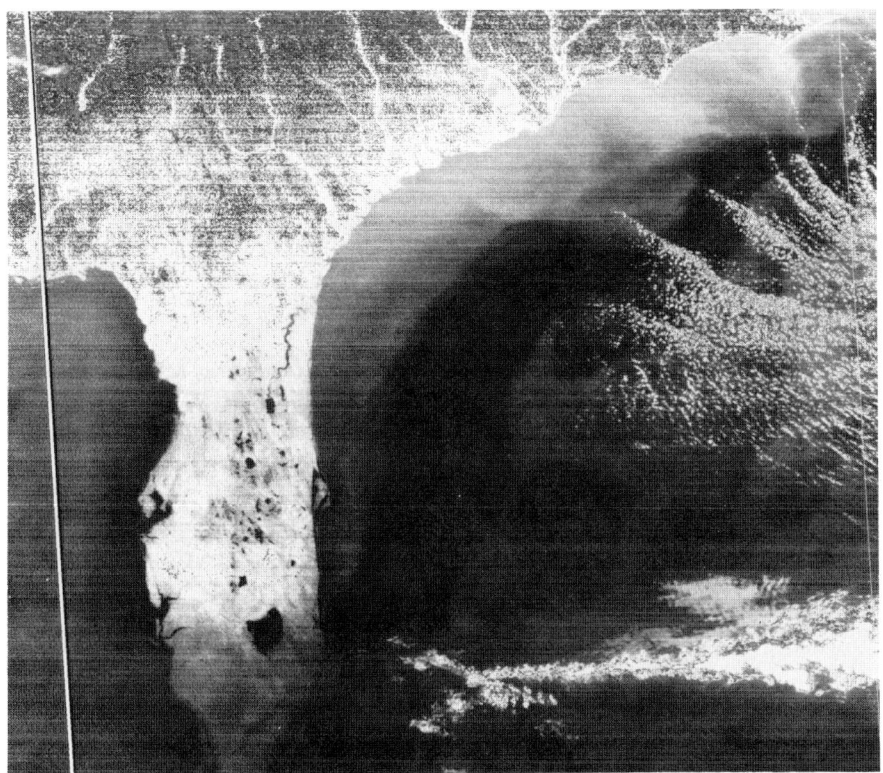

FIGURE 1: NOAA-3 infrared image of Gulf Stream and the southern part of the Florida peninsula, 4 April 1975.

linear function of temperature in the spectral region of the sensor, the temperature observed by the satellite can be written

$$T_o = \tau T_s + (1-\tau)\,\theta \qquad (2)$$

where T_s is the surface temperature, θ is the water-vapor-weighted average atmospheric temperature, and τ is the transmittance of the atmosphere. Assuming that the atmosphere is constant across the area of interest, satellite and surface temperature differences are related by:

$$\Delta T_o = \tau \Delta T_s$$

or

$$\tau = \frac{\Delta T_o}{\Delta T_s} \qquad (3)$$

It should be noted that equations 1, 2 and 3 hold for black-body emission (ε = 1.0). These approximations (eq. 2 and 3) are therefore valid for SST's where the emissivity (ε) is very nearly 1 but not for land temperatures, where ε is significantly less than 1.

Equation 1 can be evaluated for specific atmospheres if transmission as a function of atmospheric absorbers is known. The LOWTRAN computer compilation (Selby et al., 1978) is such a model. The results of using LOWTRAN to calculate temperatures observed by the NOAA-5 satellite radiometer through six model atmospheres (Table 1) are shown in Figure 2. According to equation 3, the slopes of these curves are the atmospheric transmittances. The relationship between the calculated transmittance and total precipitable water (TPW) is shown in Figure 3. The linear fit, forcing the transmittance to be unity at 0 cm of TPW, can be written in the form

$$\tau = 1 - KW \tag{4}$$

where W is the total precipitable water and K = 0.131.

TABLE 1: Model Atmospheres

Atmosphere	TPW (cm)	θ (κ)
Tropical	4.19	289.0
Midlatitude Summer	2.98	284.6
Midlatitude Winter	0.87	264.1
Subarctic Summer	2.12	275.9
Subarctic Winter	0.42	253.4
1962 U.S. Standard	1.44	274.6

A field experiment to gather data to compare with this theoretical relationship has been conducted by the author and his associates. Surface temperatures for two adjacent distinct water

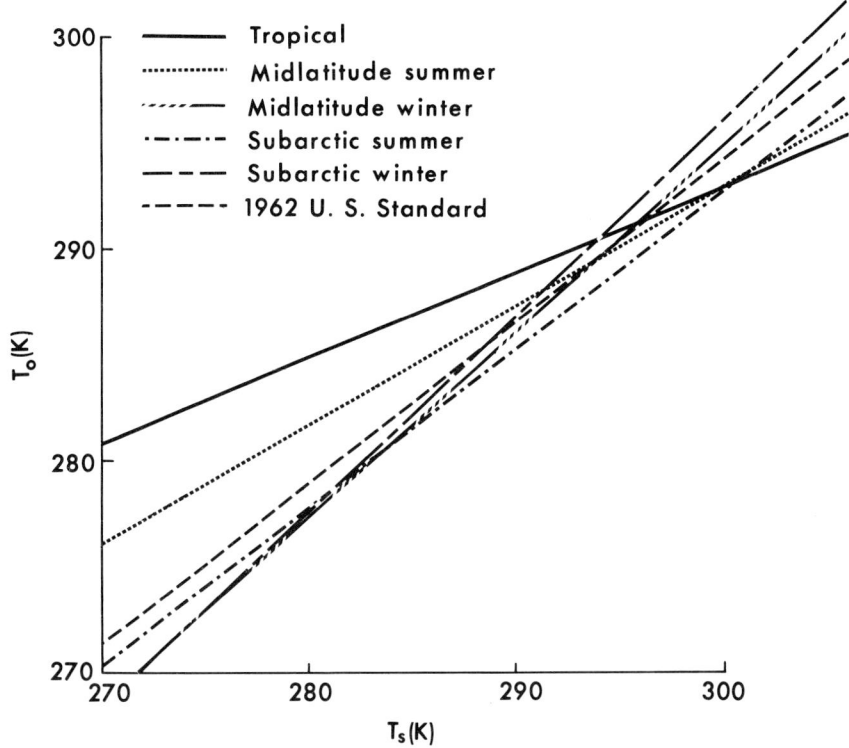

FIGURE 2: LOWTRAN-4 predictions of temperatures observed by the
NOAA-5 IR-VHRR for six model atmospheres.

masses of different temperatures were measured under clear-sky
conditions by surface thermometry and satellite radiometry. Total
precipitable water was calculated from data gathered by radiosondes
launched coincident with the satellite overpass. The results of
these measurements are shown in Figure 3 by the triangles.

    With only five data sets obtained under relatively dry condi-
tions, not much can be said above how a full curve of measured
transmittances would be plotted. Some initial observations are
appropriate, however. Most obvious of all is that the measurements
do not agree with the theoretical curve. Below approximately 1 cm
of total precipitable water the transmittance was found to be
unity. In this range temperature differences were accurately
measured by satellite, though the satellite values of SST were

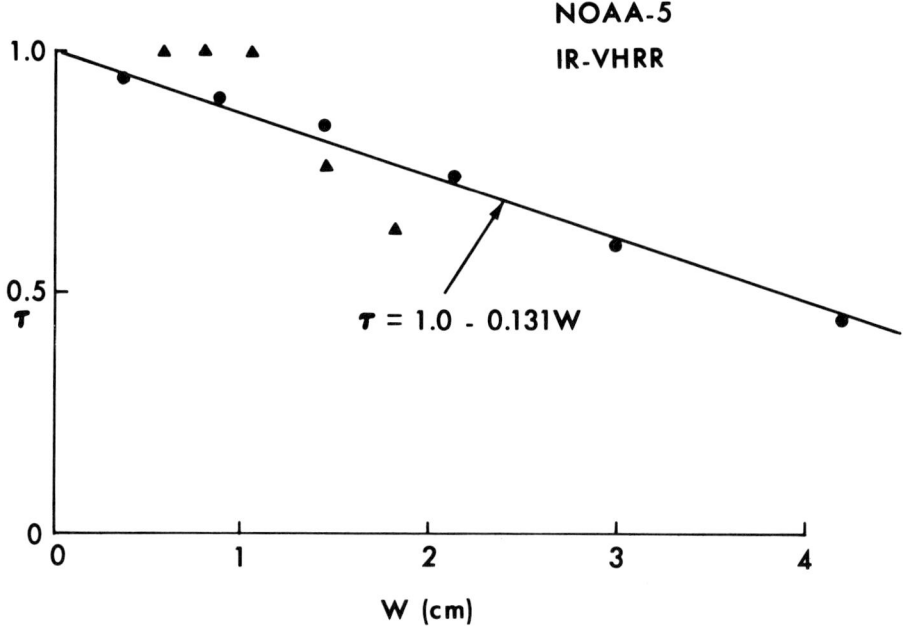

FIGURE 3: Atmospheric infrared transmittance (τ) for the NOAA-5 IR-VHRR as a function of the precipitable water in the atmosphere (W). Triangles represent experimental measurements and dots represent prediction from LOWTRAN-4.

lower than those indicated by surface measurements. For water content greater than 1 cm the transmission coefficient decreased, but further conclusions must await the collection of more data. A relationship similar to that shown in Figure 3 has been determined for the Geostationary Satellite infrared sensor by Maul et al. (1978).

The decrease in transmittance as atmospheric water vapor content increases results in a loss of thermal contrast in satellite-sensed temperature fields. In terms of the digitizations of the temperature data the decrease in transmission results in an effective increase (Υ) in the quantization interval of the sensor (Fig. 4), that is

$$D_E = \Upsilon D. \tag{5}$$

where D is the satellite quantization interval and D_E is the effective quantization interval. For the NOAA-5 infrared sensor model

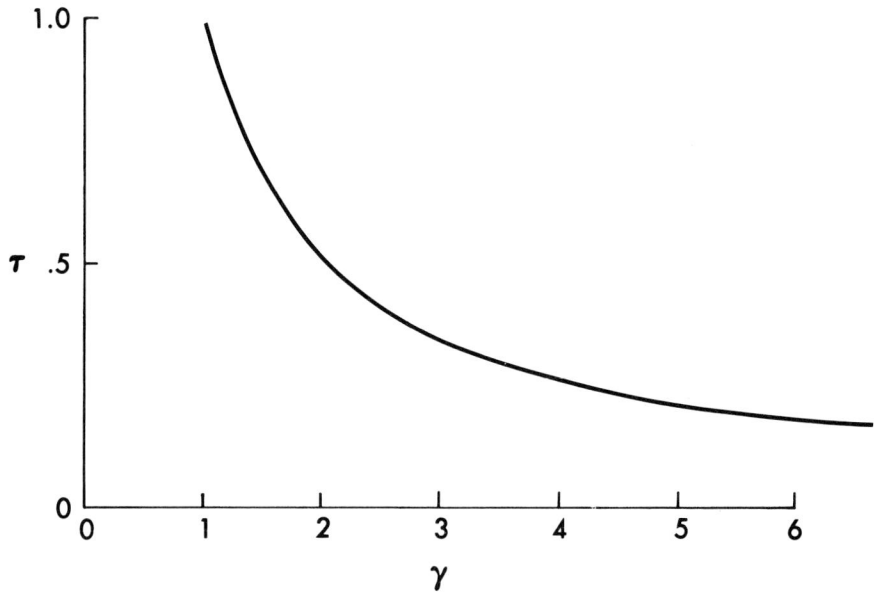

FIGURE 4: Effective quantization coefficient (Υ) as a function of atmospheric infrared transmittance (τ).

value of Υ is two when the TPW is 3.8 cm. This value of TPW is not abnormally high for cloud-free maritime atmospheres, especially in the temperate and tropical regions in the summer months. For these atmospheres ocean thermal fronts with temperature differences of less than 1°C will not be detected, while those fronts that are detected will have a temperature difference twice that indicated by the satellite data.

The danger of looking at the sea through holes in the cloud cover should also be apparent. In the extreme case surface-emitted radiation may be completely masked by emission by atmospheric water vapor so that a water vapor weighted atmospheric temperature is being measured rather than the temperature of the sea surface.

On the other hand, optimum conditions for the detection of SST differences occur when the TPW is less than 1 cm. In the temperate regions this condition occurs over the ocean in the winter months following the outbreak of polar air from the eastern sides of the continents.

The replacement of the NOAA-5 satellite by the TIROS-N series offers the opportunity to improve the accurate mapping of SST features. The first improvement will be based on narrower sensor resolution. The NOAA-5 radiometer was sensitive in the range 10.5 to 12.5 μm, whereas channel 4 of the TIROS-N radiometer is sensitive between 10.5 and 11.5 μ. This reduction in band width results in the reduction in sensitivity to atmospheric water vapor since the higher wavelengths lie near the edge of a water absorption band. Using the LOWTRAN model, calculations of atmospheric transmittance were made for the six standard atmospheres with the response curves for the two infrared channels of TIROS-N. Writing the transmittance in the form of equation 4 yields

$$K = 0.0481$$

for channel 3 (3.5-3.9 μm) and

$$K = 0.0877$$

for channel 4 (10.5-11.5 μm). The improvement from the NOAA-5 sensor to the TIROS-N channel 4 sensor is not very great. At 1.5 cm of TPW the transmittance is increased from 0.80 to 0.87 or, equivalently, the effective quantization (Υ) decreases from 1.26 to 1.18. A further improvement is found in the TIROS-N channel 3. For this sensor the transmittance is above 0.8 until TPW is greater than 4 cm. This channel is, however, contaminated by reflected solar radiation during daytime passes.

The greatest improvement in the determination of SST will result from the method initially proposed by Anding and Kauth (1970) based on the fact that measured temperatures in the two "window" channels will be different. This difference results from different absorptions and implies some information about properties of the intervening atmosphere. Surface temperatures may then be written as a function of the two sensed temperatures without the necessity of any information about the composition of the atmosphere. Further information about this method can be found in Prabhakara et al. (1974), McMillin (1975), and Becker (1978).

ACKNOWLEDGEMENTS

This research was sponsored by the Geography Programs of the Office of Naval Research, Arlington, Virginia, under a contract with Coastal Studies Institute, Louisiana State University. Special appreciation is extended to Oscar K. Huh for many stimulating discussions about satellite radiometry.

REFERENCES

Anding, D., and R. Kauth. 1970. Estimation of Sea Surface Temperature from Space, Remote Sensing Environ., 1: 217-220.

Becker, F. 1978. Temperature Superficielle de la Mer et Transmission Atmospherique par Radiometrie Differentielle. Comptes-Rendus des Journees CNES-CNEXO, Brest, Publ Sci. Techn. CNEXO Actes Collog. No. 5, 141-160.

Maul, G. A., P. W. DeWitt, A. Yanaway, and S. R. Baig. 1978. Geostationary Satellite Observations of Gulf Stream Meanders: Infrared Measurements and Time Series Analysis, J. Geophys. Res., 83 6123-6135.

Maul, G. A., and M. Sidran. 1973. Atmospheric Effects on Ocean Surface Temperature Sensing from the NOAA Satellite Scanning Radiometer, J. Geophys. Res., 78, 1909-1916.

McMillin, L. M. 1975. Estimation of Sea Surface Temperatures from Two Infrared Window Measurements with Different Absorption, J. Geophys. Res., 80, 5113-5117.

Prabhakara, C., G. Dalu, and V. G. Knude. 1974. Estimation of Sea Surface Temperature from Remote Sensing in the 11- 13 μm Window Region, J. Geophys. Res., 79, 5039-5044.

Saunders, P. M. 1967. Aerial Measurement of Sea Surface Temperature in the Infrared, J. Geophys. Res., 72, 4109-4117.

Selby, J. E. A., F. X. Kneizys, J. H. Chetwynd, and R. A. McClatchey. 1978. Atmospheric Transmittance/Radiance: Computer Code LOWTRAN-4. Air Force Geophysical Lab. Tech. Rep-78-0053.

Active Radiation Probing of the Sea Surface Using HF Radar

Dennis B. Trizna

ABSTRACT

A review of the various kinds of experiments involving inferring parameters of the sea surface from high frequency radar data
are discussed. These include mapping of wind direction fields of
weather systems at sea, mapping of ocean currents, measurement of
directional ocean wave spectra using first order radar spectrum
measurements, and inferring directional wave spectrum parameters
using second order radar spectrum features. Comments are also made
on some proposed measurement systems as well as the state of the
theory of HF radar backscatter interpretation.

INTRODUCTION

Radio waves transmitted in the high-frequency (HF) radio band, between 3 and 30 megahertz, are useful for remote sensing at beyond line-of-sight distances because of two propagation effects which appear at these frequencies. These two effects in turn define two modes of over-the-horizon (OTH) propagation, each with different ranges of coverage in distance. The first of these effects is the refraction of radio waves by the ionosphere, effectively coherently reflecting transmitted radio energy back to earth. This energy is scattered by the earth's surface, leaving a characteristic modulation imprint on the radio signal. A fraction of this energy returns via another ionospheric reflection to a second antenna and radar receiver for processing analysis. As the radio frequency is tuned less than thirty megahertz, the refracting properties of the ionosphere cause the radar energy to illuminate shorter ranges. As the frequency is lowered to the order of 3 megahertz, ionospheric absorption at D-region layers prevents propagation to the higher refracting E and F layers and this mode of sky-wave propagation is cut off. Typical ranges made available by this OTH mode under daytime ionospheric conditions range from 500 to 2000 nautical miles from the radar transmission site. A comprehensive review of the development of the HF radar technology at the Naval Research Laboratory has been published recently (Headrick and Skolnik, 1974), as well as a description of applications to sea surface sensing (Barrick, 1978). The first application of HF radar to mapping of primary direction of sea surface waves, and inferred driving winds, over a large area of the North Atlantic Ocean was performed at the Naval Research Laboratory (Long and Trizna, 1973). Since then further developments have taken place which indicate that remote determination and mapping of significant wave height, primary wave direction and possibly the measurement of the directional wave spectrum itself appear feasible using sky-wave OTH radar (Barrick, et al., 1974; Ahearn et al., 1974).

The second effect on wave propagation at HF radio frequencies to be considered is concerned with the low losses encountered when radiation is launched from a vertically-polarized transmitting antenna located at the shoreline of the sea. At the frequencies in the HF band the salt water sea surface appears as a low-loss conducting surface. Because of the boundary conditions associated with Maxwell's Equations of electromagnetism, the radio energy effectively is trapped on the sea surface and propagates again beyond line of sight via the surface-wave propagation mode to OTH distances. Figure 1 shows an example of both HF OTH propagation modes. The maximum range now is no longer determined by the geometry of a reflecting surface, as in the case of ionospheric OTH propagation, but by the intensity of the radio energy transmitted and the propagation losses of the surface. For a surface-wave radar with a transmitting power, antenna, and processing gain product of 10^{10} Joules (100 dBJ), ranges as far as 200 nautical miles have been achieved using low HF radar frequencies. Propagation losses increase with increasing frequency in this case.

Using one of the two OTH propagation modes just defined, the measurement made for remote sensing use is a spectral analysis of reflected radio energy transmitted in a pulsed format. In the most basic format a shaped pulse train is transmitted, coherent on a pulse-to-pulse basis, whose corresponding frequency spectrum is very narrow, with low-frequency side lobes. The amplitude of the reflected pulse train is sampled a number of times in the period between consecutive transmitted pulses over a series of time increments nearly equal to the pulse length. Each of these time increments defines a range bin. A time history of reflected amplitudes is stored for each of the range bins. When a time series is sufficiently long to provide useful Doppler frequency resolution, say from 25 to 100 seconds, the time series is transformed to yield a power frequency spectrum for each defined range bin. The features of this spectrum are then analyzed for information about the scattering surface. An example Doppler spectrum is shown in Figure 2.

If there were only stationary scatters in a given range cell, the frequency spectrum of the reflected energy would be identical to that transmitted, with no Doppler shifts superimposed. In the case of reflections from the sea surface, the reflections are shifted in frequency because of the motions of the sea surface. To first order, these reflections are coherent returns from a specific set of wave components of the directional wave number sea surface

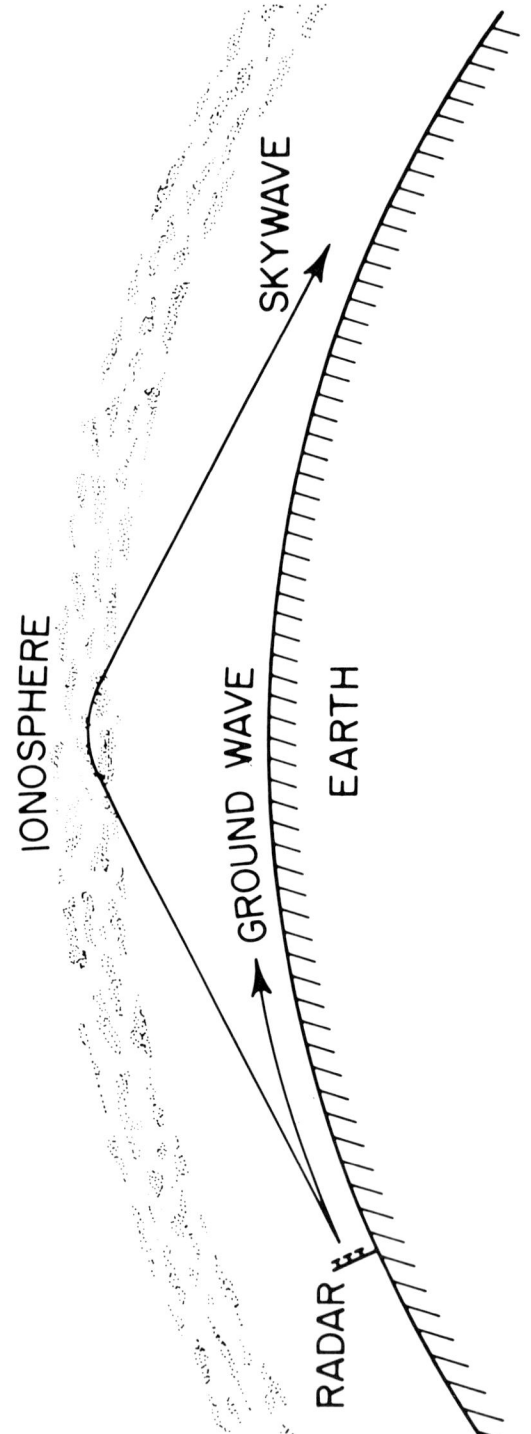

FIGURE 1: HF Radar energy propagates beyond the horizon by ionospheric refraction and by ground wave diffraction.

energy spectrum. That set of components must have wavelengths of one-half the radio wavelength, and must be propagating parallel and anti-parallel to the transmitted radio wave vector, as was first discovered by Crombie (1955). The Doppler shifts imposed by these two wave components are just those due to a target moving with the phase velocity of the surface-wave component predicted by linear gravity wave theory:

$$v_p = \pm \Omega/K = \pm (gL/2\pi)^{\frac{1}{2}} \tag{1}$$

The resulting shifted amplitudes are called first-order Bragg lines.

Theoretical analysis of the scatter mechanism involved indicates that the radar spectral amplitudes are directly proportional to the pair of sea surface wave number spectral amplitudes of the directional energy spectrum designated earlier (Barrick, 1972). Hence a well-calibrated measurement of the first-order Bragg lines as a function of frequency and radar azimuthal look angle can provide a direct measurement of the directional sea surface energy spectrum and its behavior with time. Wave growth and wave response to changing wind direction can be measured with higher dynamic ranges that that ever achieved with traditional oceanographic techniques. Dynamic range is defined in radar terminology as the ability to measure a small amplitude signal at one frequency in the radar power spectrum, in the presence of a much larger signal at a second frequency. As an example, typical maximum ratios of approach-to-recede Bragg lines for developed sea conditions are of the order of 24 dB for upwind-downwind radar aspect. Hence one is measuring downwind sea spectral amplitudes of the order of 0.4% of the upwind value.

In addition to the first order Bragg lines, there is a significant amount of information in the higher order structure to be found in the Doppler spectrum. This is shown in Figure 2, and will be discussed in a later section. First, the SEA ECHO radar, a typical but sophisticated pulse-Doppler radar, is discussed.

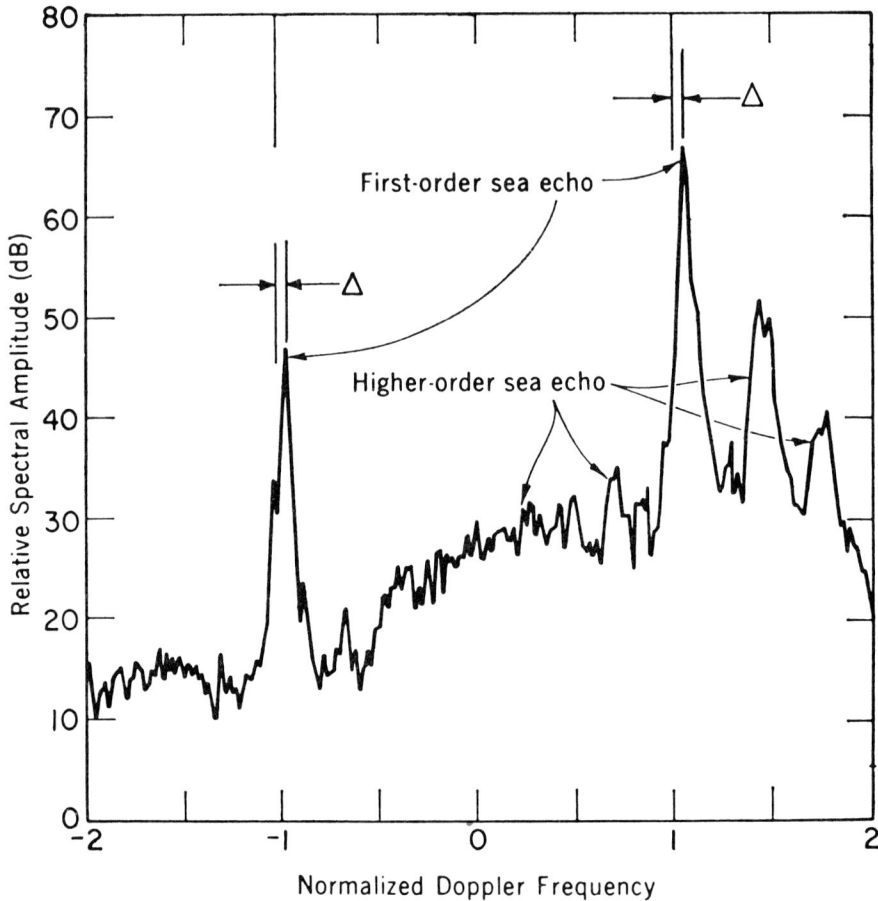

FIGURE 2: A typical Doppler spectrum of sea backscatter is shown, dominated by the two Bragg lines of first order at normalized frequencies, ±1. A shift in their position by an amount, Δ, is seen, due to ocean surface current. Higher order continuum and resonant spikes are also present, and are discussed in the text.

SEA ECHO RADAR

The Sea Echo HF Radar on San Clemente Island is a most versatile and powerful research tool. The flexibility of programming operating parameters to do simultaneous tasks allows experiments to be conducted in oceanography and ionospheric diagnostics which can be done with virtually no other kind of research instrument. These

include measurements made both in the surface-wave and OTH modes of
operation.

As with any new large research facility, such as a high-energy
particle accelerator or a radio or optical telescope, those experiments which can be done for the first time because of the unique
characteristics of the facility are the most interesting. The
unique feature of the Sea Echo radar most useful in this regard is
the ability to operate at an interleaved series of radar frequencies and beam positions on a pulse-to-pulse basis, allowing in
effect simultaneous operation for all of these parameters, typically to within one part in two thousand. This can be achieved
over more than a decade of operating bandwidth, 2 to 25 MHz, and
with 2° azimuthal resolution for frequencies 16 MHZ and above over
the 80° azimuthal coverage of the antenna. The beam width spreads
for frequencies less than 16 MHz inversely with frequency. The
capability exists to eliminate elements from the antenna array so
as to compensate for this effect, allowing near identical beam
widths to be synthesized over the entire operating bandwidth, again
on a pulse-to-pulse basis.

Pulse lengths available for operation include 20, 50, and 100
microseconds, as well as an FM chirp format. In this latter mode
up to a ten-millisecond FM pulse is transmitted and deramped to as
short as ten microseconds. This chirp format is generally useful
only in the OTH mode, and allows additional gain to be achieved,
equal to the ratio of the transmit pulse length to the dechirped
pulse length. When transmitting the ten-millisecond pulse, signals
from the group delay within this transmit period cannot be received; hence only returns from between 1500 km and the time delay
up to the next transmitted pulse are available for remote sensing.

With regard to radar hardware, the antenna design has two
particular features which distinguish it from previous HF antennas
and cut cost significantly. The first feature is the transmitter
system, in which separate 1.5-kW modules are placed at the base of
each of 25 transmit/receive elements and drive the elements
directly without the need for high power duplexing for receiver
protection. This results in greater simplicity of operation and

maintenance, with substantial cost savings as well. The transmitters arranged in such a manner develop 40-kW peak power. A
model of the antenna is shown in Figure 3.

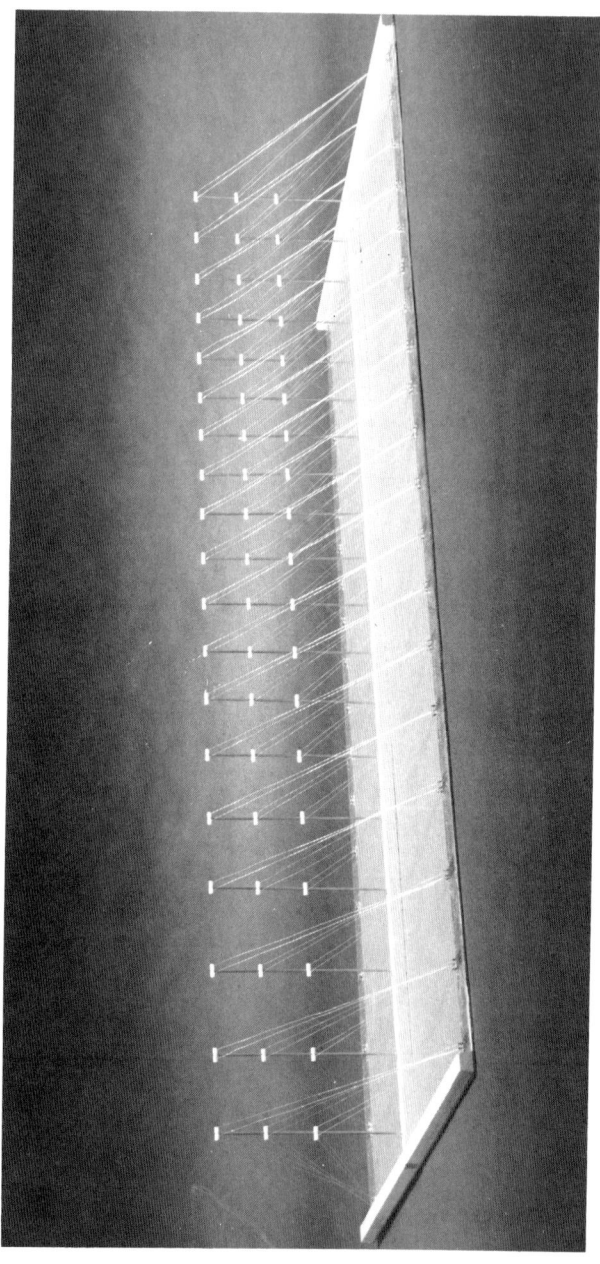

FIGURE 3: A model is shown of the Sea Echo transmit/receiver HF radar antenna. A single element is an inverted half rhombic, the full rhombic created by the image in the ground plane. Three different rhombics can be chosen, to optimize antenna gain for three different ranges of the HF band.

The second distinguishing feature of the antenna is the elements themselves, vertically-polarized half-rhombics. They have an
excellent elevation angle pattern for both surface-wave and over-
the-horizon propagation, with most of the energy confined to low-
elevation angles of transmission. Again low cost is a feature of
the construction. An element consists of a single 150-ft tower
with standard four-guy supports connected at a single anchor point
on the ground, and at different heights on the tower. In this case
one of the top three guys is the driven half-rhombic element, a
selection of one of the three being made for different frequency
bands within the 2 to 25 MHz range. The guy lines are electrically
driven at the near anchor point and terminated at the other anchor
using 200 ohm resistors, with radial ground screen wires running
from that point several hundred feet to the sea. The entire area
beneath the antenna structure is covered with a large mesh ground
screen also. The twenty-five elements of the array are spatially
tapered for better sidelobe control. More details are found in Ma
and Tveten (1975).

The two-way peak gain of the array is typically 55 dB over
most of the operating bandwidth. It is steerable in two-degree
increments over eighty degrees in azimuth centered at a boresight
of 320° true bearing.

The data are sampled, digitized, and taped for later playback
and analysis on the Texas Instruments Advanced Scientific Computer
at NRL. As many as ten different ranges can be sampled when the
radar is run in the multi-frequency and azimuth mode using 50
channels, so that 500 samples are taken for one scan or pulse
repetition frequency. When interleaved in such a manner one can
choose variable integration times with which to process the data
depending on the experiment at hand. Periods of up to 204.8
seconds are typically chosen for those cases when spectral resolution is of interest, or of the order of a few seconds when statistics on the variability of the Bragg peaks (and ocean wave resonant
components) are studied.

OTHER HF RADAR SYSTEMS

Other radar sites being used for radar experiments in the U.S. include the Naval Research Laboratory MADRE research radar, which illuminates the Atlantic Ocean, the Stanford Research Institute International Wide Aperture Research Facility in California, which covers the Pacific Ocean, and the Office of Naval Research Whitehouse field site, also on the east coast. All of these radars are suitable for OTH operation via skywave propagation. The ONR and SRI radars have angular resolution of 1 and 0.5 deg at 15 MHz, respectively. The ONR receiver array is currently being used in a program at NRL to map ocean currents associated with ocean eddies and the Gulf Stream. The program is in a very preliminary stage at this time, and initial results appear very promising.

A new development in the area of remote sensing at HF frequencies is occuring at the National Oceanic and Atmospheric Administration Wave Propagation Laboratory under the direction of Dr. Donald Barrick. He is developing portable radars which can be set up on the shoreline and map offshore wave parameters. Two different systems are under development. One measures the current field offshore, up to 75 km distant, on a range-azimuth grid basis. Using just a single log periodic vertical monopole array of four elements, and two citizen band whip antennas for receiving, he is able to resolve currents of five cm/s with 6 km spatial resolution. Two of such instruments must be used in tandem to resolve currents unambiguously, since a single system measures the radial component of the current.

A second technique also under development by Barrick is another portable system which will attempt to measure directional wave spectra over a similar area as is done by the current measuring radar. The transmitting antenna is a monopole this time, and the receiving system consists of three antennas, two crossed loops and a single monopole. Three separate signals are received and digitized for each antenna. The results are combined after the fact in processing to form a single beam which can be steered in computer processing. Since the loops each have a cosine theta pattern, and the monopole is omni-directional, one can sum the three outputs using the following relationship:

$$1 + \cos \theta \sin \emptyset + \sin \theta \cos \emptyset = 2 \cos^2 ((\theta-\emptyset)/2) \qquad (2)$$

where θ is the angle relative to the pointing direction of one of the loop antennas, and $\emptyset$ is the variable steer angle introduced in the computer processing of the three signals. The monopole signal is normalized to the loop outputs by attenuation and is represented by unity in the equation above. Hence one has a steerable beam pattern represented by the right hand side of the equation. The system is currently under development at NOAA WPL in Boulder, Colorado, using a single radar frequency. The analysis requires inversion of second order components of the Doppler spectrum, which shall be discussed later. It is noted that a similar system could be designed with a broadband transmitter operating pulse to pulse in a multi-frequency mode as discussed in the section describing the Sea Echo radar. The analysis would then consist of a simple measurement of the first order Bragg lines as a function of ocean wave frequency and direction.

A final type of HF radar application is a synthetic aperture system which has been used to measure the directional distribution of 75 m waves operating at a radar frequency of 1.9 MHz. This work was done by Stanford University and Scripps Institute of Oceanography scientists in an experiemnt on Wake Island (Tyler et al., 1974). They made use of the transmission of a LORAN C signal, normally used for navigation, and received it on a van which was driven along a straight line on an unused airfield. The first order Bragg lines were due to scatter from a wide selection of angles because of the broad beam of the cardiod pattern antenna which was used. However, the Bragg lines were broadened due to the selective additional Doppler shift imposed as a function of direc- tion by the motion of the van. Energy received from a direction perpendicular to the path of the van had no additional Doppler imposed and appeared at the expected frequency. Positive and negative shifts were added by the van motion to that energy re- ceived off angle to the perpendicular, a greater Doppler shift the greater the angle. This spread in the first order Bragg line could then be analyzed to determine the spreading of wave energy with angle. The results were compared with a pitch and roll buoy and were quite similar. Such an ingenious technique could be applied to a number of radar frequencies simultaneously to obtain a true directional spectrum of ocean waves versus an equivalent number of wave frequencies. For reference purposes, the span of ocean wave frequencies scanned by a radar operating from 1 to 24 MHz is 0.1 to 0.5 Hz.

REMOTE SENSING CAPABILITIES OF HF RADAR

We shall now report on some recent experimental results which have been achieved at NRL and other research organizations in the U.S. These are not meant to be a comprehensive listing of research in the U.S., but representative of what can be accomplished with HF radar, both in the skywave and surface wave modes of propagation.

Wind Field Mapping Using OTH HF Radar

The capability for remote sensing of wind direction fields has come a long way since the first map derived by Long and Trizna, (1971), with the biggest strides made using higher spatial resolution. The work by Long and Trizna was conducted at the Naval Research Laboratory using the MRL MADRE HF radar, with a beam width of 12 deg at 13.5 MHz. Operated monostratically, the beamwidth is halved. With this beamwidth and a pulse length of 250 μ-sec, a cell remotely sensed at 2,000 km is 75 km deep and 200 km wide, enough to show gross features of weather systems but not the details of a wind reversal along a front, for example. The SRII WARF radar, with a ½ deg beamwidth, operating bistatically with a broader beam for transmit, and a 100 μ-sec pulse length, forms a cell 15 km by 33 km at 2,000 km range, and has been used to show such features. Maresca and Carlson (1978) have also used the high spatial resolution features of the system to accurately track the position of hurricances in the Gulf of Mexico to an order equal to or better than that achieved with aircraft or satellite data.

The technique of wind field mapping involves measuring the approach and recede Bragg line amplitudes, taking their ratio, assuming a form for the directional spectrum at the wave frequency probed by the radar, then determining an angle between the radar look angle and the wind and wave direction. At the shorter wavelengths sensed by a skywave radar operating at 15 MHz, 10 m, the ocean wave components respond on the order of ten minutes or so in terms of their directional distribution. If one assumes a form for the direction spectrum given by

$$S\ (K,\ \theta) = S'\ (K)\ F\ (\theta) \tag{3}$$

where $S'(K)$ is the omni-directional spectrum measured by a buoy integrated over all angles, then $F(\theta)$ is the angular spreading function governing the distribution of wave energy with angle, θ, to the wind direction. Apparently the waves respond sufficiently rapidly to show a change in $F(\theta)$ which can be used to define a wind reversal along a front. The technique is implemented in the following way. Since the first order Bragg lines in the radar spectrum are proportional to $S(K,\ \theta)$ and $S(K,\ \theta + \pi)$, when one takes the ratio of amplitudes of the Bragg lines, one has a measure of $F(\theta)/F(\theta + \pi)$. Assuming a model for $F(\theta)$, one can then calculate an angle which the wind is blowing relative to the radar look angle. Two models have been used in work done using HF radars. Both allow for energy to exist over a full 360 deg in angle, typically assumed symmetric about the wind direction.

The first is a model derived by Long and Trizna (1973) from a set of radar data collected over a wind field operating in the sky-wave mode. The data were collected using an 18 MHz radar frequency, probing ocean wavelengths of 8.5 m, and were probably fully developed for the reported winds of the storm. Analysis of the radar amplitudes for the approach and recede Bragg lines showed a relationship for the angular apreading function for energy in the rear half plane of the form

$$F(\theta) = A + \cos\ \theta \quad \text{for} \quad \theta < \pi/2 \tag{4}$$
$$F(\theta + \pi/2) = \text{const.}/F(\theta) \quad \text{for} \quad \theta < \pi/2$$

The constant is set to $F^2(\pi/2)$ for continuity at $\theta = \pi/2$. Since the energy spread in the front half plane is thought to be related in some way to the driving wind field, this says that the energy in the rear half plane is inversely proportional to that in the forward half plane, and implies some non-linear mixing between wave components of the same frequency but opposite in direction of propagation. Wind directions inferred from this model agreed quite well with those observed at sea where comparisons could be made.

A second model used by Tyler et al. (1974), is one of the form

$$F(\theta) = A + \cos^S (\theta/2) \qquad (5)$$

The pedestal, A, is necessary to allow for wave energy in a direction opposite that of the wind. This model has been used in oceanographic circles in analysis of pitch and roll buoy data (see, for example, Ewing, 1969), where the angular filter describing the buoy response is also assumed to be of the same form as Equation 4. In the analysis, it appears that this is for sake of computational purposes rather than from first principles. It becomes convenient in the analysis to assume a similar form for the ocean wave spreading then also. The model is useful in that it is monotonically decreasing with angle, for even values of S it can be expanded in a finite cosine series, and parametrized for narrowness with increasing S. Neither model is derived from first principles and both give about equal accuracies in wind direction estimation.

As a final word on the subject, one could imagine how angular spreading could result from a consideration of the coherence of the wind field. If one had a perfectly coherent wind with no fluctuations in the horizontal coordinates, then one might expect a very narrow wave spectrum to be developed. Its narrowness would depend upon the size of the wind system itself. The spread in K space in angle for a given wavelength would be inversely related to the horizontal size of the wind field system, by a Fourier transform analysis. If one now allows the wind field to exhibit fluctuations in space and time, as exhibited by a turbulent wind, then the spreading in angle of the wave energy would be related to the

coherence sizes of the cells of wind for a given wavelength. A very
turbulent wind system would be expected to generate a very broad
angular spread in wave energy using this concept. In addition to
broadening due to the source, wave-wave interactions between ocean
wave components are expected to broaden the spectrum with time, and
is probably responsible for the typical cosine-squared spectra
which ahve been reported by many observers. Trizna et al. (1980)
have reviewed directional spectrum measurements to date and have
found little work done in this area for a broad range of wave
frequencies.

Ocean Current Measurements

A second very straightforward measurement available from an HF
radar is that of near surface currents. It has been found by a
series of experiments using drifter buoys that currents affect the
radar measurement by shifting the position of the Bragg line in the
Doppler spectrum by an amount due to the radial velocity of the
current. That is, a given ocean wave Fourier component with phase
velocity given by Equation 1, has an additional velocity component
added to the phase veolcity which is equal to the component of the
current velocity which is radial to the radar bearing. The radar
then observed both Bragg lines, approach and recede, Doppler
shifted an equal amount in the same direction due to the surface
current. It appears from multi-frequency measurements made simultaneously by Barrick et al. (1974), that the radial current measured is actually at a depth of the order of the wavelength of the
ocean wave, since they observed a change in direction of the current with radar frequency, a current shear with depth, an actual
change in sign of the current. This technique would appear to have
applications both using surface wave and over-the-horizon propagation modes, and we shall describe examples of both.

A new technique has been developed which makes use of a com-
pact and portable surface wave radar operating at a single frequency, 25 MHz. A relatively broad beam receive antenna is used (a
small array of 4 whips) with a log periodic transmit antenna on a
beach, and data is collected for a Doppler spectrum from a wide

selection of angles. If one has a current travelling parallel to the shore, the Bragg lines are broadened due to the fact that the radar is averaging over approaching and receding components of the current. If one uses a very short pulse with two tandem systems, one can create a matrix of range-azimuth cells with unambiguous current vectors determined for each of the cells 3 km on a side. Ranges out to 60 km have been reported, beyond which surface wave losses preclude propagation (Barrick et al. 1977).

A second current measurement application involves OTH propagation. If one uses a radar with a sufficiently large aperture so as to allow one degree angular resolution or less, one can measure large scale ocean currents such as are associated with the Gulf Stream or ocean eddies. If one limits operation to the very stable E-layer propagation, allowing maximum ranges of the order of 2,000 km, one can achieve sufficient Doppler resolution to measure currents of the order of tens of cm/sec. If one uses 100 sec integration times, so as to achieve 0.01 Hz resolution in Doppler frequency, then one can achieve current radial speed resolution of the order of 10 cm/sec with a 15 MHz radar frequency. Experiments are currently being conducted by the Naval Research Laboratory using the ONR Chapel Bell receive array described earlier to attempt to map the Gulf Stream in the Atlantic. Early results appear encouraging. Figure 4 shows an example of one of the experiments. The position of the center of each range cell, defined by the pulse length and beam width of the receive array, is identified by a small circle. There are two lines associated with each range-azimuth cell which correspond to the Doppler shifts of the approach and recede Bragg lines, respectively. The lengths of the lines indicate the radial speeds of the currents which would produce the observed shifts. The data are shown in pairs to verify the consistency of each of the two measurements. Although the objective of the experiments is simply to identify the quality of propagation available by the E layer for such measurements, with no attempt at confirming surface truth data at this time, some positive statements can be made about the observed Doppler shifts being due to surface effects, and not ionospheric effects. These data were

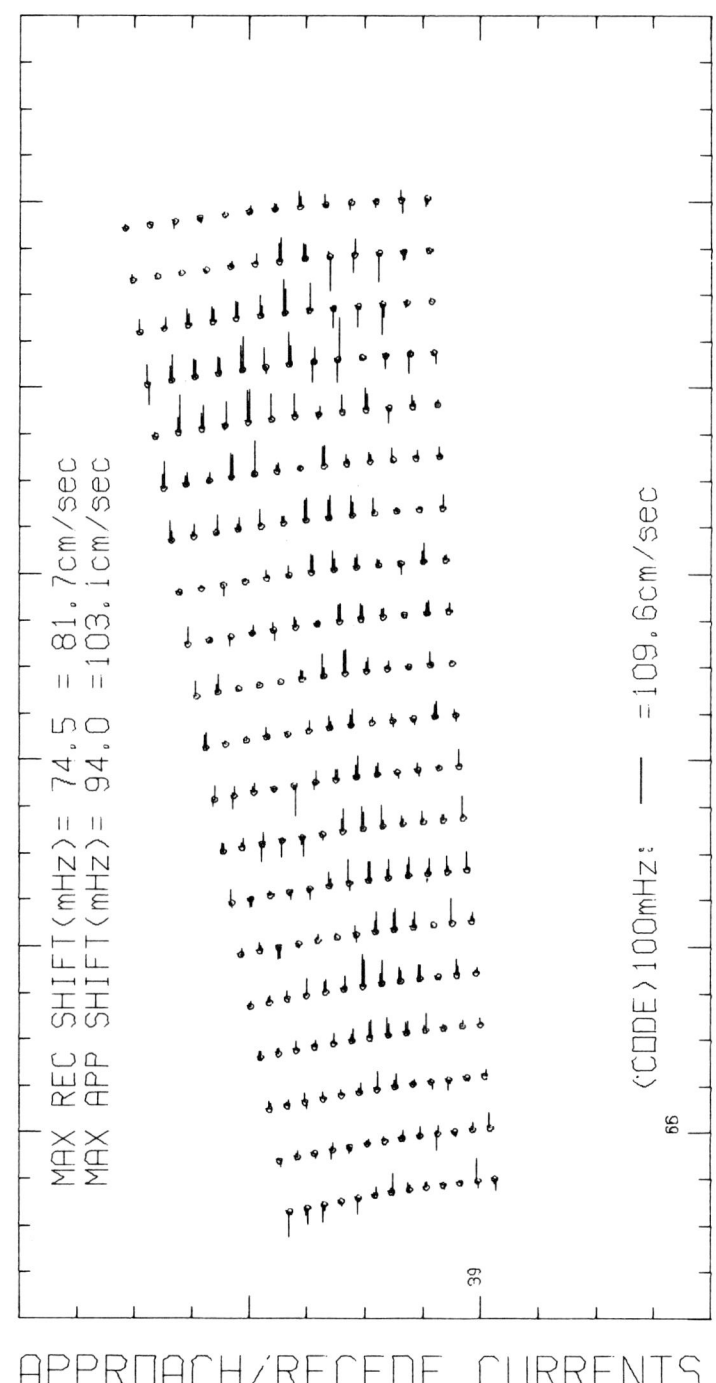

FIGURE 4: An example from a recent experiment of ocean surface currents measured by an HF radar is shown, as described in the text. The measurement is that of the shift in Bragg lines, Δ, shown in Figure 2, for a series of range-azimuth radar cells, each of the order of 6 km on a side. The shifts appear to be associated with the Gulf Stream, based on ocean analysis chart comparisons.

collected using a seven minute dwell at a given azimuth, before moving to the next adjacent beam position. Hence, counting time needed to change parameters, roughly two hours are required for the entire data set to be collected. There are several groupings of shifts which can be interpreted as currents, with similar magnitudes, extending both in the range and azimuth coordinates. An argument that these effects are ionospheric in origin is not realistic since any Doppler shifts due to the ionosphere must be associated with traveling ionospheric disturbances (TID's), where the electron density changes in such manner as to effectively cause the ionosphere to be shaken like a sheet, causing a change in effective height. Such a wave has been observed and the effect upon the Doppler spectrum is to cause the entire spectrum to be shifted higher, then lower in frequency, in a sinusoidal fashion, over a few minutes in time. The effect on this type of data display would be to broaden the average four spectra of 102.4 seconds of the order of half a Hertz or more. In addition, the scale sizes of such ionospheric disturbances are tens of kilometers in both horizontal coordinates, so that features which are correlated from one azimuth to the next, collected 30 minutes apart for a set of four, for example, could not be produced by a TID.

Several experiments were conducted to determine the time period over which the ionosphere remains reasonably stable by considering the characteristics of the amplitude of the backscatter versus range and how these change with time. It was found that F-layer modes became more prevalent after noontime, and that the entire morning would exhibit stable E-layer propagation modes with maximum clutter returns from 750 to 2,000 km, the maximum range associated with zero launch angle and a 100 km E-later altitude. These conditions would prevail from the time the ionosphere stabilized after sunrise to local noon, at which time F-layer scatter from beyond 2,000 km would become apparent and quite often dominate the E-layer scatter in amplitude. Although the phenomenology for this behavior is beyond the scope of this small data review, suffice it to say that E-layer propagation sufficiently stable to allow location of ocean currents is available in the morning periods before local noon a large fraction of the time.

Directional Ocean Wave Parameters From
Second Order Doppler Contributions

One of the most promising capabilities of HF radar is possible inference of directional ocean wave characteristics from second order contributions to the radar Doppler spectrum. If one identifies the two Bragg lines as first order scatter, i.e., single scatter from ocean wave components with wavelengths one half the radar wavelength, then all other lower amplitude contributions to the Doppler spectrum are typically 15 dB and further below the weaker of the two Bragg lines, and hence, are classified as second order or higher. Three mechanisms contributed to this higher order scatter:

A. Single scatter from ocean wave components with wave-number, K_n, satisfying the momentum equation

$$n \; K_n = 2k \tag{6}$$

B. Double scatter from pairs of ocean wave components, K_1 and K_2, which satisfy the momentum conservation equation:

$$K_1 + K_2 = 2k \tag{7}$$

C. And, finally, first order scatter from non-linear contributions to the ocean wave spectrum due to interactions between wave components which also satisfy Equation 7.

All three of these types are of second order or higher in the sense that they occur proportional to $(hK_n)^n$, or as products, $(h_1 K_1)(h_2 K_2)$, where h_1 and h_2 are the waveheights of the spectral components with wavenumbers K_1 and K_2, respectively.

Considering first single scatter from multiple integral wavelength components of the Bragg components, reflections from these are also Bragg scatter, the analogues of the higher order diffraction maxima from a grating. In this case, the angle is kept fixed and the "gratings" are changed, with spacings integrally related to the first. If, to gain physical insight into the problem, one performs an analysis of the problem using the Rayleigh approximation, one generates a set of solutions in Bessel functions of Mth order. The first order Bessel function represents the first order Bragg scatter, was the appropriate Doppler shift associated with it, and when expanded in a Taylor series, the first term is equal to that obtained from a linear perturbation analysis. Bessel functions of Mth order represent scatter from the integrally related wavelength components. The Doppler frequencies associated with these contributions are at square-root of M times the first order Bragg frequency. Again, if one expands the Bessel function of second order, the first term, when added to the second term of the Bessel function of first order, is identical to the result of a perturbation analysis to second order; it does not appear in the first order analysis. The same holds true for the additional higher order terms. In all of these cases, the ocean wave components have directions of propagation toward and away from the radar.

Whereas the above contributions are at discrete Doppler frequencies in the spectrum, the double scatter mechanism allows contributions at a continuum of Doppler frequencies extending from $-2^{3/4}$ to $+2^{3/4}$ times the First order Bragg frequency. One must consider all possible combinations of wavenumbers, K_1 and K_2, which satisfy Equation 7, which is basically a continuum of components, and which generate a continuum of amplitudes in the Doppler spectrum. These are again first order scatter, occuring two times. The first scatter is not in the direction back to the radar, but occurs off a component traveling non-parallel to the radar bearing, and reflects off the wave at an angle to the ocean wave vector which is equal and opposite to the incident angle (i.e., specular in the plane of ocean waves). A second scatter occurs similarly to the first, but from a second wave component, K_2, which reflects the radar energy back to the radar. The net Doppler shift is the sum of the Doppler shifts from each of the two scatters, each equal to the phase velocity of the scattering ocean wave times the cosine of the angle between the radar wave vector and the ocean wave component wave vector. In addition, other scattering occurs which does not return energy to the radar, and which satisfies a different conservation of momentum equation than given by Equation 7, but which is not necessary to consider since it is not received by the radar. This is the reasoning behind the theory presented by the author in 1977, (Trizna et al. 1977), and is slightly different from the theory of Barrick (1977). Although it is too intricate a matter to deal with here, a few comments can be made. In his derivation of the resonant coupling coefficient for the electromagnetic term, Γ_{em}, Barrick allows wavenumbers other than those satisfying Equation 7 to contribute, leading to the resonant nature of Γ_{em}. Allowing only contributions satisfying Equation 7 effectively treats the resonance as a delta function in evaluating the integral for the Doppler spectrum amplitude, so the difference is really not great between the two approaches. The off resonant contributions must carry some momentum in the z direction perpendicular to the surface, and these are either damped or evanescent waves, depending upon the sign of z, the perpendicular co-ordinate. Those terms which are identically resonant and satisfy Equation 7 are perpendicular to one another, and are sometime called the "corner reflector" terms.

The third contribution, the hydrodynamic term of Barrick, is disputed at this time to contribute as significantly as he has calculated. Applying the calculation of Valenzuela (1974), which includes both electromagnetic and hydrodynamic terms, to HF frequencies shows that the hydrodynamic terms contribute very little below 30 MHz. Barrick finds this term to be most dominant near the Bragg lines, away from zero Doppler. However, he has not included viscocus damping term in his calculation as did Valenzuela, and this would cause the hydrodynamic term to be very large near the Bragg line, and, indeed, become infinite as one approaches the Bragg line, when sufficiently long wavelengths are present in the ocean spectrum to contribute in this Doppler region. It must be stressed that the techniques of remote sensing of ocean waves do not depend upon one or the other of the viewpoints discussed above, but only the accuracy of the output of their application.

The ocean wave parameters which can be inferred from second order Doppler structure in the radar spectrum include dominant wavelength and R.M.S. waveheight, and, if some assumptions are made about the spreading of wave energy with angle, a directional wave spectrum. If a series of look angles with reasonably high angular resolution are available to the radar, then the assumption of a model for the spreading is not necessary, and the directional spectrum can be actually measured, using the Doppler spectra for each look angle. Some assumptions are necessary about the slope of the ocean wave spectrum in the region of Bragg components. Because the radar second order structure is proportional to $h_1 h_2$, and h_1 is near the Bragg ocean wave component and varies slowly as one moves away from the first order Bragg peak, this assumption is necessary in order to retrieve h_2, which is in the long wavelength region of the spectrum. Barrick makes the simplest assumption, that h_1 is just the Bragg waveheight and is fixed as one moves thorough the Doppler spectrum on a point by point basis. A more accurate approximation would be to assume that the spectrum varies as f^{-5} in this region. Using the corner reflector model, for a given radar spectral amplitude at a given discrete Doppler shift, the wave-number and frequency of the ocean wave are specified by Equation 7, and the waveheight of the component is retrieved, using equations specified in the author's work (Trizna et al., 1977). Using Barrick's model, one must perform an integral inversion for each point in the radar Doppler spectrum, including terms representing the corner reflector, evanescent and damped E.M. waves, and the hydrodynamic contributions. In any case, actual ocean wave ampli-tudes for specific angles can be determined. Dominant ocean wave-lengths can be estimated very rapidly without integral inversion. One simply measures the differences in Hertz between a Bragg peak and the frequency at which the second order continuum falls sharply either side of the Bragg peak. Since in this region one of the interacting wave components is very nearly the Bragg ocean wave component, this difference frequency is very nearly the ocean wave frequency of the dominant frequency of the ocean wave spectrum. R.M.S. waveheight is determined by integrating the area under the

continuum, and then dividing by the energy of the strongest Bragg
line, a self-normalizing effect as first pointed out by Barrick.

Directional Ocean Wave Spectra
From First Order Bragg Lines

Scientists at the Naval Research Laboratory (Trizna et al.
1980), using the SEA ECHO radar in the surface wave mode of operation, measured directional ocean wave spectra by a simpler technique than that discussed in the last section. The radar was run
at ten different frequencies and ten azimuths and the first order
Bragg line amplitudes measured. This allowed the Bragg ocean wave
components to be measured for each of the corresponding ocean wave
frequencies and angles and allowed an estimate to be made of the
directional ocean wave spectrum for this range of parameters.
Radar frequencies spanned the range 2.0 to 11.65, with corresponding ocean wave frequencies of 0.144 to 0.348 Hz, 7 second down to 3
second waves. The range of bearings ran from 285 to 359 degrees
true. The conditions under which the data were collected were most
interesting in that they included a bi-directional spectrum, i.e.,
two wave systems, which were created locally as well as far at sea.
The omni-directional spectrum for the periods of radar data collection are shown in Figure 5. The broad low frequency region represents wave energy which was created some 200 miles west of San
Clemente Island in the Pacific and which propagated into the
region. The higher frequency components which are higher in amplitude and which are seen to grow with time are due to winds of the
order of 12 to 14 knots which began to blow very early on the day
of observation and continued on throughout the day. The direction
of the wind was primarily from 310 to 320 degrees, in contrast to
the 270 to 290 degrees direction of propagation of open ocean wind
driven seas. The directional spreading of wave energy was measured
by the radar through the day and its development showed some very
interesting results. The longest wavelength ocean waves are the
slowest to respond to winds so that the simultaneous measurements
over the selection of wavelengths showed, in effect, a snapshot of
directional angular spreading in different stages of development.
The 0.144 Hz waves showed a distribution very nearly cosine-squared

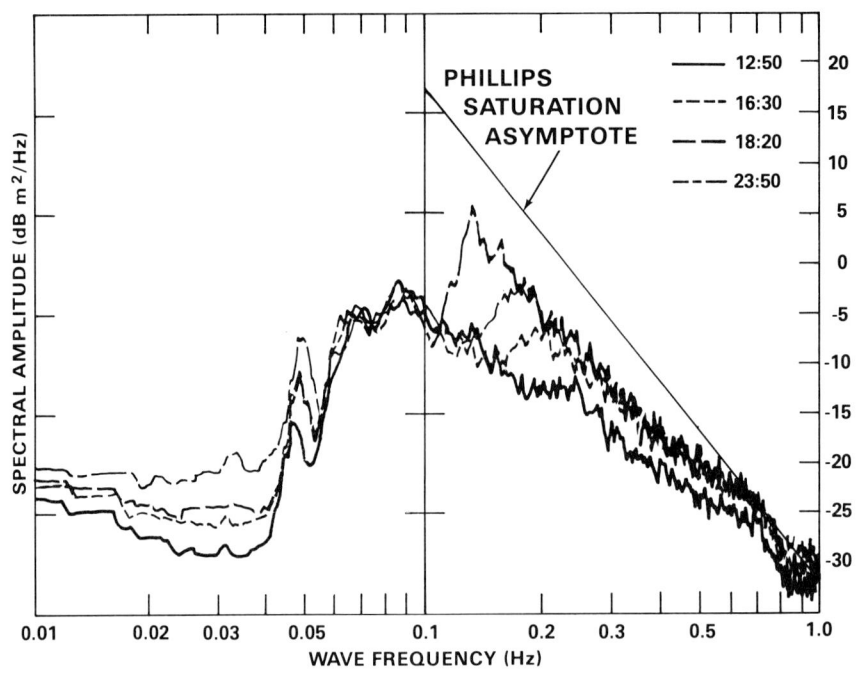

FIGURE 5: Ocean wave spectra are plotted for four successive time periods on a log-amplitude, log-frequency basis. The energy extending down to 0.1 Hz is due to a storm some two hundred miles west of the region. Energy which is observed growing with time down to 0.14 Hz is due to local winds.

in angular dependence, similar to that reported by many others in the literature for open ocean, long fetch conditions, and showed no response to local winds for the first measurement time period. As one goes to higher wave frequencies, the response to the local winds become more dominant, and show different stages of development, with a maximum observed at the direction of the wind for the highest frequencies, as well as at two discrete angles either side of the wind direction. These discrete angles are found to be frequency dependent and the peaks wave been predicted by Phillips (1957), but never observed in as great a detail as was possible in this experiment. They are the manifestation of the so-called Phillips resonance mechanism for generation of wind waves, and are worthwhile discussing to some extent.

The most effecient mechanism for the generation of wind waves cannot interact with a perfectively smooth surface, since the surface presents no frictional reaction to the wind to allow momentum transfer from the wind to the waves. However, if one has a turbulent windfield, turbulence components of this wind system can deliver momentum to the waves. Since the turbulence travels at the same speed as the average wind, it appears frozen from a reference system moving with the wind, and can deliver energy to the water surface selectively, with a given component of turbulence interacting with the ocean wave component which has its phase velocity just equal to that of the wind, since it remains "in phase" with the ocean wave component in the moving reference frame. In addition, other turbulence components can deliver energy to shorter ocean wave components at angles either side of the wind direction, which have phase velocities satisfying the equation:

$$u = v_\omega \cos \theta \qquad (8)$$

where v_ω is the wind speed, and u is the phase velocity defined in Equation 1. Peaks are predicted to occur either side of the wind direction very early in the wave generation process, should be excited linearly with time, and should soon be overcome by nonlinear wave generation processes which develop energy primarily along the wind direction.

Figures 6 and 7 show data for two different radar frequencies, and, hence, two different wave frequencies, each showing three different time periods. Figure 6 shows the twin peaks of the Phillips resonance with very little or no energy along the wind direction for the first time period plotted as the lowest of the three. With time, energy is developed along the wind direction, and dominates the Phillips spikes as is seen in the later time measurements, in which the residues of the spikes are still seen, however. Figure 7 shows the same time periods for a higher ocean wave frequency. Here the first time period shows the presence of energy along the wind direction already. The Phillips spikes are still present, at slightly different angles, as identified by the arrows for their predicted positions based on first identification

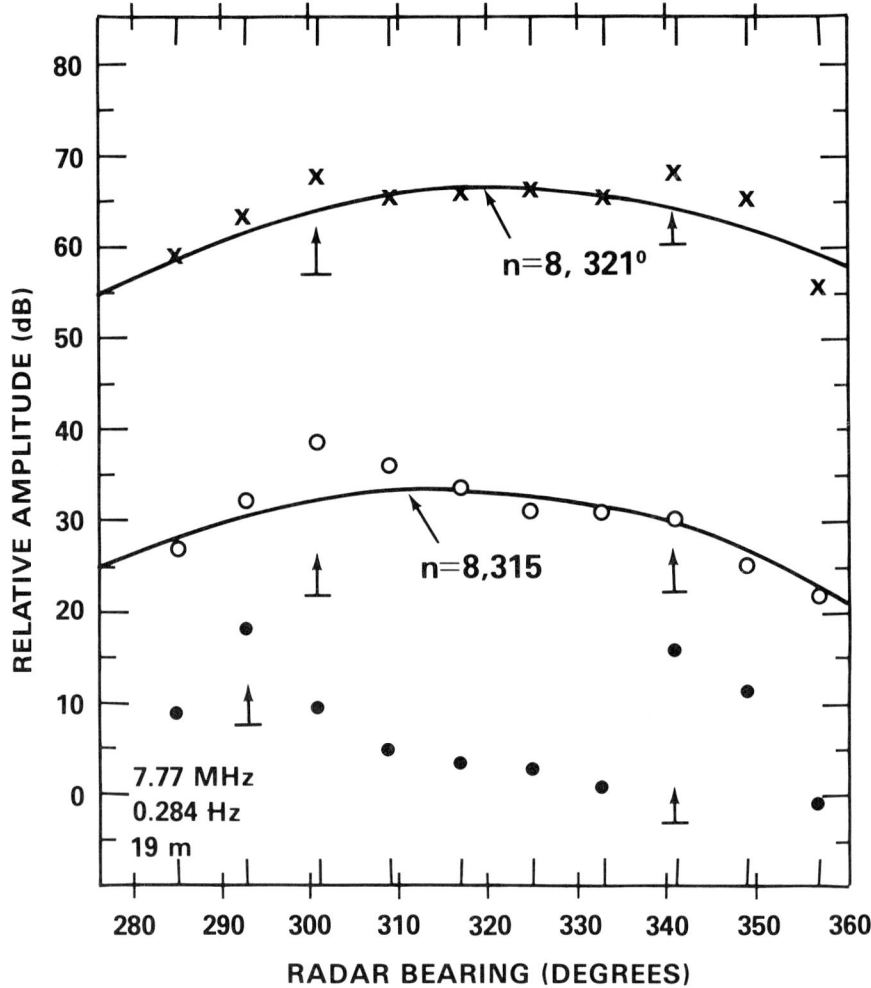

FIGURE 6: Ocean wave energy for 0.284 Hz waves is plotted in log amplitude versus angle, on a relative amplitude basis, for three different time periods roughly six hours apart. Plots from one time period to the next are not related in amplitude. The curved lines are cosine to the N power curves drawn for comparison purposes, centered at the indicated bearings. The two Phillips resonance spikes described in the text are marked by arrows. They dominate the angular spreading for the first time period and are still observable for the later time periods as dominant contributions.

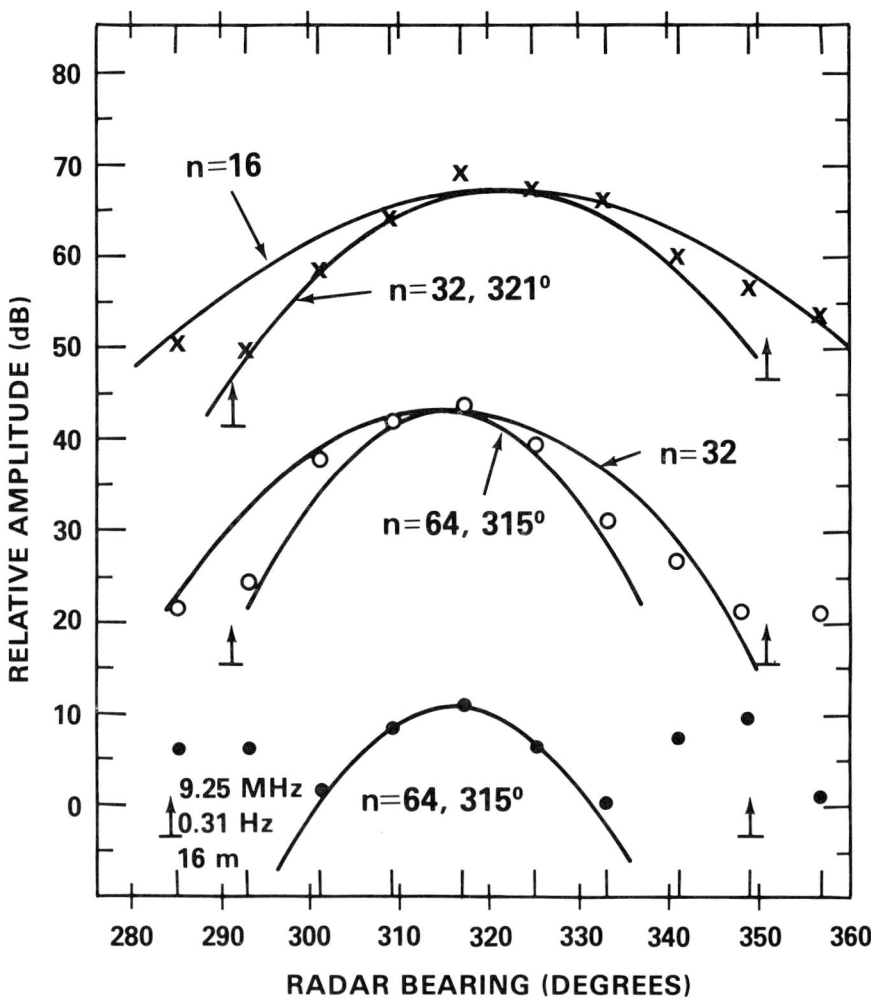

FIGURE 7: Data as described for Figure 6, but for 0.310 Hz waves. At this higher frequency, other wave generation mechanisms along the wind direction are much more prevalent for the first time period, and fully dominate the Phillips resonance spikes for later time periods.

at the 7 MHz frequency. The wind speed required to generate the spikes at these angles is of the order of 12 knots, in very good agreement with local observations. All four wave frequencies higher than the 0.284 Hz waves showed the Phillips spikes at their predicted values, while the only other wave frequency component

which could have, for the 12 knot wind, the 0.265 Hz component,
appeared very confused and did not show the spikes. The four
frequencies with phase velocities greater than the wind speed were
not expected to exhibit the spikes and did not. The widths of the
spikes agreed quite well with that predicted by the calculations of
Stewart and Manton (1971), who related them to the coherence times
of the ocean wave components. This experiment is the first in
which the Phillips resonances were observed simultaneously at a
series of different wave frequencies, from their initial excitation
to periods when they were dominated by other wave generation
mechanisms.

This set of data show the potential of observing the development of ocean wave spectra remotely, in a single homogeneous area,
for extended periods, to a degree not achievable by traditional
oceanographic techniques. Operated in conjunction with other
surface truth instrumentation, wind speed and turbulence sensors
and sea and air temperature measurements at a series of heights to
characterize the boundary layer, the study of development of ocean
wave directional spectra could be studied to a degree of detail and
accuracy never before achieved.

SUMMARY

A general outline has been given of the various systems and
techniques for remote sensing of ocean waves and currents done
using high frequency radar, both in the sky wave and surface wave
mode of operation, with a general introduction into just how such
data are collected and processed. The fundamental relationships
between the processed radar data and the ocean surface have been
outlined insofar as they are understood at the present, with the
uncertainties in this understanding discussed when they exist.
Although not a comprehensive study of all of the work going on in
the field, the references provided will allow the reader to access
the relevant work in which he is interested. It is hoped that this
paper will be of use to the PRIMARS participants for use in this
spirit.

REFERENCES

Ahearn, J. L., S. R. Curley, J. M. Headrick and D. B. Trizna.
    1974. Tests of remote sky-wave measurement of ocean surface
    conditions. Proc. IEEE, 62: 681-687.
Barrick, D. E., M. W. Evans, and B. L. Weber. 1977. Ocean surface
    currents mapped by radar. Science, 198: 138-144.
Barrick, D. E., J. M. Headrick, R. W. Bogle, D. D. Crombie. 1974.
    Sea backscatter at HF, interpretation and utilization of the
    echo. Proc. IEEE, 62: 673-680.
Barrick, E. D. 1972. First order theory and analysis of MF/HF/VHF
    scatter from the sea. IEEE Trans., AP-20: 2-10.
Barrick, D. E. 1972. HF radio oceanography - a review. Boundary
    Layer Meterology, 13: 23-43.
Crombie, D. E. 1955. Doppler spectrum of sea echo at 13.56 MHz.
    Nature, 175: 681-682.
Ewing, J. A. 1969. Some measurements of the directional wave
    spectrum. J. Mar. Res., 27: 163-171.
Headrick, J. M. and M. I. Skolnik. 1974. Over-the-horizon radar
    in the HF band. Proc. IEEE, 62: 664-673.
Long, A. E. and D. B. Trizna. 1973. Mapping of North Atlantic
    winds by HF radar sea backscatter interpretation. IEEE Trans.
    AP-21: 680-685.
Ma, M. T., and L. H. Tveten. 1975. A broadband Antenna Array for
    Sea Backscatter Measurements. IEEE Trans., AP-24: 340-347.
Maresca, J. W., Jr., and C. T. Carlson. 1978. HF skywave measure-
    ments of hurricane winds and waves. Proc. 16th conf. Coastal
    Eng., Hamburg, 190-208.
Phillips, O. M. 1957. On the generation of waves by turbulent
    wind. J. Fluid Mech, 2: 417-445.
Stewart, R. W., and M. J. Manton. 1971. Generation of wind waves
    by advected pressure fluctuations. Geophys. Fluid Dyn., 2:
    263-272.
Trizna, D. B., R. W. Bogle, J. C. Moore, C. M. Howe. 1980. Obser-
    vation by HF radar of the Phillips resonance mechanism for
    generation of wind waves. Journal Geophysical Research Vol
    C9, 4946-4956. Also NRL Report #8386, 1980.
Trizna, D. B., J. M. Headrick, R. W. Bogle and J. C. Moore. DATE.
    Directional sea spectrum determination using HF Doppler radar
    techniques, IEEE Trans. AP-25: 4-11.
Tyler, G. L., G. C. Teague, R. H. Stewart, A. M. Peterson, W. H.
    Munk, J. W. Joy. 1974. Wave directional spectra from synthe-
    tic aperture observations of radio scatter, 1974. Deep Sea
    Res. 21: 989-1016.
Valenzuela, G. R. 1974. The effect of capillarity and resonant
    interactions on the second-order Doppler spectrum of radar sea
    echo, Journal Geophysical Research 79: 5031-5037.

Processes in Marine Remote Sensing-Radar Altimetry of the Sea Surface

D.E. Cartwright

Microwave sensors, both active and passive, carried on artificial satellites are particularly valuable in monitoring the sea
surface because of the cloud penetrating properties of microwaves
and their day-night all-weather capability. Among the microwave
sensors tried on such pioneer experiments as Skylab, GEOS-3, and
Seasat, the most outstanding success has been the radar altimeter.
The principle is simple, that of timing the return from a transmitted pulse reflected from the sea surface, but high class engineering has refined the precision down to better than a nanosecond,
corresponding to an order 0.1 m resolution of the mean sea surface,
after atmospheric and other corrections have been applied. The
area of the "footprint" or area of sea surface which each pulse
sees is of order 10 km^2, but averages are usually taken over
several consecutive pulses at 0.01 second intervals, corresponding
to about 70 m of sub-satellite track.

As well as giving a measure of mean surface elevation, the
"spread" of the return pulse has been shown to give a useful measure of mean wave height to about 0.5 m precision. This is the
height of the real waves as felt by ships, as distinct from the
ripple-roughness on which scattered radiation sensors such as the
scatterometer depend. The altimeter in fact also gives a measure
of ripple-roughness through its monitoring of the intensity of the
return pulse. These features obviously open interesting opportunity for those who wish to collect synoptic statistics of wave
height and roughness over the global oceans, although they do not
provide wavelength information as in, say, synthetic aperture radar
imagery.

To return to the more novel and challenging measure of mean
sea surface elevation afforded by the radar altimeter, one of the
difficulties in interpretation is referring the elevations to
geocentric co-ordinates with comparable precision. In effect, one
has to subtract the altimeter height from the geocentric height of
the satellite itself, and this is extremely demanding of current
methods of orbit determination. Precision therefore depends on
extensive networks of satellite-to-ground ranging stations. Laser
ranging is the most accurate method, but requires cloud-free skies;
more regular tracking is usually obtained from the less precise but
more numerous Doppler stations. At present orbital determination
can at best be brought to the order of 1 m accuracy. The most
meaningful measure of sea surface elevation therefore also has
about 1 m precision at best. Even so, remarkable profiles of the
geoidal undulations, with amplitudes up to 100 m, were clearly
demonstrated in the Skylab experiments of 1972-73, and highly
detailed maps of the "sea-surface geoid" have since been made to
0.5 m accuracy by averaging over the altimetry of the thousands of
recorded orbits performed by GEOS-3. The larger wavelength features of the geoid have, of course, been known for some time from
studies of perturbations of satellite orbits, but the altimeter
enables the short wavelengths to be determined with unprecedented
accuracy.

Although orbital accuracy limits geocentric height determination to about a meter, small wavelength features of the surface can be detected (effectively as "slopes") almost independently of orbital precision. Thus, small deformations of order 1 m amplitude and 100 km horizontal scale, due to the gravitational distortion of seamounts and trenches and to concentrated current-jets such as the Gulf Stream have been clearly identified in altimeter traces. Most recently, scientists have identified cold-ring eddies associated with the Gulf Stream from Seasat altimetry. The synoptic monitoring of such eddies promises to be of great value to modern studies of circulation and climate.

Ideally, oceanographers would like to use the altimetry of surface slopes to monitor the main circulation patterns in the world's oceans. This, however, is an extremely demanding requirement. The main difficulty is that the observed slopes must be assessed not only relative to possible errors in the gradient of the estimated orbit but also relative to the true geopotential gradient of the geoid. Further, in this context the geoid referred to should not be merely the mean ocean surface but the surface of equal gravitational potential which should ideally be determined by purely gravitational measures. For this, the gravitational field at the ocean surface should be known to better than 0.1 milligal at every part of the region to be studied, and this is not the case at present. The alternative and more practical approach is to make do with a reference surface obtained by averaging the altimetric sea surface over a time span of a few years. This would differ from the gravitational geoid only by the effects of the long-term mean circulation, which would therefore be undetermined. However, the surface would be adequate as a reference datum for all time-varying events in the circulation, and these are known to be of major oceanographic interest even on climatic time scales.

Another hope is that satellite altimetry will eventually enable us finally to determine the oceanic tides. At present this is largely prevented by the errors in orbital determination, and the current approach is to use estimates of tidal elevation from computer models as "corrections" to the altimetry. This may be tolerable at current standards of accuracy, but in fact no computer model reproduces the tides with consistent global accuracy better than about 0.5 m and there are many geophysical reasons for wanting to determine the tides with better precision than that. Since the tides at any given place have fixed relationships to the known motions of the Moon and the Sun, it should be possible in principle to determine these relationships from a sufficiently long series of altimetric heights against any reasonably low noise level, without any dependence on reference levels. An rms noise error of 1 m imposed by orbit determination, however, requires many hundred satellite crossings, of say, a 2 degree square of sea surface in order to resolve tidal detail in the 0.1 m range. As with currents, one may more easily extract information on tidal slopes, or differences across a few adjacent squares. When combined with known tidal elevations in certain well-measured areas of ocean, such procedure enables the useful tidal definition to be extended. Tidal slopes are particularly easy to detect in shallow seas, some of which have important unsolved dynamic problems.

The demise of Seasat cut short a number of interesting experiments based on its altimetery which was evidently of a high order of accuracy. Nevertheless, its finite data set is a challenge for assessments of the feasibility of future altimeter-bearing satellites. Initial results from American evaluations are extremely promising, while in Europe a loosely-coordinated group of laboratories known as SURGE are pooling their resources and expertise to evaluate Seasat altimetry over the North Sea and northeast Atlantic ocean. The European philosophy is not to try to learn anything about oceanography from these data, but to use the known tides and wind-induced deformation of the sea surface to combine accurate knowledge of the local gravitational geoid, tracking data from a dense array of Doppler and laser stations and high precision orbital calculations to evaluate the precision of the altimeter signals and their applied corrections. If results are consistent, it may be possible to deduce geoidal and other oceanographic features in sea areas where they are less well known. In any case, a great deal of useful experience will be learned for application to the next generation of altimeter-satellites, promised for the mid-1980's.

REFERENCES

There are many papers and articles describing the principles of the satellite altimeter. A general survey of principles and applications is given in Chapter 3 of:

Active Microwave Workshop Report (Ed. R. E. Matthews), N.A.S.A., 1975, U.S. Government Printing Office, 502 pp.

Several papers on the results from GEOS-3 altimetry are given in Vol. 84 (B8), 1979, of J. Geophysical Research.

Whitecapping, A Manifestation of Air-Sea Interaction with Implications for Remote Sensing

Edward C. Monahan

ABSTRACT

Analysis of photographs taken during the 1978 JASIN experiment lends support to the previously published suggestion (Monahan, 1971) that the relationship between the wind speed at 10-m elevation, U, and W, the maximum percentage whitecap coverage for a particular wind speed, can be described by the following expression: $W = 0.00135U^{3.4}$. Since whitecaps markedly affect the microwave emissivity of the sea surface, and thus apparent brightness temperature of the sea, it should be feasible to determine from space the whitecap coverage of the ocean from the multiple microwave brightness temperatures recorded by Scanning Multichannel Microwave Radiometers, such as the one aboard Nimbus-7. Combining remotely sensed whitecap coverage values with the equation relating U to W, it should be possible to estimate the near surface wind speeds over those portions of the earth covered by water.

INTRODUCTION

All who put to sea are aware that as the wind and sea build up
a stage is reached where whitecaps (or whitehorses) start to form.
Indeed, this has long been reflected in the description of the
various Beaufort sea states. Since whitecaps present marked visible and microwave signatures their detection provides a potential
basis for the remote determination of sea state, and less directly,
of low elevation wind speeds.

PHYSICAL DESCRIPTION

When a wave becomes so steep as to be unstable its crest curls
over or spills and a volume of air is thereby entrapped which
quickly breaks up into many small bubbles (Blanchard and Woodcock,
1957; Monahan and Zietlow, 1969). This cloud of bubbles is carried
down to a depth on the order of the wave height (Donelan, 1978),
and then the bubbles begin to rise slowly to the surface, with the
larger bubbles rising more rapidly (Blanchard and Woodcock, 1957).
There is little or no coalescence of these bubbles as they rise
(Scott, 1975), but the smallest bubbles are subject to absorption,
albeit the sea water is saturated with air (Blanchard and Woodcock,
1957). When these bubbles reach the sea surface they immediately
break (Blanchard and Woodcock, 1957), unless the surface is contaminated with a layer of organic material, in which case the foam
may persist for a protracted period (Abe et al., 1963; Abe, 1963).
Again, no significant bubble coalescence occurs on the sea surface
(Blanchard and Woodcock, 1957). The term whitecap is usually
applied to the foam patches and individual bubbles upon the sea
surface, taken together with the readily visible cloud of submerged
bubbles.

As more and more of the bubbles in the original cloud rise and
break the whitecap begins to decay, with its surface area decreasing in an exponential manner (Monahan and Zietlow, 1969). The time

constant characterising this exponential decay varies with the salinity of the water, which affects the bubble spectrum (Monahan, 1965; Monahan and Zietlow, 1969), and with the water temperature (Miyake and Abe, 1948; Abe, 1955; Blanchard, 1971). As the whitecap thus dies out, it does not do so uniformly. Reflecting the scales of turbulence present, an older whitecap usually breaks up into many smaller patches.

As each of the many bubbles reaches the surface its upper film drains and then bursts producing hundreds of film droplets (Blanchard, 1963). The collapse of the resulting cavity in the sea surface results in the formation of a vertical jet which usually injects several jet droplets into the air (Kientzler et al., 1954; Blanchard 1958; Blanchard 1963).

Given the turbulence associated with a whitecap (Donelan, 1978), it is to be expected that even after the last bubble of a particular whitecap has reached the sea surface and burst a tem-perature scar may persist reflecting the locally enhanced vertical mixing of the near surface layer.

At still higher wind speeds than those associated with the onset of whitecapping a supplementary droplet and foam producing mechanism comes into play. This process is the mechanical shearing off of the wave crests (Blanchard, 1974). Many of the resulting drops strike the sea surface upwind of the crest from which they were torn, producing additional droplets (Harlow and Shannon, 1967) and bubbles. The streaks of foam, or spume, so formed tend to lie parallel to the wind.

Whitecap Coverage Versus Wind Speed

For the equilibrium, fully developed sea associated, in the case of a neutrally stable atmosphere, with a particular 10 m elevation wind speed, we would expect to have a particular percent-age of the sea surface covered by whitecaps. While recognizing that many, perhaps most, of the simplifying assumptions implicit in the preceding sentence are not satisfied by the conditions usually encountered at sea, most investigators nonetheless have attempted to relate their observed whitecap coverage values to wind speed.

Munk (1947), determining the number of foam patches per unit area, concluded that foam patches only appeared when the wind was greater than 7 m sec^{-1}, and identified this as a critical velocity for air-sea interaction. Blanchard (1963) analyzed five photographs from a naval aviation manual (anonymous, 1952) and deduced that there were no whitecaps when the wind was less than 3 m sec^{-1}, and whitecap coverage at higher speeds was proportional to the square of the wind speed (see Table 1 and Figures 1 and 2, from Monahan, 1979). Gathman and Trent (1968) studied a set of photographs (Table 1) and concluded, in agreement with Blanchard's result, that whitecaps were totally absent when the wind was less than 3 m sec^{-1}. Monahan (1969) observed whitecaps on the Laurentian (North American) Great Lakes (Table 1) and concluded that for the same value of deck height wind speed greater whitecap coverage is encountered when the atmosphere is thermally unstable than when it is stable. Ross et al. (1970) presented graphically a whitecap coverage versus wind speed relationship extracted from an unpublished report of Murphy (1968), and a comparable curve based on the semi-empirical calculations of Cardone (1969). Cardone's semi-empirical relation was based on the essentially linear relationship (Equation 1) he had

$$W(F) = 0.0185 + 0.893 \times 10^3 \text{ E} \tag{1}$$

obtained between the fresh water whitecap coverage values, W(F), reported by Monahan (1969) and the rate of energy transfer, E, "to the fully developed portion of the (wave) spectrum" Cardone had calculated for the conditions under which Monahan had taken his whitecap photographs.

Monahan (1971) analyzed a large set of oceanic whitecap photographs (Table 1 and filled circles of Figure 3, from Monahan, 1979), many taken during the BOMEX experiment, and concluded that for 10 m elevation wind speeds, U, between 4 and 10 m sec^{-1}, a curve (Equation 2) could be defined that coincides

$$W = 0.00135 \text{ U}^{3.4} \tag{2}$$

closely with the highest whitecap percentage coverage values, W, at the various wind speeds, and which thus described an envelope over all the data points in this wind range (curve B on Figures 1 and 2,

TABLE 1: Summary of Photographic Studies of Oceanic Whitecap Coverage.

Citation	All Wind Speeds		U_J 4m sec^{-1}		U_J 10m sec^{-1}		Source
	Intervals	Photos	Intervals	Photos	Intervals	Photos	
Munk, 1947[a]	8	-	6	-	4	-	
Blanchard, 1963	5	5	5	5	2	2	Anon., 1952
Gathman & Trent, 1968[b]	-	44	-	-	-	-	
Monahan, 1969[c]	20	292	14	211	5	97	
Ross, et al., 1970	-	-	-	-	-	-	Murphy, 1968 (unpublished)
Monahan, 1971	70	432	60	320	8	40	
Nordberg, et al., 1971	6	-	5	-	4	-	
Toba & Chaen, 1973	41	164	32	128	6	24	
Ross & Cardone, 1974	13	427	13	427	13	427	
Tang, 1974	-	-	-	-	-	-	Stogryn, 1972 (unpublished)
Lai & Shemdin, 1974[d]	10	-	10	-	10	-	Shemdin et al., 1972 (unpublished)
Monahan, 1979	21	217	21	217	7	67	

a   Numerical foam patch concentration
b   Only presence or absence of whitecaps reported
c   Fresh water observations
d   Laboratory tank study

dashed line on Figure 3). Taking Cardone's (1969) semi-empirical
relation based on Monahan's (1969) freshwater whitecap observations, adjusting the curve so that the W(F) values were in terms of
wind speeds at 10 m elevation rather than 19.5m elevation, and
multiplying the resulting W(F) values by 1.51 to account for the
fact that the lifetime of a saltwater whitecap is 51% greater than
that of a freshwater one (Monahan and Zietlow, 1969), Monahan
(1971) obtained Equation 3. This function, referred to

$$W = 0.0012 \ U^{3.3} \tag{3}$$

as C-M-Z-10 m in Monahan (1971), appears as curve C in Figures 1
and 2.

Nordberg et al. (1971), analyzed photographs taken from a
Convair 990 over the Atlantic Ocean west of Ireland and over the
North Sea (Table 1). The analysis, involving the use of a digital
densitometer, yielded whitecap coverage and foam streak or spume
coverage separately. The results for whitecaps only (open circles)
and for whitecaps plus spume (crosses) are depicted on Figures 1
and 2. Toba and Chaen (1973), analyzing whitecap photographs taken
in the East China Sea and Japanese coastal waters using the technique described in Monahan (1969), obtained results "of the same
order with Monahan's". Their data points appear as open squares on
Figures 1 and 2.

Ross and Cardone (1974), expanding upon the data base presented in Nordberg et al. (1971), and describing in more detail the
technique of analysis which involved the use of a scanning false
color densitometer to identify individually whitecap and spume
areas, obtained the results presented as open diamonds (whitecaps
only) and filled diamonds (whitecaps plus spume) in Figures 1 and
2.

Tang (1974) presented some of the whitecap and spume observations of Nordberg et al. (1971), and quoted an expression relating
W to U (Equation 4) from

$$W = 7.75 \ x \ 10^{-4} \ U^{3.231} \tag{4}$$

an unpublished report by Stogryn (1972b). Tang stated that this
expression, which appears as curve D on Figures 1 and 2, was the

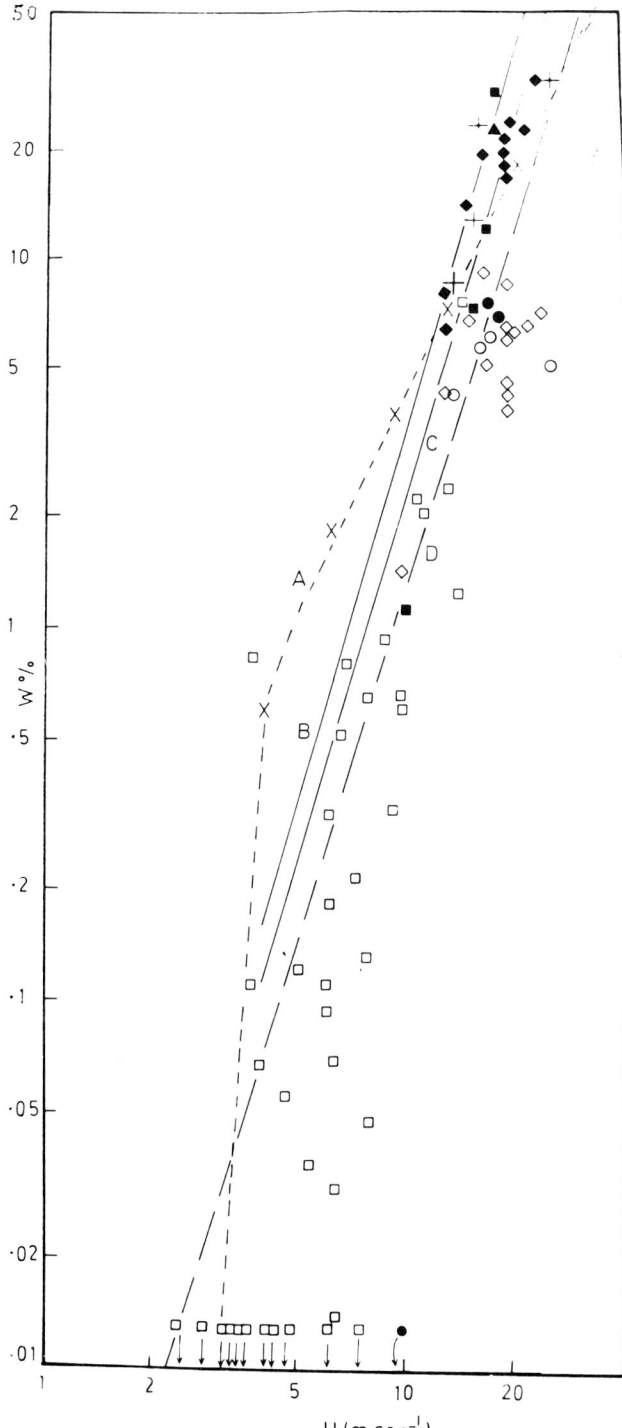

FIGURE 1: Whitecap coverage in percent versus wind speeds, in m sec^{-1}. X's and dashed line A (Blanchard, 1963). Open squares, "percentage of breaking area" (Toba and Chaen, 1973). Open diamonds, whitecap coverage only (Ross and Cardone, 1974). Filled diamonds, whitecaps plus streaks (Ross and Cardone, 1974). Open circles, whitecap coverage only (Nordberg et al., 1971). Crosses, whitecaps plus streaks, (Nordberg, et al., 1971). Whitecap coverage at 24.4 m fetch in saltwater tank (Lai and Shemdin, 1974) with no swell, filled squares; with 5 cm swell, filled circles; with 15 cm swell, filled triangle. Points of Toba and Chaen (1973), and of Ross and Cardone (1974), plotted against 10 m elevation wind speeds. Line B represents maximum whitecap coverage envelope given by Monahan (1971), here extrapolated beyond 10 m sec^{-1} winds. Line C is the C-M-Z-10m curve defined in Monahan (1971), which is also extrapolated to winds above 10m sec^{-1}. Line D is the power law relation of Stogryn (1972) as quoted by Tang (1974). (From Monahan, 1979).

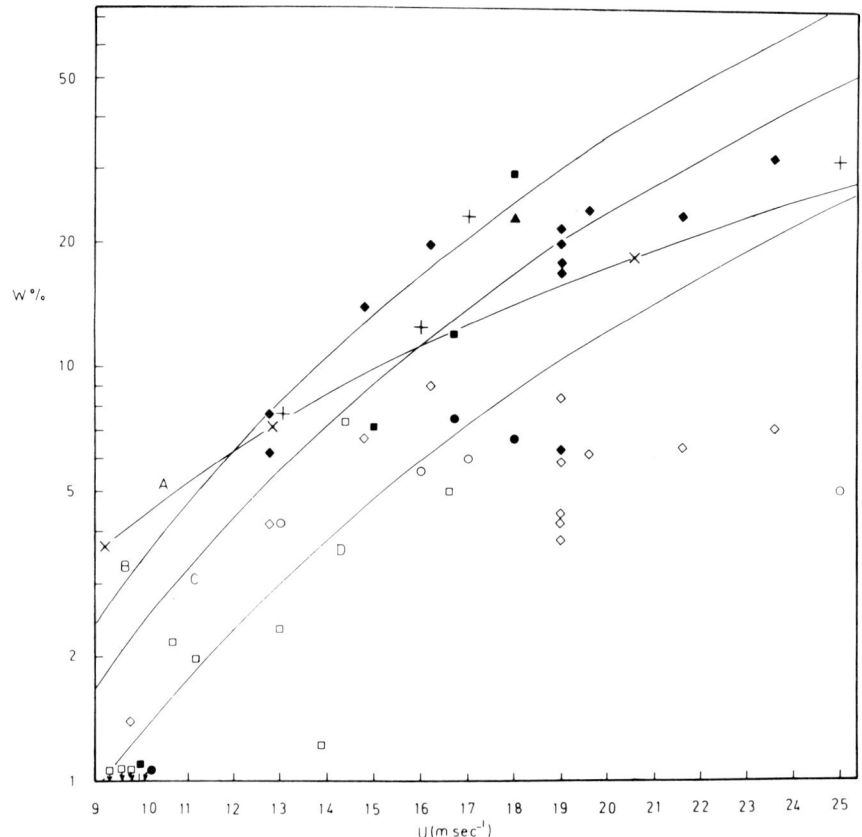

FIGURE 2: Whitecap coverage in percent versus wind speeds, in m sec^{-1}, for wind speeds above 10 m sec^{-1}. Symbols, and curves A, B, and C, and D, as defined in legend of Figure 1. (From Monahan, 1979).

result of a least squares fit to the observations of Murphy (1968), Williams (1970), Rooth and Williams (1970), and Monahan (1969, 1971). It perhaps should be noted that the only quantitative statement in Williams (1970) was to the effect that he had analyzed "a copy of the Naval Squadron photographs, (Murphy, 1968), and arrived at foam coverages about 30% lower than Blanchard's in the range considered here, and higher than Blanchard's at higher velocities." It would appear from this statement, and from a comment in Blanchard (1971), that the analyses in Blanchard (1963), Murphy (1968), and Williams (1970), were all carried out on the same data base (anonymous, 1952).

The results of Shemdin, et al. (1972) reported in Lai and Shemdin (1974), which were obtained in a wind flume at a fetch of 24.4 m and may be of somewhat limited applicability, are shown as filled squares, circles, and triangles in Figures 1 and 2. The effect of the presence of swell can be noted in these results.

Whitecap observations obtained during the 1978 JASIN experiment (Table 1 and Monahan, 1979) are presented as open circles on Figure 3. It is to be noted that only one of these points representing the mean whitecap coverage during the various JASIN observation intervals is above the data envelope described in Monahan (1971) (Equation 2) by more than one standard error, and that point is less than two standard errors above the line. These results thus reinforce the finding (Monahan, 1971) that W values determined from Equation 2 coincide with the highest whitecap coverage at the various wind speeds, and suggest that the range of applicability of this expression extends well beyond the 10 m sec^{-1} quoted in Monahan (1971). When the whitecap coverage calculated from Equation 2 (curve B) is compared with the observations of other authors (Figures 1 and 2), it appears that Equation 2 is valid out to wind speeds in excess of 17 m sec^{-1}. But in making this comparison, it must be noted that the apparent agreement is with the observations of Nordberg, et al. (1971) and Ross and Cardone (1974) of whitecaps plus spume streaks, and not with those authors' observations of whitecaps alone. In suggesting that these authors may have had difficulty in differentiating whitecaps from spume streaks it might be pointed out that they used false color television imagery at one stage of their analysis (Ross and Cardone, 1947). Case D of Nordberg, et al. (1974), corresponding to the first observation listed in Table 2 of Ross and Cardone (1974), is a specific example which might be called into question. Nordberg, et al. (1971) found for this case, where the wind speed was 16 m sec^{-1}, that 6.9% of the sea surface was covered by spume streaks. This represented 55% of the total foam cover determined for case D. Yet Figure 4a of Nordberg, et al. (1971), which is a photograph of the sea surface identified with case D, shows few if any linear features, and corresponds with what many observers would charac-

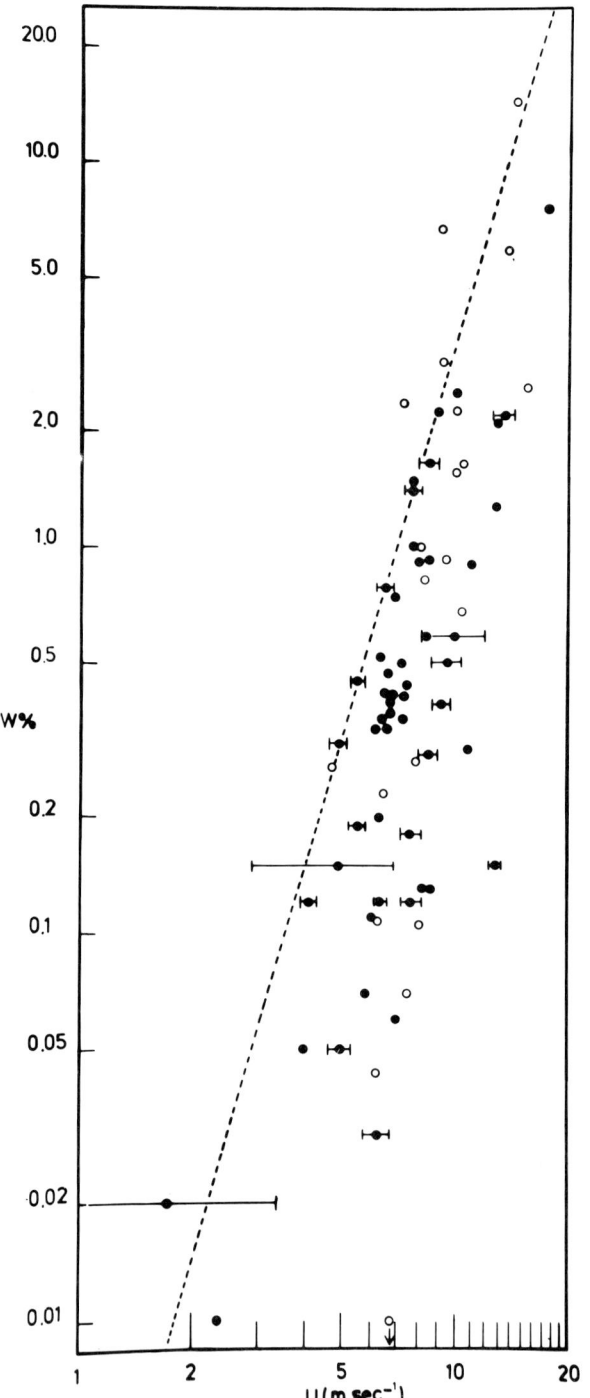

FIGURE 3: Whitecap coverage in percent versus 10 m elevation wind speeds, in m sec^{-1}. Open circles, JASIN results. Filled circles, results from Monahan, (1971). Horizontal bars associated with some filled circles span range of wind speeds measured during observation periods. Dashed line represents maximum whitecap coverage envelope given in Monahan (1971) (From Monahan, 1979).

terize as typical conditions for the sea surface at the stipulated wind speed.

The whitecap photographs taken during the JASIN experiment (Monahan, 1979) were taken aboard the R. R. S. CHALLENGER, which carried comprehensive meteorological instrumentation. A preliminary comparison of the whitecap observations with the friction velocity, U*, determined from stability dependent bulk aerodynamic formulae (Monahan and Davidson, 1979) indicates a close correlation (Figure 4), and has led to the tentative suggestion that U*, or the surface stress, might be more appropriate than U for scaling W.

IMPLICATIONS FOR REMOTE SENSING

It has long been recognized by practitioners of passive microwave remote sensing (e.g., Williams, 1969) that the apparent microwave brightness temperature of the sea surface was a function of the low elevation wind speed, and that in this relationship lay the seeds of a technique for the remote determination of sea surface wind speeds from satellite microwave radiometric measurements. The role of whitecaps or sea foam in increasing the emissivity of the sea surface, and of thereby increasing the apparent sea surface microwave brightness temperature in the range of frequencies from 13.4 to 37 GHz has been summarized by Stogryn (1972a). A simple physical model for the high microwave emissivity of sea foam has been presented by Droppleman (1970). Webster et al. (1976), have put forth the physical argument that the bubbles in a whitecap produce cusps on the surface of the sea, and that this cusped sea surface represents a varying dielectric layer which leads to increased emissivity and hence to a higher apparent sea surface microwave brightness temperature. The emissivity of a whitecap covered sea surface has been calculated as a function of viewing angle, polarization, and microwave wavelength (Webster et al. 1976). Using a 1.55 cm (19.35 GHz) microwave radiometer looking vertically downward Ross et al. (1970) found that the apparent brightness temperature of the sea surface increased with wind speed, for winds between 17 and 25 m sec^{-1}, at a rate of 1°K/m

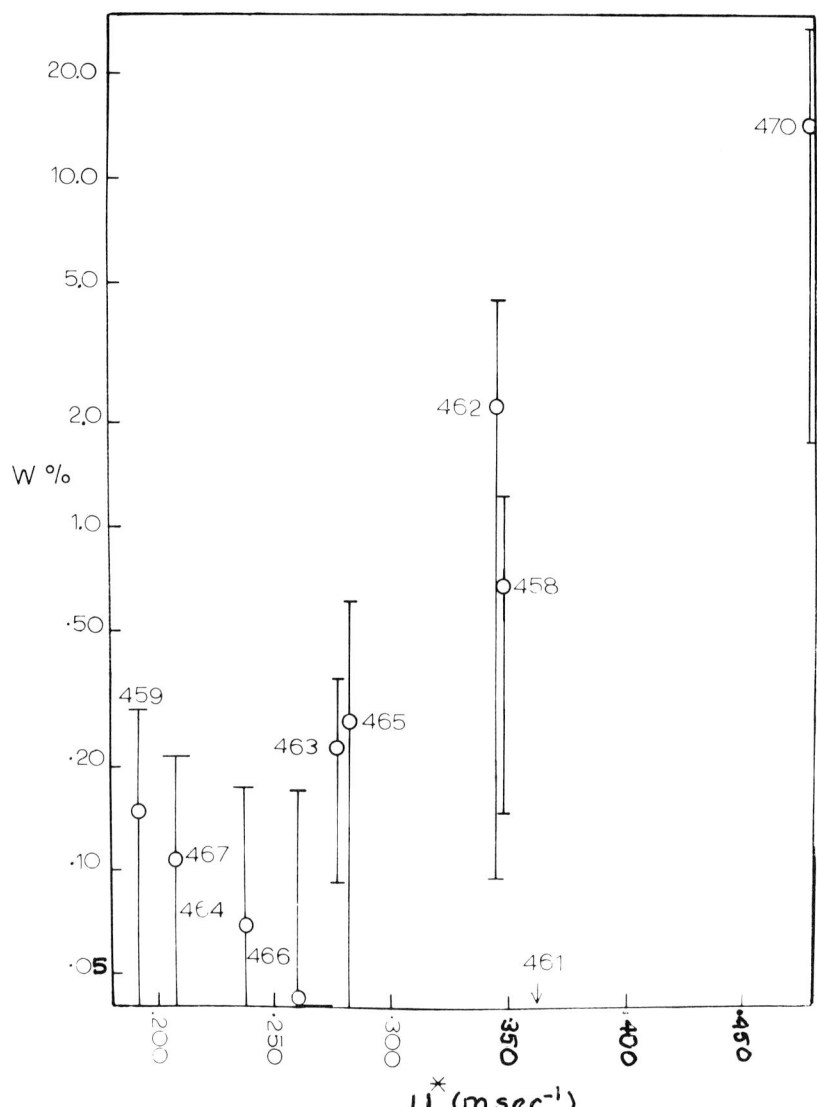

FIGURE 4: Oceanic whitecap coverage, in percent, versus friction velocity, U^*, in m sec^{-1}. Mean values, with indicated standard deviations, for those JASIN whitecap observation intervals for which U^* estimates, obtained from stability dependent bulk aero-dynamic formulae, were immediately available. (From Monahan and Davidson, 1979).

sec^{-1}. Using the same system, Nordberg et al. (1971), reported that the apparent brightness temperature increased over the wind speed range of 7 to 25 m sec^{-1} at a rate of 1.2°K/m sec^{-1} when viewing in the nadir direction, and at a rate of 1.8°K/m sec^{-1} when viewing 70° off the nadir direction, and stated that these increases were due primarily to white water cover. The measured rates of apparent sea surface brightness temperature increase with wind speed, as a function of viewing angle, polarization, and microwave wavelength have been presented by Webster et al. (1976) and Gloersen and Barath (1977). Ross et al. (1970), have shown clearly the effect of individual whitecaps in raising the apparent microwave temperature of the sea surface as measured from low flying aircraft. Spikes representing an increase of as much as 70°K above the background brightness temperature were observed in the aircraft radiometer record when a big whitecap all but filled the antenna beam.

The Scanning Multichannel Microwave Radiometer (SMMR), which was carried aboard SeaSat-A and which is also included as part of the instrumentation aboard Nimbus-G (7), is designed to measure the apparent brightness temperature of the sea surface at five microwave wavelengths (0.8, 1.4, 1.7, 2.8, and 4.6 cm) and both polarizations. Since the influence of whitecaps on the brightness temperature varies with microwave wavelength, polarization, and viewing angle (Webster et al. 1976) it is expected that the surface wind speed to within 2 m sec^{-1} can be deduced, as well as the sea surface temperature, from the apparent brightness temperatures measured at the 2.8 and 4.6 cm wavelengths (Gloersen and Barath, 1977; Sherman, 1979).

The bursting of whitecap bubbles is a major source of the water droplets found in the air just above the sea (Blanchard and Woodcock, 1957; Blanchard, 1963; Monahan, 1968), and the concentration of particularly the larger of these droplets will increase markedly, as will whitecap coverage, with increasing wind speed (Monahan, 1968; Blanchard and Syzdek, 1972). That these droplets make a significant contribution to the transfer of energy across the air-sea interface has been pointed out by Wang and Street (1978), and by Ling and Kao (1979). Specifically, the water vapour, or humidity, as well as temperature, fields just above the surface of the ocean are strongly dependent on the strength of the spray droplet flux (Ling and Kao, 1976; 1979; Ling et al. 1979). Wu (1974) deduced from tank studies that spray droplets play the paramount role in "sea surface evaporation" when the winds are in excess of 15 m sec^{-1}. The resulting vertical distribution of these droplets in the air over the sea has been treated theoretically by Toba (1965) and by Ling and Kao (1979). This vertical droplet distribution has been measured over a laboratory wind wave tank by Wu (1973), and over the open ocean by Chaen (1973) and by Ling and Kao (1979).

Now the presence of water droplets, and indeed water vapor, influences the apparent microwave brightness temperature of the sea surface. Specifically, the spray droplets create an inhomogeneous transition layer immediately above the sea surface which "first increases the air-sea interface brightness temperature and then attenuates and re-emits radiation to make significant contributions to the measured sea surface brightness temperature" (Tang, 1974). It follows therefore that by using the apparent sea surface bright-ness temperatures measured by the SMMR at the wavelengths other than those used in the determination of low elevation wind speed and sea surface temperature, it should be possible to determine the liquid water, and water vapor, content of the air overlying the ocean (Gloersen and Barath, 1977; Sherman, 1979).

While the remarks to this point have dealt with passive micro-wave systems, Valenzuela and Laing (1970) have stated that the noisiness of the spectra and other aspects of the radar sea echo obtained with a multi-frequency (0.43, 1.23, 4.46, and 8.91 GHz) coherent pulsed radar indicate the possible influence of sea spray.

The sea surface radiance measurements made in the various visible and infra-red wavelength bands by the Coastal Zone Color Scanner aboard Nimbus-G(7) is also subject to significant whitecap influence, since at high wind speeds upwards of 10% of the sea surface can be covered by whitecaps (Fig. 1, 2 and 3).

SUMMARY

Since the algorithms now used to determine near surface wind

speed, U, from apparent sea surface microwave brightness temperatures implicitly assume that there is a simple relationship between
whitecap coverage, W, and U, it is desirable that further studies
should be carried out to define in more detail the dependence of W
on U. If whitecaps have longer lifetimes in cold water than in
warm water, then latitudinal and seasonal variations in the relationship between W and U are to be expected. Likewise, other
geographical variations in the W-U relationship should be looked
for since certain oceanic areas, such as the interior of the major
oceanic gyres, are less organically rich than other portions of the
world oceans, such as coastal upwelling regions, and the concentration of organic material on the sea surface influences the whitecapping process.

It is fortunate that the whitecap observations taken as part
of the JASIN experiment (Monahan, 1979) are available, along with
the other JASIN data, as sea truth for the SeaSat-A measurements.

Using the whitecap coverage and surface wind values recovered
from the apparent sea surface brightness temperatures measured on
several of the SMMR channels in conjunction with the models of
Toba (1965), Ling and Kao (1979), and others, it should be possible
to calculate the liquid water and water vapor content to be
expected in the air just above the sea, and thereby check the
values of these atmospheric parameters which are extracted from the
additional apparent sea surface brightness temperatures which are
recorded on the other SMMR channels.

In attempting to relate whitecap coverage, determined from the
identification and measurement of individual whitecaps on photographic transparencies of the sea surface, to some integrated
measure of the sea surface such as apparent microwave brightness
temperature, the effect of optical resolution on the former determination must not be neglected. Specifically, the relationship
between photographic resolution and apparent whitecap coverage
needs to be more thoroughly investigated.

ACKNOWLEDGEMENTS

Our study of the influence of whitecaps on the marine atmosphere is sponsored by the U.S. Office of Naval Research (N00014-78-G-0052). The recent shipboard whitecap photography and the extensive analyses of the resulting photographs were made possible by the assistance provided the author by E. Murphy and F. Cunningham. We gratefully acknowledge the assistance of D. Ellett and A. Edwards of the Scottish Marine Biological Association and the aid of the personnel of the R. R. S. CHALLENGER during the 1978 JASIN cruises.

REFERENCES

Abe, T. 1955. A study on the foaming of sea water: on the mechanism of the decay of bubbles and their size distribution in foam layer of sea water. Papers Meteor. Geophys., 5, pp. 240-247.

Abe, T. 1963. In situ formation of stable foam in sea water to cause salty wind damage. Papers Meteor. Geophys., 14, pp. 93-108.

Abe, T., T. Ono, and M. Kishino. 1963. A fundamental study on the prevention of the salty damages due to the foaming of Sea water (preliminary report). Journal Oceanogr. Soc. Japan, 18, pp. 185-192.

Anonymous. 1952. Wind estimations from aerial observations of sea conditions. Unpublished report, Weather Squadron Two (VJ-2), Jacksonville, Fla.

Blanchard, D. C. 1958. Electrically charged drops from bubbles in sea water and their meteorological significance. Journal of Meteorology, 15, pp. 383-396.

Blanchard, D. C. 1963. The electrification of the atmosphere by particles from bubbles in the sea. Progress in Oceanography, 1, pp. 71-202. Pergamon Press, New York.

Blanchard, D. C. 1971. Whitecaps at sea. Journal of the Atmospheric Sciences, 28, p. 645.

Blanchard, D. C. 1974. International Symposium on the Chemistry of Sea/Air Particulate Exchange Processes: Summary and Recommendations. Journal de Recherches Atmosphériques, 8, pp. 509-513.

Blanchard, D. C., and L. Syzdek. 1972. Variations in Aitken and Giant Nuclei in Marine Air. Journal of Physical Oceanography, 2, pp. 255-262.

Blanchard, D. C. and A. H. Woodcock. 1957. Bubble formation and modification in the sea and its meteorological significance. Tellus, 9, pp. 145-158.

Cardone, V. J. 1969. Specification of the wind distribution in the marine boundary layer for wave forecasting. Tech. Report GSL-69-1, New York University, pp. 1-131.

Chaen, M. 1973. Studies of the Production of Sea-Salt Particles on the Sea Surface. Memoirs of the Faculty of Fisheries, Kagoshima University, 22, No. 2., pp. 49-107.

Donelan, M. A. 1978. Whitecaps and momentum transfer, pp. 273-287, in A. Favre and K. Hasselmann, Eds., Turbulent fluxes through the sea surface, wave dynamics, and prediction, NATO Conference Series: V, Air-Sea Interactions, vol. 1, pp. 1-677, Plenum Press, New York.

Droppleman, J. D. 1970. Apparent Microwave Emissivity of Sea Foam, Journal of Geophysical Research, 75, pp 696-698.

Gathman, S. and E. M. Trent. 1968. Space charge over the open ocean. Journal of the Atmospheric Sciences, 25, pp. 1075-1079.

Gloersen, P., and F. T. Barath. 1977. A Scanning Multichannel Microwave Radiometer for Nimbus-G and SeaSat-A. IEEE Journal of Oceanic Engineering, OE-2, pp. 172-178.

Harlow, F. H., and J. P. Shannon. 1967. Distortion of a Splashing Liquid Drop. Science, 157, pp. 547-550.

Kientzler, C. F., A. B. Arons, D. C. Blanchard, and A. H. Woodcock. 1954. Photographic investigation of the projection of droplets by bubbles bursting at a water surface. Tellus, 6, pp. 1-7.

Lai, R. J. and O. H. Shemdin. 1974. Laboratory Study of the Generation of Spray Over Water. Journal of Geophysical Research, 79, pp. 3055-3063.

Ling, S. C. and T. W. Kao. 1976. Parameterization of the Moisture and Heat Transfer Process over the Ocean under Whitecap Sea States. Journal of Physical Oceanography, 6, pp. 306-315.

Ling, S. C. and T. W. Kao. 1979. Multiphase Fluxes over North Atlantic Ocean - JASIN 1978 experiment, pp. 1-35. Separate from JASIN Data Display Meeting, Woods Hole, 21-25 May, 1979.

Ling, S. C., T. W. Kao, and A. I. Saad. 1979. The Role of Micro Water Droplets on the Transport of Heat and Moisture from the Ocean and Lake, J.A.S.C.E (In press).

Miyake, Y., and T. Abe. 1948. A study on the foaming of sea water. Part 1. J. Marine Res., 7, pp. 67-73.

Monahan, E. C. 1965. A Field Study of Sea Spray and its Relationship to Low Elevation Wind Speed-Preliminary Results, Woods Hole Oceanographic Institution, Massachusetts, pp. 1-147.

Monahan, E. C. 1968. Sea Spray as a Function of Low Elevation Wind Speed. Journal of Geophysical Research, 73, pp. 1127-1137.

Monahan, E. C. 1969. Fresh Water Whitecaps. Journal of the Atmospheric Sciences, 26, pp. 1026-1029.

Monahan, E. C. 1971. Oceanic Whitecaps. Journal of Physical Oceanography, 1, pp. 139-144.

Monahan, E. C. 1979. The dependence of oceanic whitecap coverage on wind speed. (MS in prep.).

Monahan, E. C., and K. L. Davidson. 1979. Preliminary Intercomparisons of JASIN Wind, Whitecap, and Aerosol Observations. Separate distributed at JASIN Data Display Meeting, Woods Hole Oceanographic Institution, 21-25 May, 1979, pp. 1-11.

Monahan, E. C., and C. R. Zietlow. 1969. Laboratory Comparisons of Fresh-Water and Salt-Water Whitecaps, Journal of Geophysical Research, 74, pp. 6961-6966.

Munk, W. H. 1947. A critical wind speed for air-sea boundary processes. Journal of Marine Research, 6, pp. 203-218.

Murphy, H. 1968. Percentage foam vs. wind velocity. Internal report, University of Miami, Coral Gables.

Nordberg, W., J. Conaway, D. B. Ross, and T. Wilheit. 1971. Measurements of Microwave Emission from a Foam-Covered, Wind-Driven Sea. Journal of the Atmospheric Sciences, 28, pp. 429-435.

Rooth, C., and G. Williams. 1970. Microwave radiometry of the ocean. Quarterly report, University of Miami, Coral Gables.

Ross, D. B. and V. Cardone. 1974. Observations of Oceanic Whitecaps and their Relation to Remote Measurements of Surface Wind Speed. Journal of Geophysical Research, 79, pp. 444-452.

Ross, D. B., V. J. Cardone, and J. W. Conaway, Jr. 1970. Lasar and Microwave Observations of Sea-Surface Condition for Fetch-Limited 17- to 25- m/s winds. IEEE Transactions on Geoscience Electronics, GE-8, No. 4, pp. 326-336.

Scott, J. C. 1975. The preparation of water for surface-clean fluid mechanics. Journal of Fluid Mechanics, 69, pp. 339-351.

Shemdin, O. H., R. J. Lai, A. Reece, and G. Tober. 1972. Laboratory investigations of whitecaps, sprays and capillary waves. Tech. Rep. 11, Dept. of Coastal and Oceanogr. Eng., Univ. of Florida, Gainesville.

Sherman, J. W. 1979. SEASAT applications to oceanic monitoring, pp. 1-50. Separate from Seminar/Workshop on Ocean Products and IGOSS Data Processing and Services System, Moscow, 2-11 April 1979, sponsored by I.O.C. and W.M.O.

Stogryn, A. 1972a. The Emissivity of Sea Foam at Microwave Frequencies. Journal of Geophysical Research, 77, pp. 1658-1666.

Stogryn, A. 1972b. A study of radiometric emission from a rough sea surface. NASA Contract Report CR-2088.

Tang, C. C. H. 1974. The Effect of Droplets in the Air-Sea Transition Zone on the Sea Brightness Temperature. Journal of Physical Oceanography, 4, pp. 579-593.

Toba, Y. 1965. On the giant sea-salt particles in the atmosphere. II. Theory of the Vertical Distribution in the 10 m Layer Over the Ocean. Tellus, 17, pp. 365-382.

Toba, Y. and M. Chaen. 1973. Quantitative Expression of the Breaking of Wind Waves on the Sea Surface. Records of Oceanographic Works in Japan, 12, No. 1, pp. 1-11.

Valenzuela, G. R. and M. B. Laing. 1970. Study of Doppler Spectra of Radar Sea Echo. Journal of Geophysical Research, 75, pp. 551-563.

Wang, C. S. and R. L. Street. 1978. Transfers Across an Air-Water Interface at High Wind Speeds: The Effect of Spray. Journal of Geophysical Research, 83, pp. 2959-2969.

Webster, W. J., T. T. Wilheit, D. B. Ross, and P. Gloersen. 1976. Spectral characteristics of the microwave emission from a wind-driven foam-covered sea. Journal of Geophysical Research, 81, pp. 3095-3099.

Williams, G. F. 1969. Microwave Radiometry of the Ocean and the Possibility of Marine Wind Velocity Determination from Satellite Observations. Journal of Geophysical Research, 74, pp. 4591-4594.

Williams, G. F. 1970. Comments on "Fresh water whitecaps" Journal of the Atmospheric Sciences, 27, p. 1220.

Wu, J. 1973. Spray in the Atmospheric Surface Layer: Laboratory Study. Journal of Geophysical Research, 78, pp. 511-519.

Wu, J. 1974. Evaporation due to spray. Journal of Geophysical Research, 79, pp. 4107-4109.

Marine Lidar for Subsurface Sensing

Ronald L. Schwiesow

ABSTRACT

Oceanographic lidar (an acronym for light detecting and ranging) has a number of applications in sensing subsurface features, including bathymetry, determination of turbidity profiles, fluorosensing of composition, and measurement of temperature profiles. We first consider in this tutorial review the type of return signals expected from a practical lidar and the lidar equation. Then the types of actual and projected oceanographic lidar applications are classified and interrelated from the physical interactions involved in the scattering of laser light. Bathymetry and fluorosensing have been done from aircraft, and temperature profiles have been measured from a surface platform.

133

PERSPECTIVE

Remote sensing of the ocean can involve passive (radiometric) measurement of upwelling radiance and active (backscatter) detection of a transmitted pulse of electromagnetic radiation. Both active and passive sensing can be done in spectral regions ranging from radio waves to the ultraviolet. Examples of radiometric sensing include ocean color scanning in the visible, sea-surface temperature sensing in the infrared, and microwave radiometry for surface temperature and sea state. Backscatter remote sensing techniques are represented by high frequency radar for waveheight and direction, radar scatterometry for sea state and altimetry, and synthetic aperture radar.

The purpose of this tutorial review is to bring active optical backscatter techniques to the attention of the marine remote sensing community. Active optical sensing is commonly called lidar, an acronym for light detecting and ranging. Oceanographic lidar has been used for bathymetry, sensing of organic materials in the ocean, and temperature profiling. We will summarize these applications and refer the interested reader to more detailed discussions in the literature.

The advantage of active sensing in the blue-green part of the electromagnetic spectrum is that such light penetrates the surface to make measurements of subsurface properties or objects. Microwave and infrared techniques penetrate at most a few millimeters into the sea surface. Surface-layer and bulk-average values of temperature and salinity, for example, can be quite different. In contrast to acoustic remote sensing, where the transmitter and receiver must be in contact with the water for efficient operation, light couples effectively into and outward from the water.

Because oceanographic lidar can work effectively from an airborne platform, it is possible to cover wide areas of ocean rapidly. Scanning transverse to the flight path allows good horizontal resolution without flying a pattern spacing equal to the desired resolution. In the near future, limitations on practical laser power and receiver telescope size, together with the inverse square power dependence of signal on range, mean that oceanographic lidar will be used with aircraft rather than spacecraft.

LIDAR SIGNALS

Ambient Light

The lidar receiver will generally receive radiant energy from various extraneous sources in addition to subsurface scattering. Sunlight reflects from the surface to generate sea glint. This phenomenon has been studied as an indicator of spatial spectra of waves and other surface properties of the ocean (Guinn et al., 1979). Sea glint, although covering a broad spectral region, is bright enough to swamp the lidar return and must be avoided by proper sensing geometry.

Ocean color differences result from differential scattering and absorption of sunlight by constituents in the sea, such as chlorophyll. For recent work see Cracknell (this volume) and Kim et al., (1979). Radiance from scattered sunlight is not as strong as sea glint and is spread over the entire visible spectrum. Compared to marine lidar, ocean color scanning has the advantages that it does not require a laser transmitter and is operable from satellite altitudes. Interpretation of ocean color differences, especially for quantitative information, is difficult because of interfering effects. Atmospheric scattering by aerosols or thin cirrus affects the spectral and spatial distribution of the incident solar radiation. The atmosphere also produces spectrally dependent attenuation of the sea scatter return and an overlapping upwelling radiance of its own. It can be difficult to separate various contributions to the ocean scattering itself, including many different types of sediment, chlorophyll, algae, other plants, and foam.

For certain applications, oceanographic lidar provides data specific to the parameter of interest. The effects of interfering signals can be reduced by using the pulsed nature and/or narrow spectral bandwidth radiation of the laser transmitter.

Lidar Transmitter

Lasers used as lidar transmitters are spatially coherent. This means that a single-transverse-mode laser in principle can be focused at the surface to a diffraction-limited spot of less than 3-mm diameter with a 30-cm beam-expanding telescope working at an altitude of 500 m. Such tight focusing reduces the variation in angle of incidence with roughness of the sea surface (Prettyman and Cermak, 1969) and in turn leads to maximum flux density from scattered light at the receiver focus. Atmospheric refractive index inhomogeneities can induce beam spreading (Clifford, 1978; Schwiesow and Calfee, 1979; Yura, 1979), but the magnitude of this effect in the marine environment is not yet well established. A useful way of viewing atmospheric effects is to consider a (range-dependent) maximum coherent aperture (effective aperture) diameter permitted by the atmosphere and compare this to the actual transmitter telescope aperture. The spot size at the air-water interface for a diffraction-limited beam varies inversely with effective aperture diameter and directly with range.

For some applications such as bathymetry (bottom depth measurement), the received flux density is not as important as knowing the beam-average angle of incidence so that an accurate range-to-depth transformation can be made. The spot size at the surface can be made larger than the diffraction limit by collimating rather than focusing the transmitted beam.

Transmitter lasers are also more or less temporally coherent. This means that the laser output is spectrally narrow, or that the linewidth is a very small fraction of the center wavelength. Typical values of monochromaticity range from a width of 0.5 nm at 500-nm wavelength to 1 MHz at 600 THz (600×10^{12} Hz) in frequency units. The principal virtue of a narrow linewidth is the ability to discriminate between narrowband scattered return signal and broadband interference from solar sources. Some lidar remote sensing techniques, such as Brillouin and Doppler spectroscopy, require narrow laser linewidths to measure properly the return spectra in high resolution.

Types of lasers used for oceanographic lidar include argon-ion (usually cavity-dumped), dye, frequency-doubled Nd:YAG, and pulsed neon. The argon-ion laser is more spatially and temporally coherent than the others, but less efficient on an optical power out to electrical power than the others, but less efficient on an optical power out to electrical power in basis. The two major Ar+ laser lines are at 488 nm and 514.3 nm. Dye lasers have the advantage of operating at any chosen wavelength in the visible part of the spectrum. They are more energy efficient than Ar+, but tend to be optically and mechanically the most complex systems. Doubled Nd:YAG lasers are comparatively coherent and efficient, but require a sensitive nonlinear crystal for doubling. They operate at 530 nm. Pulsed neon lasers at 540 nm have the advantage of simplicity and a short 3-ns pulsewidth, but exhibit poor spatial coherence. These four types of lasers are well developed; detailed data are available from a variety of commercial suppliers.

Newly developed lasers that may be appropriate for oceanographic applications are copper vapor at 510 and 578 nm (Anderson et al., 1975) and excimer at 505 nm (Eden et al., 1979). These are comparatively highpower, pulsed gas laser systems with efficiencies near 1%.

The Lidar Equation

Range information is available from timing the return signal with respect to a reference point. The depth corresponding to a signal at time t is

$$d = ct/2n \qquad (1)$$

where c is the speed of light in vacuum, t is the time after reference, and n is the refractive index of the sea water. A convenient reference point is the large spike in received radiant power from the diffuse reflectance of the sea surface. Depth d is then measured with respect to the local surface as a reference level. In round figures, the range-time conversion is approximately 110 m/μs. A 3-ns pulse leads to a depth uncertainty of less than 40 cm on a single-shot basis from timing considerations alone. Wave slope refraction may cause larger errors. Pulse averaging can reduce the depth uncertainty if the signal-to-noise is large enough.

The returned signal can be approximated by an adaption of the lidar equation (Sizgoric and Carswell, 1975). The instantaneous peak received radiant power ϕ_s is

$$\phi_s = \phi_0 \, A \, \eta \, T_1 \, \exp[- \int_0^d \alpha_1(z,\lambda)dz] \, \beta \, \exp[\int_d^0 \alpha_2(z,\lambda)dz] \, T_2$$
$$\cdot \, \Delta d \, (R + d/n)^{-2} + B(\lambda) \, \Delta\lambda \, \Delta\Omega \, (R\Delta\theta)^2. \qquad (2)$$

The power depends on peak transmitted power ϕ_0, receiver telescope area A, overall optical efficiency η, transmission through the air to water interface T_1 and water to air interface T_2, and attenuation coefficient for downward-propagating light $\alpha_1(z,\lambda)$ and upward-propagating $\alpha_2(z,\lambda)$ at wavelength λ. Variable z is the line-of-sight propagation distance below the surface. The volume backscatter coefficient of the water β is the ratio of backscattered to incident optical power per length of the scattering region and per solid angle of the receiver telescope as seen from the scatterers; it has units of $[m^{-1} \, sr^{-1}]$. The lidar pulse length Δd is set by the greater of the laser pulse time and the receiver system time resolution Δt; the lidar depth resolution can be no better than Δd. The height of the lidar above the surface is R. The background spectral radiance $B(\lambda)$ depends on the wavelength received, atmospheric and sea state, and ambient lighting conditions. It has units of optical power per unit area of radiating surface, per solid angle of the receiver telescope, and per receiver spectral bandpass $\Delta\lambda$. The solid angle of the receiver telescope as seen from the scattering region is $\Delta\Omega = A(R + d/n)^{-2}$. For simplicity we assume all background radiance to come from the sea surface, so the area of background radiator contributing to the background signal is $(R\Delta\theta)^2$, where $\Delta\theta$ is the angular field of view of the lidar system.

The lidar equation shows a number of characteristics of marine lidar performance. The spectral volume attenuation coefficient $\alpha(z,\lambda)$ ranges between approximately 0.03 m^{-1} and 1 m^{-1} (Duntley, 1963). The attenuation length (or optical depth) is α^{-1}, the depth at which incident light is attenuated to 1/e of its original flux.

The lidar penetration depth and receiver dynamic range are determined by the attenuation, and lidar signal decreases exponentially
with depth of the sensing volume.

In applying marine lidar to a specific problem, a scientist
has control over a number of parameters in equation (2). Of course
the laser power, telescope area, and optical efficiency should be
as large as practical, and the lidar platform as close to the sea
surface as permitted by the application. The range cell Δd should
be no smaller than necessary. Noise from the background radiance
varies as the square root of the background radiant power. To
minimize this noise, the lidar must operate with a narrow spectral
bandwidth and small field of view. These two parameters are ulti-
mately limited by the temporal and spatial coherence of the laser
in the system, although the scattering mechanism used for the
sensing can require a wider receiver bandwidth than the laser
linewidth, and atmospheric effects can degrade the lidar field of
view.

The state of polarization of the return signal depends on the
type of scattering involved. Scattering from suspended particulates small with respect to the wavelength of light, from spherical
particles (Mie scatter), and from molecules preserves the polarization state of the incident laser light. Scattering from hard
targets or the bottom tends to be depolarized; that is, such signals consist of unpolarized light with approximately equal power in
polarization directions parallel and perpendicular to the incident
polarization. Using a receiver sensitive to the perpendicular
polarization discriminates against particulate scattering in favor
of target scattering (Sizgoric and Carswell, 1975). Multiple
scattering in turbid water, where the scattered signal contains
some light that has undergone more than one scattering interaction,
is partly depolarized (Granatstein et al., 1972).

This review can only mention several oceanographic optical
topics that bear on the performance of marine lidar. Problems such
as beam spread by underwater scattering, single-scatter and
multiple-scatter attenuation lengths, and volume backscatter co-

efficient values are topics of active research. Additional information is available from standard works such as Duntley (1963),
Jerlov (1976), Preisendorfer (1976), and Tyler (1977).

REPRESENTATIVE APPLICATIONS

Marine lidar for subsurface sensing is a developing field
where much of the literature is in conference proceedings and
contract reports. A useful overview of applications was given by
Kim and Ryan (1975) and by Gordon (1980).

Marine lidar applications can be divided into four main
classes. Bathymetry (depth mapping) senses the range difference
between surface and bottom returns. Turbidity measurements determine aspects of suspended particulate loading. Constituent detection of chlorophyll, oil, dye, etc., usually depends on
fluorescence. Temperature profiling relies on the temperature
dependence of various types of scattering mechanisms.

Bathymetry

Depth measurement involves determining the time delay between
signal returns from the surface and the bottom (or other submerged
target). The advantage of airborne lidar bathymetry over surface-
borne data collection techniques is that large areas may be covered
rapidly. Optical attenuation is the depth-limiting factor in the
lidar approach.

The instrumental limitations to accuracy include the laser
pulselength and receiver dynamic range. Refraction at nonhorizon-
tal wave slopes causes departures from the projected beam angle to
vertical, which are reflected as errors in conversion between lidar
range and depth.

Hickman and Hogg (1969) gave a convincing feasibility demon-
stration and detailed discussion of marine lidar for bathymetry.
Their system, which used a 3-ns pulsed neon laser, achieved a
resolution (not necessarily accuracy) of ± 0.5 m to a depth of
approximately four attenuation lengths (40 m in the test case).
One significant problem is the strong surface reflection, which can
be minimized by working at non-normal incidence for a smooth sur-

face, or ignoring data taken when rough-water wave surfaces give specular reflection. Other alternatives include using photomultiplier gating, which does not prevent photocathode overload, and taking advantage of beam spreading with depth to use spatial filtering (Levis et al., 1975).

Sizgoric and Carswell (1973, 1975) performed bathymetry with a rapidly pulsed, cavity-dumped argon-ion laser. This transmitter has the advantage of high average power and very narrow spectral linewidth.

More recent work with a frequency-doubled Nd:YAG laser transmitter showed a resolution of ± 0.25 m at approximately 1.7 attenuation lengths (Kim, 1977). Although studies have provided only line scans of bottom depth along the flight path, scanning perpendicular to the ground track will provide better coverage and a denser data set. The effective attenuation length for a receiver system that has a field of view large enough to accept the forward-scattered peak from multiple scattering can be as much as five times larger than the attenuation length for a narrow field of view system (Kim, 1977), depending on the details of the scattering mechanism. This advantage is balanced against the background noise contribution from equation (2), so that the lidar field of view must be chosen correctly for each differing application. Kim (1977) found in practice that polarization discrimination suppressed the surface return by a factor of 4, but that the depolarization by bottom scattering produced a signal with a crossed analyzer only 30% less than the no-polarizer signal.

Lidar bathymetry is developed to the point that it can be usefully employed for bottom mapping. Continuing research is directed toward improving data processing techniques and extending our understanding of sea roughness on the return. New applications include the location and assessment of fish stocks, particularly those varieties that come closer to the surface at night and are not easily spotted by visual means. These studies have been conducted at the Chalmers University of Technology (S. Svanberg, personal communication) and the National Defence Research Institute in Sweden. Continuing bathymetry work is being done at the

Canadian Centre for Remote Sensing in Ottawa and at the U.S.
National Aeronautics and Space Administration Wallops Flight
Center, where a system named the Airborne Oceanographic Lidar has
been tested.

Turbidity

The accuracy of lidar-inferred turbidity measurements has not
yet been clearly determined. At least two different principles may
be used.

One technique is based on a measure of the volume backscatter
coefficient as a function of depth $\beta(z)$ and the very crucial
assumption of a constant ratio of backscatter-to-attenuation for
the experiment. In this case, the return power is expressed as

$$\phi_s(z) \propto \beta(z) \ \{\exp[- \int_0^z \alpha_1(z')dz' + \int_z^0 \alpha_2(z')dz']\}, \tag{3}$$

where the single-scatter attenuation $\alpha_1(z')$ and multiple-scatter
(wide angle) attenuation $\alpha_2(z')$ are presumed known from the $\beta(z')$
already determined at shallower depths. Sizgoric and Carswell
(1975) discussed a test of this approach that used volumetric water
scattering data.

The alternative way of measuring tubidity is to use a scattering interaction with a presumed known and constant scattering
cross section. In this case, $\beta(z)$ is known, and $\alpha_1(z') + \alpha_2(z')$
can be determined from $\phi_s(z)$ by a simple inversion process. This
measurement technique is the same in principle as determining the
backscattered optical power from a white target as it is lowered
through the water. If the attenuation coefficients are constant in
depth, they can be directly evaluated as the slope of a plot of the
log of the range-corrected received power vs. depth (Sizgoric and
Carswell, 1975).

Two scattering mechanisms have been proposed to provide the
constant, known backscatter cross section for turbidity measurements. Liquid water exhibits a broad Raman scattering signal
shifted from the exciting laser incident light by 3230 cm^{-1} to 3639
cm^{-1}. Leonard et al. (1979) have reported on attenuation coefficients determined by the Raman technique. Brillouin scattering
lines are shifted much less than Raman (about 0.3 cm^{-1}) and have

been suggested by Hirschberg et al. (1975) for turbidity measurements. Analysis of the accuracy of these two approaches continues.
The Raman results have been tested in the field and corrected for
the change in attenuation coefficient of pure water over the width
of the Raman band.

Constituents

Chlorophyll, algae, oil, and dye tracers are examples of
constituents in the ocean that fluoresce when illuminated. A lidar
transmitter is an ideal pulsed illuminator for fluorescence studies
because much of the optical background can be eliminated by time
gating the receiver at the proper platform-to-surface range.
Because fluorescence involves the absorption of an incident photon
and the emission of a photon of a lower energy, the fluorescent
signal is completely depolarized.

Fluorescence spectroscopy involves both an emission and an
excitation spectrum. The emission spectra of various refined
petroleum products differ enough from each other to provide some
measure of identification (Houston et al., 1975). The emission
peak for refined products is at a shorter wavelength than for crude
oils. Fish oils also display characteristic emission spectra.
Most of the hydrocarbon fluorescence emission spectra have been
taken with 337.1-nm excitation where water penetration is not as
good as in the blue-green. If the fluorescing material is assumed
to be concentrated at the surface, then the time delay and pulse
shape of the fluorescence are other distinguishing characteristics.

The excitation spectra of algae and chlorophyll are much more
characteristic of different constituents in this class than are
emission spectra. Most of the phytoplankton fluoresce in a narrow
region within approximately 25 nm of a 685-nm band center. The
response of different algae at 685-nm emission depends markedly on
the excitation wavelength. Green algae respond most strongly to
excitation at wavelengths shorter than 360 nm, whereas red algae
peak at approximately 560 nm and bluegreen algae at approximately
610 nm (O. Steinvall, personal communication). Although the
excitation spectra overlap somewhat, it is clear that constituent

identification by using fluorescence emission spectroscopy (multiple transmitter wavelengths) is possible.

Although constituent identification by using excitation, emission, and time delay spectroscopy has been demonstrated, quantitative measurements are more difficult. Backscattered power vs. phytoplankton concentration followed a power law rather than linear relationship in one experiment (Houston et al. 1975). This behavior was interpreted as depending on whether the optical thickness of the sample was dominated by absorption or scattering attenuation. Nonphytoplankton turbidity will probably also affect the dependence of backscattered power on chlorophyll or algae concentration.

Field measurements of chlorophyll-A density to fractions of a mg/m^3 sensitivity have been made by Kim (1973). This work used 590-nm dye laser excitation and monitored 685-nm fluorescence from an altitude of 30 m. Kim (1973) found that the lidar chlorophyll analysis showed the same tidal cycle as a wet-chemistry analysis and the same hourly average values. Transects across Lake Ontario showed density variations corresponding to upwelling and wind-caused effects.

Fluorescent dye tracers have been used to trace water transport. S. Svanberg and K. Fredriksson (personal communication) have traced rhodamine-G dye to 1 µg/ℓ concentration.

Research on the quantitative accuracy of lidar fluorosensing of minor ocean constituents is continuing, particularly on the effects of turbidity. Sea state does not significantly affect spectral intensity measurements as long as breaking water is avoided (Rayner et al., 1978). Sea state can cause a variable time-delay in the pulse reception time, but with only a slight effect on the pulsewidth.

Temperature Profiles

Oceanographic lidar measures depth profiles of water temperature by means of temperature-dependent scattering interactions. Types of scattering used or proposed include Raman, Brillouin, and Rayleigh scattering.

In Raman scattering, incident light is scattered with a frequency shift that is characteristic of energy level differences in the scattering molecule. In contrast to fluorescence, where the emitted wavelength is characteristic of the scatterer, the Raman-scattered frequency (photon energy) maintains a constant frequency difference from the incident optical frequency. For water the O-H bond stretch is a Raman-active mode at an energy difference of approximately 3400 cm^{-1}. This bond energy is affected by the environment of the water molecule. There are two principal components in the Raman band at 3225 cm^{-1} and 3425 cm^{-1} (Walrafen, 1967), which are attributed to the O-H bond in the environment of a water monomer or dimer. The relative intensity of the two components is a measure of the temperature and changes by approximately 1% per degree K (Schwiesow, 1973; Leonard et al., 1979). In addition, the depolarization from Raman scattering depends both on the temperature and on the part of the Raman band detected (Cunningham and Lyons, 1973).

Interfering effects must be considered. The relative intensity of the two spectral components depends on salinity as well as on temperature. The difference in water attenuation for the two components 200 cm^{-1} or 5 nm apart affects apparent temperature and must be corrected for. Any fluorescence from constituents other than water introduces an error signal. Background radiance can be significant over the spectrally broad Raman band.

Leonard et al. (1977, 1979) have made field measurements of temperature and turbidity profiles using a Raman lidar with a dye laser transmitter, which operated from a surface platform. These authors performed an extensive analysis of interference and error sources and concluded that their relative accuracy in a profile was ± 1 K. In one experiment lidar-measured and reference temperature values agreed to an rms difference of ± 1.2 K. Further research on measurement and analysis of the effect of the diffuse attenuation coefficient on the inferred temperature is planned.

Brillouin scattering occurs when incident light is scattered from density inhomogeneities in the sea, which are caused by random, thermally generated acoustic waves. From the angle and fre-

quency (wavelength) continuum of acoustic waves, lidar backscatter
selects those obeying the Bragg condition; these are acoustic waves
propagating along the lidar axis with wavelength equal to one-half
the optical wavelength. Light scattered from the propagating waves
is Doppler shifted, giving rise to a spectral doublet, symmetric
about the lidar transmitter optical frequency. The separation of
each Brillouin peak from the central Rayleigh and Tyndall (or Mie)
scattered line is approximately 0.4 cm^{-1} (12 GHz = 12 x 10^9 Hz).
Much higher resolution is required for Brillouin spectroscopy than
for water Raman spectroscopy. Essentially, measurement of the
Brillouin doublet separation measures the velocity of sound at
ultrahigh frequencies near 5 GHz (O'Connor and Schlupf, 1967).

The velocity of sound changes with temperature, salinity, and
density. The change with temperature is approximately 0.2% per
degree K. Because frequency shifts can be measured with high
accuracy, 0.2% per degree K is a reasonable sensitivity. If the
sea salinity can be assumed or inferred from other data, the depth
profile of Brillouin doublet separation gives a depth profile of
water temperature (Hirschberg et al., 1975, 1980).

Hirchberg et al. have made continuous-wave lidar (not range
resolved) Brillouin measurements from a near-surface platform.
Their results with an exploratory system showed a temperature
scatter of ± 1.5 K. Current research is directed toward instrumentation developments including the incorporation of a pulsed
laser, a Michelson interferometric beamsplitter, and a two-channel
Fabry-Perot interferometer.

The on-frequency backscatter contains light scattered from
molecules (Rayleigh scattering) and from suspended particulates
(Mie or Tyndall scattering). The molecules are in random thermal
motion and will Doppler shift the scattered light. The transla-
tional velocity of the molecules in thermal equilibrium is much
higher than the thermal motion of the heavier particulates, so by
high resolution spectroscopy it should be possible to separate
molecular and particulate contributions to the central backscatter
peak. The width of the molecular velocity probability distribution
is proportional to the square root of the absolute temperature

(Batchelor, 1970, p. 55), so a measurement of the Rayleigh scattering linewidth determines temperature. Rayleigh linewidth
measurement requires high spectral resolution because the 1/e
halfwidth is approximately 1.8 GHz (0.06 cm^{-1}) or less.

Temperature profiling by Rayleigh linewidth has been demonstrated in the atmosphere (Fiocco et al., 1971). It has been
proposed for oceanographic lidar but not yet demonstrated
(Schwiesow, 1980).

SUMMARY

Oceanographic lidars can be used to make rapid, wide-area
measurements of various physical properties below the sea surface
from airborne platforms. Such lidars include pulsed laser transmitters, usually operating in the blue-green but ranging from 337
nm to 590 nm, and sensitive optical receivers.

Lidar experiments have measured bottom depth, turbidity
(attenuation coefficient), minor constituent concentration (oil,
chlorophyll, dye), and temperature profiles. These measurements
make use of a range of scattering interactions, which include
diffuse reflectance of a surface, volume scattering from suspended
particulates, Rayleigh (elastic or on-frequency) and Raman (in-
elastic) scattering from water molecules, absorption and fluorescent emission from organic constituents, and Brillouin scattering
from thermally generated acoustic waves. Various lidar applications presently being studied range in state of development from
proposed to demonstrated in the field.

None of the lidar techniques is in final form. Improvements
in lasers and data processing are continuing, but the most active
research in oceanographic lidar concerns the scattering interactions themselves and various sources of interference such as sea
state (wave spectra) and ambient light. Attenuation of the incident laser light and the scattered light by the sea is a fundamental limitation on the range and applicability of subsurface
lidar.

A number of research applications are within the range of capabilities and limitations of oceanographic lidar. These include the possibility of economically gathering operational data on such quantities as bulk heat content and biological productivity in the first 50 m to 100 m of the ocean.

REFERENCES

Anderson, R. S., L. Springer, B. G. Bricks, and T. W. Karras. 1975. A discharge heated copper vapor laser. IEEE J. Quantum Electron. QE-11, 172-174.

Batchelor, G. K. 1970. An Introduction to Fluid Dynamics. 615 pp. Cambridge: University Press.

Clifford, S. F. 1978. The classical theory of wave propagation in a turbulent medium. In: Laser Beam Propagation in the Atmosphere. pp. 9-43. Ed. by J. W. Strohbehn. New York: Springer-Verlag.

Cunningham, K., and P. A. Lyons. 1973. Depolarization ratio studies on liquid water. J. Chem. Phys. 59, 2132-2139.

Duntley, S. Q. 1963. Light in the sea. J. Opt. Soc. Am. 53, 214-233.

Eden, G., R. Burnham, L. F. Champagne, T. Donohue, and N. Djeu. 1979. Visible and UV lasers: problems and promises. IEEE Spectrum 16, 50-59.

Fiocco, G., G. Benedetti-Michelangeli, K. Maischberger, and E. Madonna. 1971. Measurement of temperature and aerosol to molecule ratio in the troposphere by optical radar. Nature Physical Science 229, 78-79.

Granatstein, V. L., M. Rhinewine, and A. M. Levine. 1972. Depolarization of laser light scattered from turbid water. Appl. Opt. 11, 1870-1871.

Guinn, J. A., Jr., G. N. Plass, and G. W. Kattawar. 1979. Sunlight glitter on a wind-ruffled sea: further studies. Appl. Opt. 18, 842-849.

Hickman, G. D., and J. E. Hogg. 1969. Application of an airborne pulsed laser for near shore bathymetric measurements. Remote Sens. Environ. 1, 47-58.

Hirschberg, J. G., A. W. Wouters, F. N. Cooke, Jr., K. M. Simon, and J. D. Byrne. 1975. Laser application to measure vertical sea temperature and turbidity: NASA CR-139184. 120 pp. Coral Gables, Florida: University of Miami. (available NASA).

Hirschberg, J. G., A. W. Wouters, and J. D. Byrne 1980. Ocean parameters using the Brillouin effect. In: Gordon, op. cit. pp. 29-44.

Hoge, F. E., and R. N. Swift 1980. Oil film thickness measurement using airborne laser-induced water Raman backscatter. Appl. Opt. 19, 3269-3281.

Houston, W. R., D. G. Stephenson, and R. M. Measures. 1975. LIFES: Laser induced fluorescence and environmental sensing. In: Kim and Ryan, op. cit. pp. 153-169.

Jerlov, N. G. 1976. Marine Optics. 231 pp. Amsterdam: Elsevier.

Kim, H. H. 1973. New algae mapping technique by the use of an airborne fluorosensor. Appl. Opt. 12, 1454-1459.

Kim, H. H. 1977. Airborne bathymetric charting using pulsed blue-green lasers. Appl. Opt. 16, 46-56.

Kim, H. H., and P. T. Ryan. 1975. The use of lasers for hydrographic studies. NASA publication SP-375. 207 pp. Washington: NASA. (available NTIS).

Kim, H. H., C. R. McClain, and W. D. Hart. 1979. Chlorophyll gradient map from high-altitude ocean-color-scanner data. Appl. Opt. 18, 3715-3716.

Gordon, H. R. 1980. Ocean remote sensing using lasers. NOAA Technical Memorandum ERL PMEL-18. 200 pp. Seattle: NOAA Pacific Marine Environmental Laboratory. (available NOAA).

Leonard, D. A., B. Caputo, R. L. Johnson, and F. E. Hoge. 1977. Experimental remote sensing of subsurface temperature in natural ocean water. Geophys. Res. Letters 4, 279-281.

Leonard, D. A., B. Caputo, and F. E. Hoge. 1979. Remote sensing of subsurface water temperature by Raman scattering. Appl. Opt. 18, 1732-1745.

Levis, C. A., W. G. Swarner, C. Prettyman, and G. W. Reinhardt. 1975. An optical radar for airborne use over natural waters. In: Kim and Ryan, op. cit. pp. 67-80.

O'Connor, C. L., and J. P. Schlupf. 1967. Brillouin scattering in water: the Landau-Placzek ratio. J. Chem. Phys. 47, 31-38.

Preisendorfer, R. W. 1976. Hydrologic Optics. Vol. I, 218 pp. Vol. II, 400 pp. Vol. III, 246 pp. Vol. IV, 207 pp. Vol. V, 296 pp. Honolulu: U.S. Dept. of Commerce.

Prettyman, C. E., and M. D. Cermak. 1969. Time variation of the rough ocean surface and its effect on an incident laser beam. IEEE Trans. GE-7, 235-254.

Rayner, D. M., M. Lee, and A. G. Szabo. 1978. Effect of sea-state on the performance of laser fluoresensors. Appl. Opt. 17, 2730-2733.

Schwiesow, R. L. 1973. Raman scattering from pollutant gases and air-water interfaces. AIAA Journal 11, 87-90.

Schwiesow, R. L., and R. F. Calfee. 1979. Atmospheric refractive effects on coherent lidar performance at 10.6 mm. Appl. Opt. 18, 3811-3917.

Schwiesow, R. L. 1980. Lidar high-resolution spectroscopy for oceanographic measurements. In: Gordon, op. cit. pp. 11-28.

Sizgoric, S., and A. I. Carswell. 1973. Lidar as an underwater probe. Can. Aeronautics and Space J. 19, 511-513.

Sizgoric, S., and A. I. Carswell. 1975. Underwater probing with laser radar. In: Water Quality Parameters, ASTM STP 573. pp. 398-413. Philadelphia: American Society for Testing and Materials.

Tyler, J. E. (Ed.). 1977. Light in the Sea. 385 pp. Stroudsburg, Pennsylvania: Dowden, Hutchinson and Ross.

Walrafen, G. E. 1967. Raman spectral studies of the effects of temperature on water structure. J. Chem. Phys. 47, 114-126.

Yura, H. T. 1979. Signal-to-noise ratio of heterodyne lidar systems in the presence of atmospheric turbulence. Opt. Acta 26, 627-644.

Absolute Sea Surface Temperature Measurement by Remote Sensing and Atmospheric Corrections Using Differential Radiometry

F. Becker

ABSTRACT

In order to measure on line the absolute temperature of the surface of the oceans, from remote sensing data obtained from air-borne or space-borne platforms, a simple method for correcting the errors on the observed radiance produced by a clear atmosphere in the 8-14 μ window is proposed. The principle of the method consists of measuring the radiance of the ocean in two adjacent channels. The true radiance, or brightness temperature, is obtained from these measured radiances by solving a very simple linear equation which depends only on one parameter if one assumes that the Planck formula as well as the radiative transfer equations in the atmosphere can be linearized.

The only unknown parameter can be obtained from a theoretical analysis of the optical properties of the atmosphere or from empirical methods. In this paper, only three of these empirical methods are presented and discussed. Their results are compared with experimental data obtained over the Mediterranean Sea and the Rhine Valley. The accuracy is of the order of the instrumental precision (0.2 to 0.3°C). Comparison with other methods is discussed.

INTRODUCTION

It is very interesting to measure the absolute surface temperature of seas over large areas in a quasi-instantaneous way for two reasons. First, it is the only way to get a valuable representation of the distribution of these temperatures and secondly, to study their evolution with the time (Deschambenoy et al. 1977, Becker and Pontier 1978). In order to obtain these absolute temperatures, it is necessary to correct for the atmospheric perturbations which may change the observed temperature by 3 or 4 degrees (or even more).

This is a very well known problem which has already been solved in several ways. Therefore, the aim of this communication is not to present a new model, but rather to give a semi-empirical method to correct for atmospheric perturbations very quickly and with sufficient accuracy in order to provide "absolute temperature images". The method which will be presented and discussed is based on the difference of the atmospheric absorption in two adjacent atmospheric windows as it was proposed five years ago by Prabhakara et al. (1974). The results presented in this paper have been obtained in collaboration with the Laboratoire de Meteorologie Dynamique (CNRS) and already discussed in several papers mainly Becker et al. (1977) and Becker (1978).

PRINCIPLE OF THE METHOD FOR CORRECTING ATMOSPHERIC PERTURBATIONS

The radiance of the sea surface is perturbed by the aerosols and the molecules contained in the atmosphere. Since remote sensing experiments are performed when the atmosphere is rather clear, the effects of the aerosols will be neglected or at least will be dealt with empirically and only the effects of the molecules will be discussed here.

One can distinguish two kinds of such molecules:

i)   Those which are present in the atmosphere with propor-
     tions which are relatively constant both in time and in
     space, namely CO_2, N_2O, CO, CH_4, etc.

ii) Those which are present with very variable proportions,
     namely H_2O and O_3.

In this paper, we shall neglect the effect of O_3 because the main experiments which will be discussed have been made with air-borne scanners. Furthermore, if the data comes from space-born radiometers, the channels are chosen in such a way that the strong ozone absorption band at 9.6 μ is avoided.

The Figures 1-a and 1-b from McClatchey et al. (1972) show that the transmittance of the gas with a relatively constant den-sity is larger than that of the water vapor in the channel 8-12.5 μ. It is therefore a good approximation to assume that the pertur-bation produced by the atmosphere comes from the water vapor mole-cules. Furthermore, it appears in Figure 1-b that the attenuation produced by the water vapor is strongly wavelength dependent in the 8-14 μ window. Hence the principles of the proposed methods: To measure the radiance of the sea surface in two channels in which the transmittance of the water vapor molecules is sufficiently large but differentiated and in which the effects coming from the other molecules can be neglected. In the various experiments which will be discussed, the two channels,

Channel I   10.5-12.5 μ

Channel II   7.9-9.6 μ.

have been chosen. These channels are represented on Figures 1-a and 1-b. The boundary 9.6 μ (O_3) is indeed "dangerous" for satel-lite measurements, but not for airplane measurements.

Before going further in the analysis of this method, it is interesting to see whether it works. The Figure 2 shows an example of the variation of the observed radiance of the sea with the

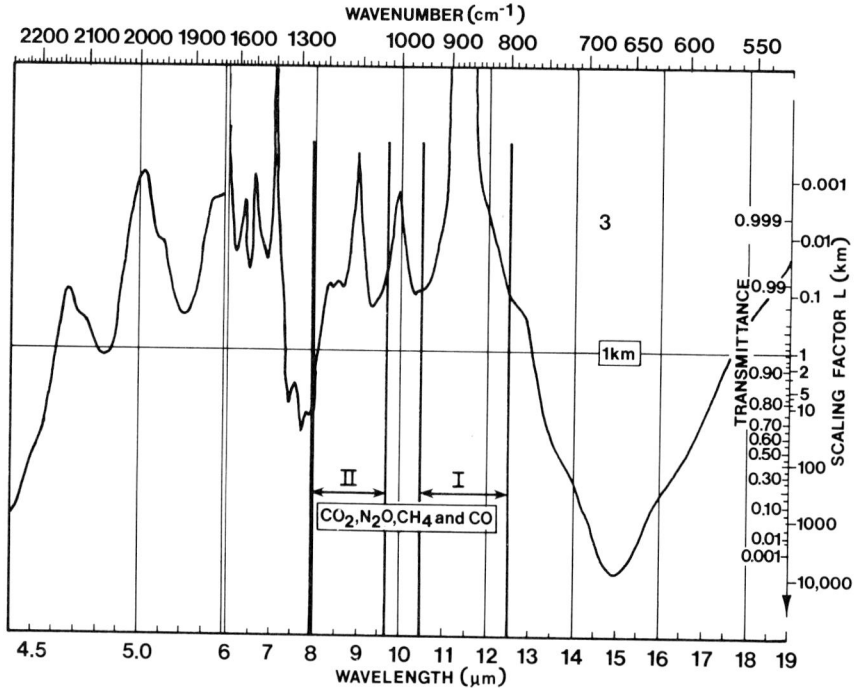

Figure 1-a: Curves to calculate the transmittance of uniformly mized gas (CO_2, N_2O, CO, CH_4), from McClatchey et al. (1972).

altitude in the two channels. The experiments have been performed by Pontier et al. with the scanner ARIES (Monge et al. 1976) flown on a DC3 airplane above the buoy BORHA II in the Mediterranean Sea. The difference between the radiances in the two channels may be as large as 1°C. It may be much larger for measurements from space-borne platforms.

LINEAR MODEL OF CORRECTION

Theoretical expression

In the line proposed by Prabhakara et al. (1974) the model discussed here is based on the linearization of both the Planck's radiation formula and the radiative transfer equation. These approximations are justified by the facts that the differences between the various temperatures observed are relatively small

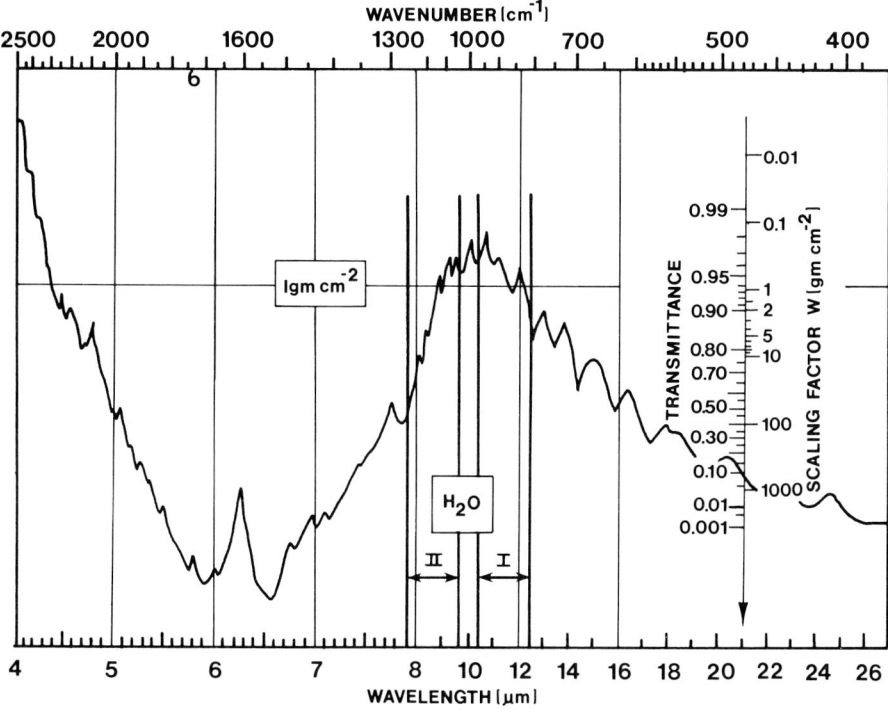

Figure 1-b: Curve to calculate the transmittance of the water vapor. It is outlined for 1 gm/cm², from McClatchey et al. (1972).

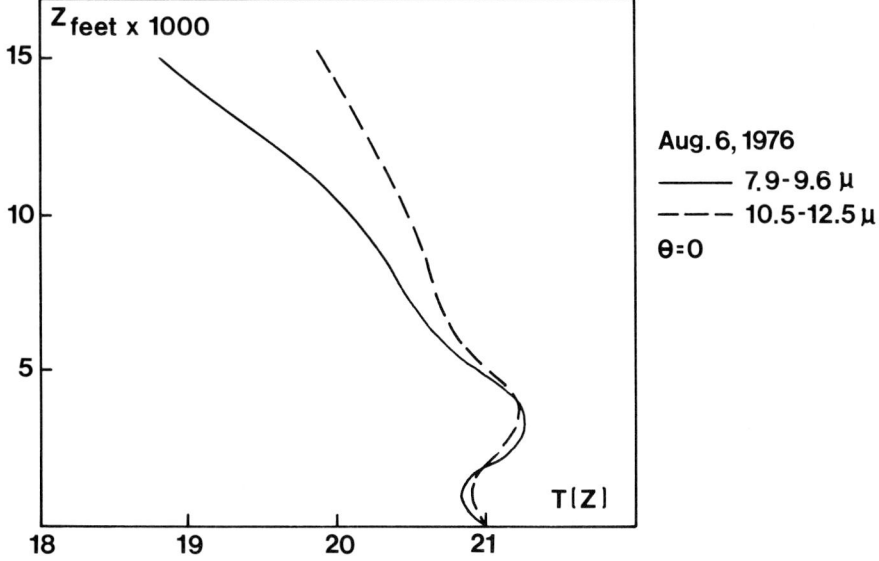

Figure 2: Experimental variation of the observed radiometric temperature in the two channels with the altitude.

$(\frac{\Delta T}{T}=1\%$ - cf. Figure 2) and the attenuation of atmosphere is not very large in the chosen spectral bands (see Figure 1a or 1b).

Neglecting the effects of the various scatterings (which are very good in the thermal infra-red bands for clear atmosphere), of the absorption by the molecules other than H_2O, and the effects of reflection by the ground ($\varepsilon \cong 0.99$), the signal recorded by the radiometer is given by

$$I = \int_{\lambda_1}^{\lambda_2} f(\lambda) \ I(\lambda) \ d\lambda \tag{1}$$

with $I(\lambda) = \varepsilon(\lambda) \ R(\lambda,T_s) \ \tau(0,h,\theta,\lambda) + \int_0^h R(\lambda,T(z)) \ \frac{\partial\tau}{\partial z} \ dz$

In these expressions, $f(\lambda)$ is the spectral response of the scanner in the band $\lambda_1-\lambda_2$, $R(\lambda,T)$ is the spectral radiance of a black body at temperature T (T_s is the surface temperature and $T(z)$ the air temperature at altitude z) and τ is the transmission coefficient given by

$$\tau(z_1,z_2,\theta,\lambda) = \exp \ \{- \int_{z_1}^{z_2} \frac{\alpha(\lambda,z)}{\cos \theta} \ \rho(z) \ dz\} \tag{2}$$

where h and θ are respectively the height and the angle of observation, α is the spectral absorption coefficient of the water vapor and $\rho(z)$ is density at altitude z.

On the other hand, if T is the radiometric equivalent temperature, the recorded signal is nothing but

$$I = \int_{\lambda_1}^{\lambda_2} f(\lambda) \ R(\lambda,T) \ d\lambda \tag{3}$$

In order to find the relationship between T and T_s we write

$$\tau = \exp\{- \int_{z_1}^{z_2} \frac{\alpha(\lambda,z)}{\cos \theta} \ \rho(z) \ dz\} \underset{\sim}{} 1 - \int_{z_1}^{z_2} \frac{\alpha(\lambda,z)}{\cos \theta} \ \rho(z) \ dz$$

and

$$R(\lambda T) \cong R(\lambda,\bar{T}_s) + (T - \bar{T}_s) \ \frac{\partial R}{\partial T}\bigg|_{\bar{T}_s}$$

Replacing τ and R by their approximate expression in equations (1) and (2) leads to

$$T(h) \underset{\sim}{} \bar{T}_s + \frac{A(\lambda) \ W(h)}{\cos \theta} \ (\tilde{T}(h) - \bar{T}_s) \tag{4}$$

where $T(h)$ is the radiometric temperature observed at altitude h with the angle of observation θ, $\bar{T}_s$ is the radiometric temperature of the sea surface $(R(\lambda,\bar{T}_s) = \varepsilon R(\lambda,T_s))$,

$$A(\lambda) \simeq \frac{\int_{\lambda_1}^{\lambda_2} d\lambda \; f(\lambda) \; \alpha(\lambda) \; \left.\frac{\partial R}{\partial T}\right|_{T_s}}{\int_{\lambda_1}^{\lambda_2} d\lambda \; f(\lambda) \; \left.\frac{\partial R}{\partial T}\right|_{T_s}} \qquad (4a)$$

is an average absorption coefficient for the water vapor,

$$W(h) = \int_0^h \frac{P_o T_o}{P(z)T(z)} \; \rho(z) \; dz \qquad (4b)$$

is the weighted precipitable water height of the atmosphere, and $\tilde{T}(z)$ is an effective air temperature given by

$$\tilde{T}(h) = \frac{\int_0^h \frac{P_o T_o}{P(z)} \rho(z) \; dz}{W(h)} \qquad (4c)$$

It has been assumed that $\alpha(\lambda,z) \simeq \alpha(\lambda) \frac{P_o T_o}{P(z)T(z)}$

Physical meaning of formula (4)

Formula (4) gives immediately the main mechanisms of atmospheric perturbations on the sea surface temperature. For a given angle of observation θ, if

i) $\tilde{T}(h) > \bar{T}_s$, the atmosphere appears to be "warmer" than the sea. It emits more than it absorbs; thus, the observed temperature $T(h)$ will be larger than the true one T_s.

ii) $\tilde{T}(h) = \bar{T}_s$, the atmosphere appears to be at the same temperature as the sea. Since it emits exactly what it absorbs, there are no corrections.

iii) $\tilde{T}(h) < \bar{T}_s$, the atmosphere appears to be "colder" than the sea. It absorbs more than it emits; thus, the observed temperature $T(h)$ will be smaller than the true one T_s.

These effects are proportional to $A(\lambda)$ for a given atmosphere, if A_1 corresponds to channel I and A_2 to channel II, and if $\tilde{T}(h) = \bar{T}_s$, the two apparent temperatures $T_1(h)$ and $T_2(h)$ in the channels I and II respectively will be equal and equal to $\overline{T}_s$. Therefore the two curves giving $T(z)$ in the two channels should crossover (if they do so) for an altitude for which $\widetilde{T}(h) = T_s$. This is actually observed experimentally on Figure 2 within 0.1°C.

Verifications of formula (4)

A "theoretical" verification of formula 4 has been made by Imbault (private communication). Using an elaborate computer code built by Scott (1974) and Chedin et al. (1977) in which the various effects of a clear atmosphere are taken into account due to a very complete absorption band library, D. Imbault has checked that

    i)   the water vapor plays the dominant role

    ii)   the difference $(T(h) - T_s)$ calculated with the complete program varies linearly with R_s according to

$$T - T_s = f(W)(\widetilde{T} - T_s) + \eta(z)$$

where $F(W)$ is a linear function of W for the small values of W:

$$f(W) = AW + \beta$$

The approximation of the formula (4) is therefore to assume that there exists a value A' of A such that

$$(AW + \beta)(\widetilde{T} - T_s) + \eta \cong A'W(\widetilde{T} - T_s)$$

In any way, formula (4) is valid only if

$$\frac{A(\lambda)W}{\cos \theta} \ll 1$$

which has been checked in all applications.

These approximations have been checked experimentally both over the Mediterranean Sea by Becker, Pontier et al. (1978) and over the Rhine River (Becker et al, 1979). During the various experiments, the apparent temperature of the sea surface has been measured at various altitudes, while ground measurements were performed. In the same time an atmosphere sounding was done giving the temperature and humidity of the air from 0 to 4 kms. In order to reduce the noise of the scanner several pixels have been averaged. In the Mediterranean Sea it has been possible to average over hundred pixels because the surface temperature was very homogeneous.

Figure 2 gives the results over the Mediterranean Sea, while Figure 3 shows, as an example, the comparison between formula (4) and the experimental data for channel I. The parameter A has been adjusted to reproduce the data at the altitude z = 15,000 feet. The experiment data are reproduced within ± 0.25°C, which is the order of magnitude of the experimental error.

The linear angular dependence on $1/\cos\theta$ predicted by the formula (4) is also in agreement with the experimental data for both channels as it is seen in Figure 4. The extrapolation to $1/\cos\theta = 0$ gives also the temperature T_s as predicted by formula (4).

Finally, the relative contrast between the various elements of the surface is predicted to be linear in W for the same altitude and the same angle of observation. If T_1 and T_2 are the observed temperatures of two pixels of surface temperature T_{s1} and T_{s2}, it is easy to show that

$$\frac{T_1 - T_2}{T_{s1} - T_{s2}} = \frac{\Delta T}{\Delta T_s} = 1 - \frac{AW(h)}{\cos\theta} \tag{5}$$

This is also in a correct agreement with the calculations for various atmospheres based on the Lowtran Model made by L. J. Rouse (private communication) for NOAA 5 as it is shown in Figure 5. In this figure, we have also reported the values obtained for A from the experiments over the Rhine River valley (Becker et al., 1979) ($A = 0.22 \pm 0.02$ cm^{-1}). It is interesting to note that the result of the simple formula (5) lies in between the best fits of L. J. Rouse using the Lowtran Model and the transmission model ARMOD.

Methods of correction based on Formula (4)

There are several ways to use formula (4):

i)   One can use the angular attenuations as was done by de Felice and Pontier (1977). Figure 4 shows that the extrapolation of the straight dependence on $1/\cos\theta$ to zero gives T_s. Nevertheless, this method is correct only if the temperature distribution is uniform over large surfaces and if the angular attenuation is sufficient.

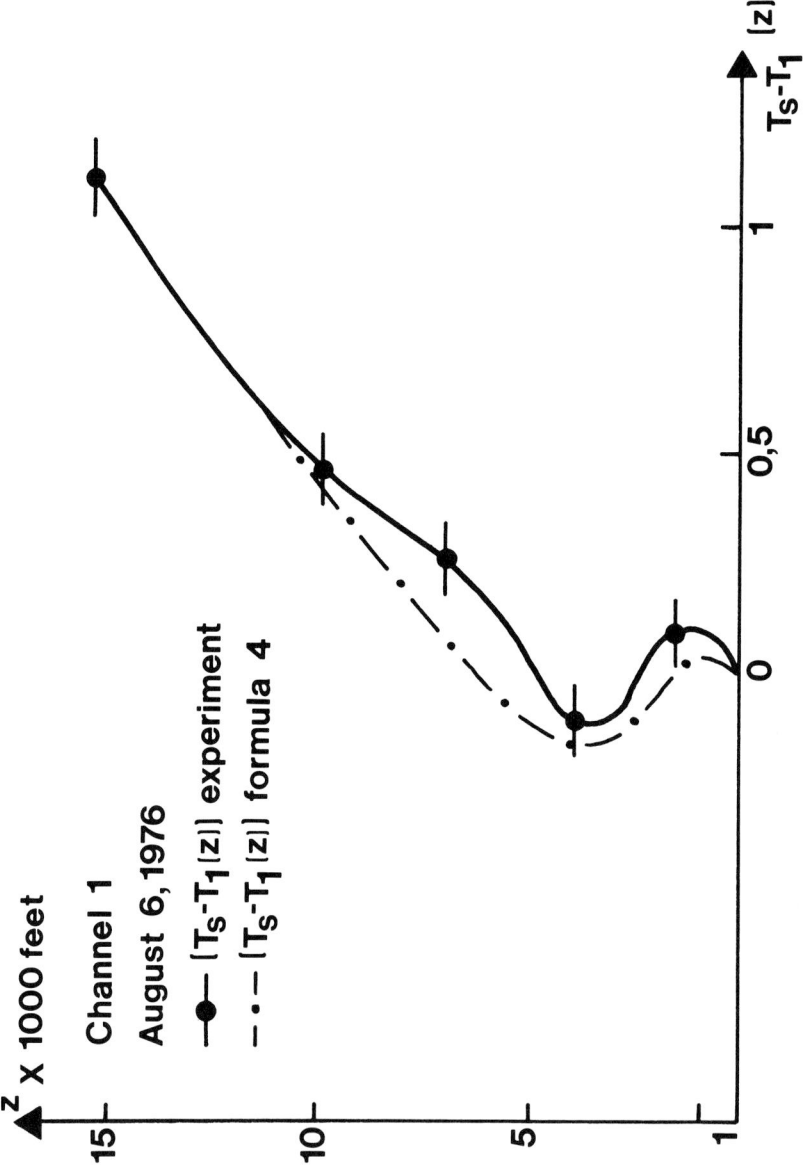

Figure 3:   -o- Experimental difference $(T_s - T_1)$ between the ground radiometric temperature and the radiometric temperature measured at altitude z as a function of the altitude
z.   .-.-.   Theoretical difference $(T_s - T_1)$ versus z given by formula (4).

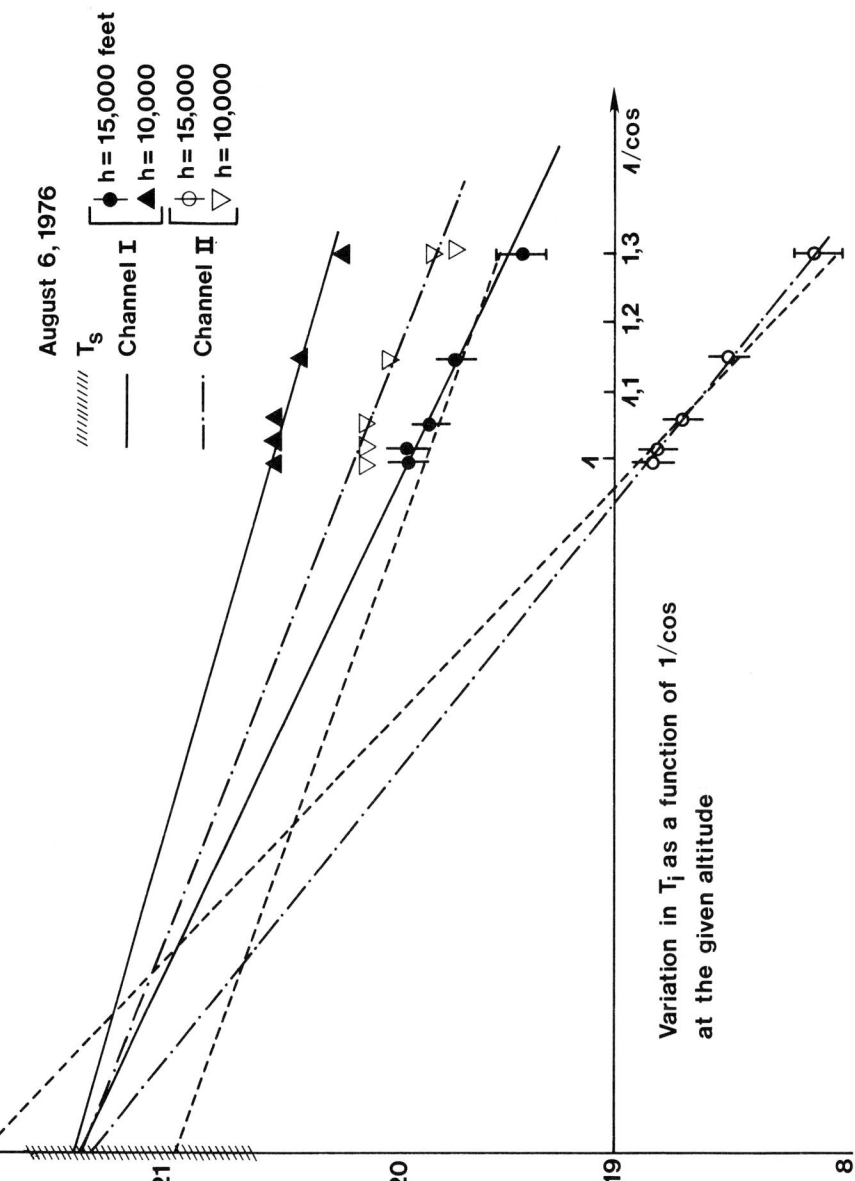

Figure 4: Variation of T_1 (θ) as a function of $1/\cos\theta$ at a given altitude. Comparison of the experimental data with the linear prediction of formula (4).

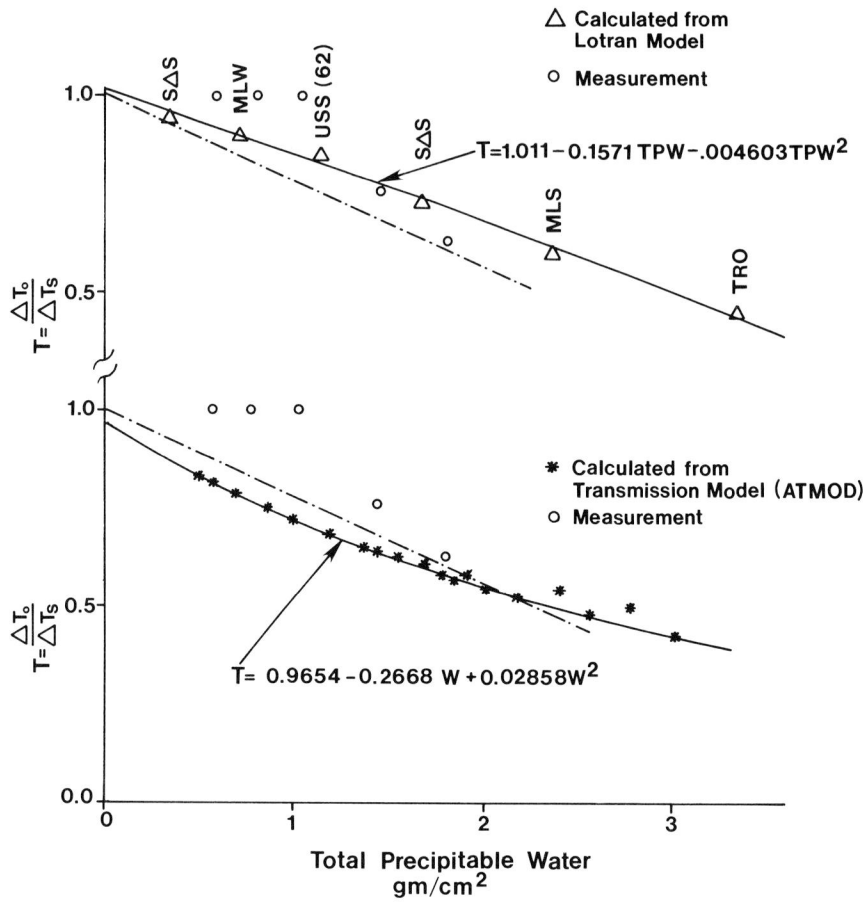

Figure 5: Variation of the contrast $\Delta T/\Delta T_S$ with precipitable water W
Δ Calculated points from Lowtran Model (from L.J. Rouse, personal communication)
*Calculated points from ATMOD (from L.J. Rouse, personal communication) Measurement
-Regression formulae of L.J. Rouse (personal communication)
As an indication the prediction of formula (5) has been plotted with the values of A obtained over the Rhine valley with 10.5-12.5 μ channel -.-.- (Becker et al. 1979)
$$\frac{\Delta T}{\Delta T_S} = 1 - 0.22 \text{ W}.$$

The error may be large because the measures can be done only between $1/\cos\theta = 1$ and $\sqrt{2}$. If the measure is done in two channels the accuracy is better because the two regression lines must converge towards the same point at $1/\cos\theta = 0$.

Formula (4) gives for this approach

$$T_s = \frac{T(\theta_1) + T(\theta_2)}{2} + \frac{\cos\theta_1 + \cos\theta_2}{2(\cos\theta_1 - \cos\theta_2)} [T(\theta_1) - T(\theta_2)] \tag{6}$$

In this formula there are no free parameters.

ii) For heterogeneous temperature distributions, it is necessary to use a method based on the altitude and angular dependences. This method has been applied by Becker et al. (1979) over the Rhine Valley: a given element on the ground of surface temperature T_s is seen successively at the two altitudes h_1 and h_2 with the two angles θ_1 and θ_2 respectively.

If the functions $T(h)$ and $W(h)$ are known, it is easy to calculate A from the two relationships

$$A = \frac{(T_1 - T_s) \cos\theta_1}{W(h_1) \ (\widetilde{T}(h_1) - T_s)} = \frac{(T_2 - T_s) \cos\theta_2}{W(h_2) \ (\widetilde{T}(h_2) - T_s)}$$

and then use formula (4). This procedure leads to a simple algorithm to produce "one line" corrected images. Actually, the accuracy obtained was better than 0.4°C in absolute magnitude.

iii) In any situation it may be possible to use the variation of the contrast between various points. It is necessary to have two points of reference on the ground. From formula (5) it is possible to calculate $\frac{AW}{\cos\theta}$ and then from formula (4) the correction for any other point.

iv) It is indeed possible to calculate A, W and $\widetilde{T}$ and to deduce T_s from formula (4).

Statistical corrections

Deschamps and Phulpin (1978) have proposed a statistical correction. From a theoretical model based on Lowtran, they calculated many couples of points $(T(h), T_s)$ for various atmospheres.

They found that the surface temperature may be related to the observed temperature for TIROS N by

$$T_s = T + 2.94 \pm 1.87 \text{ in the } 10.5 - 11.5 \text{ } \mu \text{ channel}$$

$$T_s = T + 4.9 \pm 2.5 \text{ in the } 11.5 - 12.5 \text{ } \mu \text{ channel.}$$

If one does not want very accurate results, this method gives a very simple correction procedure.

From formula (4) which can be rewritten

$$T_s = T + \frac{A(\lambda)W}{\cos \theta} (T_s - \tilde{T})$$

one sees that the approach assumes that $\frac{AW}{\cos\theta} (T_s - \tilde{T})$ is a constant over the world.

METHODS BASED ON TWO CHANNELS

Theoretical expression

The total precipitable water $W(z)$ and the effective temperature $\tilde{T}(z)$ do not depend on the frequence of the radiation; nor does T_s if one neglects the variation of emissivity in the two channels (which imposes very close channels). In the case of sea water the emissivities are respectively $\varepsilon_1 = 0.99$ and $\varepsilon_2 = 0.986$; this difference of 0.4% leads to a difference on the temperatures of $\frac{\Delta T}{T} \cong$ 0.08%, i.e. $\Delta T \cong 0.2°C$, which is in the error of the method. With these conditions the relation 4 may be applied in the two channels leading to

$$T_1(z) = T_s + A_1 \frac{W(z)}{\cos\theta}(\tilde{T}(z) - T_s) \qquad (7\text{-}a)$$

$$T_2(z) = T_s + A_2 \frac{W(z)}{\cos\theta}(\tilde{T}(z) - T_s) \qquad (7\text{-}b)$$

where T_1 and T_2 are the radiometric temperatures observed in the channels I and II respectively at the altitude z with the angle θ. Due to the linear character of the relations (7-a) and (7-b) it is possible to eliminate both $\tilde{T}(z)$ and $W(z)$ between these equations to get

$$T_s = \frac{T_1(z) + T_2(z)}{2} + \frac{A_1 + A_2}{2(A_2 - A_1)} \left| T_1(z) - T_2(z) \right| \qquad (8)$$

Expression (8) is very interesting in several respects:

i)   it is not necessary to perform an atmospheric sounding because neither $\tilde{T}(z)$ nor $W(z)$ appears in (8)

ii)   this expression does not depend, at least theoretically, on the type of atmosphere

iii) this expression does not depend on the angle of observation θ

iv)   this expression is true at any altitude.

Furthermore, it is interesting to note once again that if $T_1 = T_2$, necessarily $T_1 = T_2 = T_s$. This result can easily be explained remembering that the absorption/emission effects have been assumed to be due only to the water vapor molecules. If $T_1 = T_2$ there is no global perturbation effect and if $T_1 = T_2$, T_1 must therefore be equal to T_s. In fact in this case, second order effects have to be taken into account and one should have $T_s - T_2 = T_s - T_1 = \eta$ when $T_1 = T_2$, where η corresponds to the neglected absorption/emission phenomena. This difference is here very small as it appeared on Figure 2 and will appear on Figure 6.

Experimental verification

Formula (8) has been checked also in several ways:

i)   Using the theoretical model of Scott (1974) and Chedin et al. (1977) discussed in § III-3, D. Imbault (private communication) has calculated the dependence of T_1 as a function of T_2 for different values of T_s. The relationship obtained is not strictly linear, but the difference with a linear expression is small as it has been observed by McMillin (1975).

ii)   The experimental data obtained over the Mediterranean Sea by Pontier et al. for the channels I and II at different altitudes and different angles are on the same line at a given day, as seen in Figure 6.

The equation of the line is as follows:

$$T(z) = \frac{2a - 1}{2a + 1} T_2(z) + \frac{2}{1 + 2a} T_s \tag{9a}$$

Theoretically, "a" is independent of the atmosphere because

$$\text{"a"} = \frac{A_1 + A_2}{2(A_2 - A_1)}$$

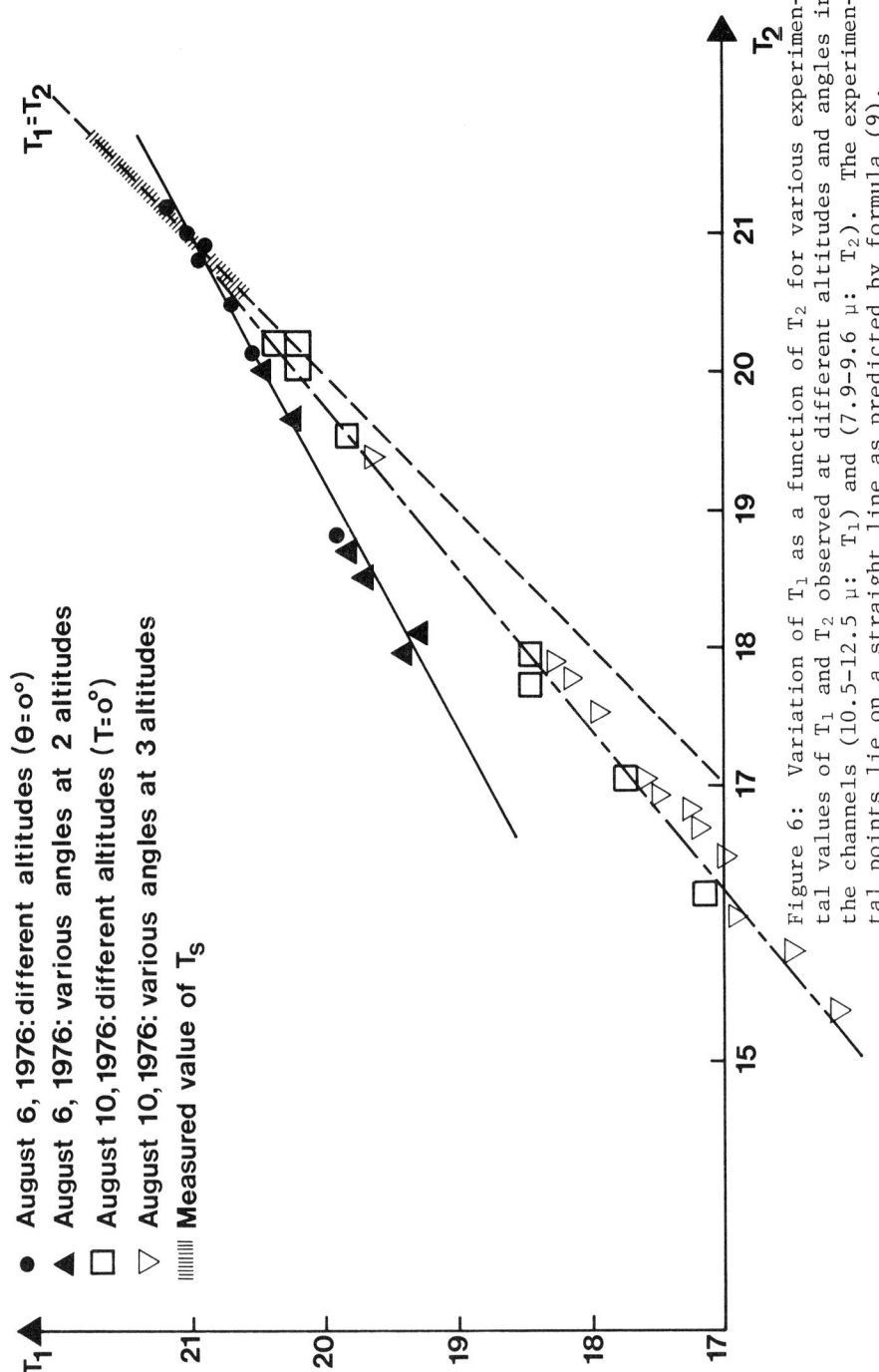

Figure 6: Variation of T_1 as a function of T_2 for various experimental values of T_1 and T_2 observed at different altitudes and angles in the channels (10.5–12.5 μ: T_1) and (7.9–9.6 μ: T_2). The experimental points lie on a straight line as predicted by formula (9).

Nevertheless, there is a systematic deviation between the two lines corresponding to the two days. This difference is not well-explained. It may correspond to second order effects neglected.

iii) A theoretical calculation made by Deschamps and Phulpin (1978) with various atmospheres has given the results of Figure 7. It turns out that the various points in a graph $(T_2 - T_s)$ as a function of $(T_1 - T_s)$ are on the same line which can be approximated by a straight line. In their regression formula which take into account large differences, the straight line does not include the origin. Nevertheless, for $T_1 - T_s < 6°$, it is possible to draw a straight line going through the origin as predicted by formula (9-a) which can be rewritten

$$(T_1 - T_s) = \frac{2a - 1}{2a + 1} (T_2 - T_s) \qquad (9\text{-}b)$$

Methods of correction based on formula (9)

The first method consists in calculating the coefficient

$$"a" = \frac{A_1 + A_2}{2(A_2 - A_1)}$$

from a theoretical knowledge of the water vapor properties and the response function of the radiometer (see formula (4-a)). This approach may not be very accurate because if the linear character of formula (9) is well established, the value of the slope is not well determined. Therefore "experimental" determinations of "a" may be more accurate.

Experimental determinations of "a" may be done in the following ways:

i) Use data at several altitudes (which is only possible for airplane platforms):

$$\frac{1}{a} = 2 \frac{T_1(z_2) + T_2(z_1) - T_2(z_2) - T_1(z_1)}{T_1(z_1) + T_2(z_1) - T_1(z_2) - T_2(z_2)} \qquad (10)$$

Numerical results are shown on Table 1. They give an idea of the dispersion of the results.

TABLE 1: Experimental data obtained in channel I (10.5-12.5 μ: T_1) and in channel II (7.9-9.6 μ: T_2) and various values of T_2 predicted by formula (8). The dispersion of the calculated points is not larger than that of the experimental points.

date	z Altitude	$T_1(z)$	$T_2(z)$	Ts predicted by (8) a=1.6.		Ts Measured
	15,000	19.9	18.8	21.11		
	10,000	20.5	20.1	20.94		$20,6<t_s<21.6$
	7,000	20.7	20.5	20.93		
6th August	4,000	21.1	21.1	21.1	21±0.1	
	1,500	20.9.	20.9	20.9		
	1,000	20.9	20.8	21.02		
	300	21	21	21		
				a=5.04		
	13,500	17.1	16.1	21.6		
	10,000	17.7	17	20.9		
	7,000	18.4	17.9	20.67		$20.6<T_s<21.6$
10th August	4,000	19.8	19.5	21.16	21.1±0.5	
	1,500	20.2	20.1	20.65		
	1,000	20.2	20.0	20.6		
	300	20.3	20.0	21.5		

ii) Use a statistical approach as proposed by Deschamps and Phulpin (1978). For instance, for the channels 11 μ and 12 μ of TIROS N, they propose the following regression formula relating the surface temperature T_s to the temperatures measured in channel 11 μ ($T_{(11)}$) and in channel 12 μ ($T_{(12)}$):

$$T_s = - 2.1795 + 3.6256 \, T_{(11)} - 2.6256 \, T_{(12)} \pm 0.511 \qquad (11)$$

which is much more accurate than the equivalent formula proposed with one channel only. The results are shown on Figure 7. We have discussed earlier the difference with formula (9-b). A simple analysis based on the data of Deschamps and Phulpin and formula (9-b) yields the values $a \cong 1.7$, i.e.

$$T_s \cong 2.2128 \ T_{(11)} - 1.2128 \ T_{(12)} \qquad (12)$$

The difference between formula (11) and (12) can be seen on Figures 7 and 8.

iii) Use the angular dependence. Since formula (9-a) has the same behaviour with respect to the angle as with the altitude one can "calculate" a from

$$\frac{1}{a} = 2 \ \frac{T_1(\theta_2) + T_2(\theta_1) - T_2(\theta_2) - T_1(\theta_1)}{T_1(\theta_1) + T_2(\theta_1) - T_1(\theta_2) - T_2(\theta_2)}$$

This method can be used with satellite having a large field of view. Comparison between the data at the center of the image and data at the edges may be very useful in this respect. Numerical examples for the experiment over the Mediterranean Sea are given on Table 2.

iv) Use good ground reference temperature giving T_{sref}. If T_{1r} and T_{2r} are the temperatures of the observed reference station, one has immediately

$$a = \frac{2 \ T_{sref} - (T_{1r} + T_{2r})}{T_{1r} - T_{2r}}$$

IMPROVEMENTS

As seen in Figures 5, 7 and 8 corresponding to calculated satellite data, the linear model proposed here is correct for small variations of the parameters. For large variations, it must be necessary to include the following second order effects:

i) In the calculation itself by including second order terms

$$\tau = 1 - \int_{z_1}^{z_2} \frac{\alpha(\lambda z)}{\cos \theta} \ \rho(z)dz + \frac{1}{2!} \left| \int_{z_1}^{z_2} \frac{\alpha(\lambda z)}{\cos \theta} \ \rho(z)dz \right|^2$$

TABLE 2: Verification of the prediction of formula (8) for the angular variation. The fluctuation of the predicted temperatures is very small and fits the experimentally measured value.

Altitude	θ	$T_1 (\theta,z)$	$T_2 (\theta,z)$	Ts predicted by (8) (a=1.6)
	0	19.9	18.8	21.1
	10	19.9	18.8	21.1
15,000	20	19.8	18.7	21
feet	30	19.7	18.5	21
	40	19.4	18.1	20.8

$$R(\lambda,T) \simeq R(\lambda,T_s) + (T - T_s) \left.\frac{\partial R}{\partial T}\right|_{T_s} + \frac{(T - T_s)^2}{2!} \left.\frac{\partial^2 R}{\partial T^2}\right|_{T_s}$$

which leads to a more complex formula.

ii) By introducing reflexion effects in formula (2). The first effect has been taken into account by Deschamps and Phulpin (1978) who show how they can be corrected for using a third channel. For instance, starting from their theoretical data for TIROS N, they show that the atmospheric correction may be done using the formula

$$T_s = 0, 14 + 1,002 \, T_{(3.7)} + 1,126 \, T_{(11)} - 1,128 \, T_{(12)} \pm 0.09°K$$

The accuracy of this formula is again much better than that of formula (11).

Finally, the last point is the elimination of the pixel including cloud average. But this is another problem which will not be discussed here.

CONCLUSIONS

The various analyses presented here show that the formulae proposed to correct for atmospheric effects, which assume a linear dependence with moisture, $1/\cos\theta$ and the equivalent atmospheric temperature, are valid if the absorption/emission is not too impor-

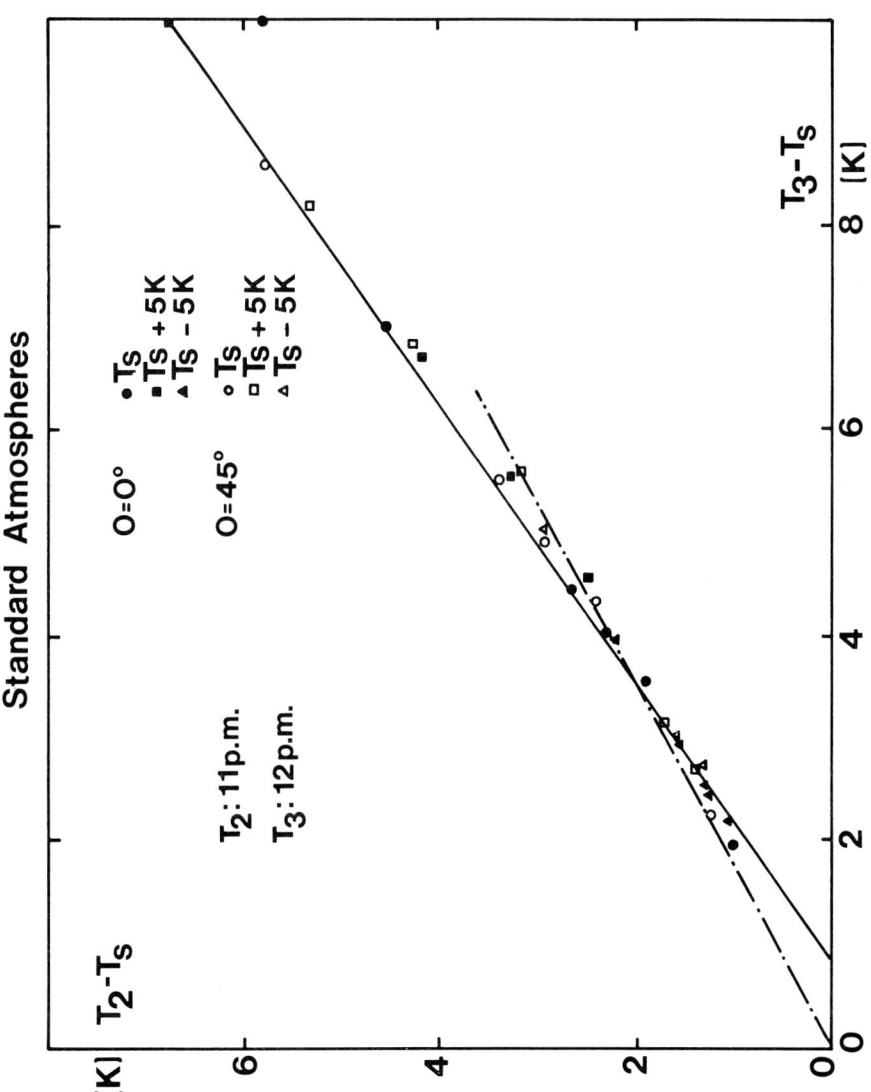

Figure 7: Calculated variations of $(T_2 - T_s)$ as a function of $(T_3 - T_s)$ for different standard atmospheres and for various angles of observation. The straight line is the regression formula given in the text (formula 11) (from Phulpin, 1978). The line -.-.- is the prediction of formula (9b) with a = 1.7. This figure may be compared with Figure 6.

tant. This leads to a simple two channel formula which depends only on one parameter, theoretically independent of the atmosphere.

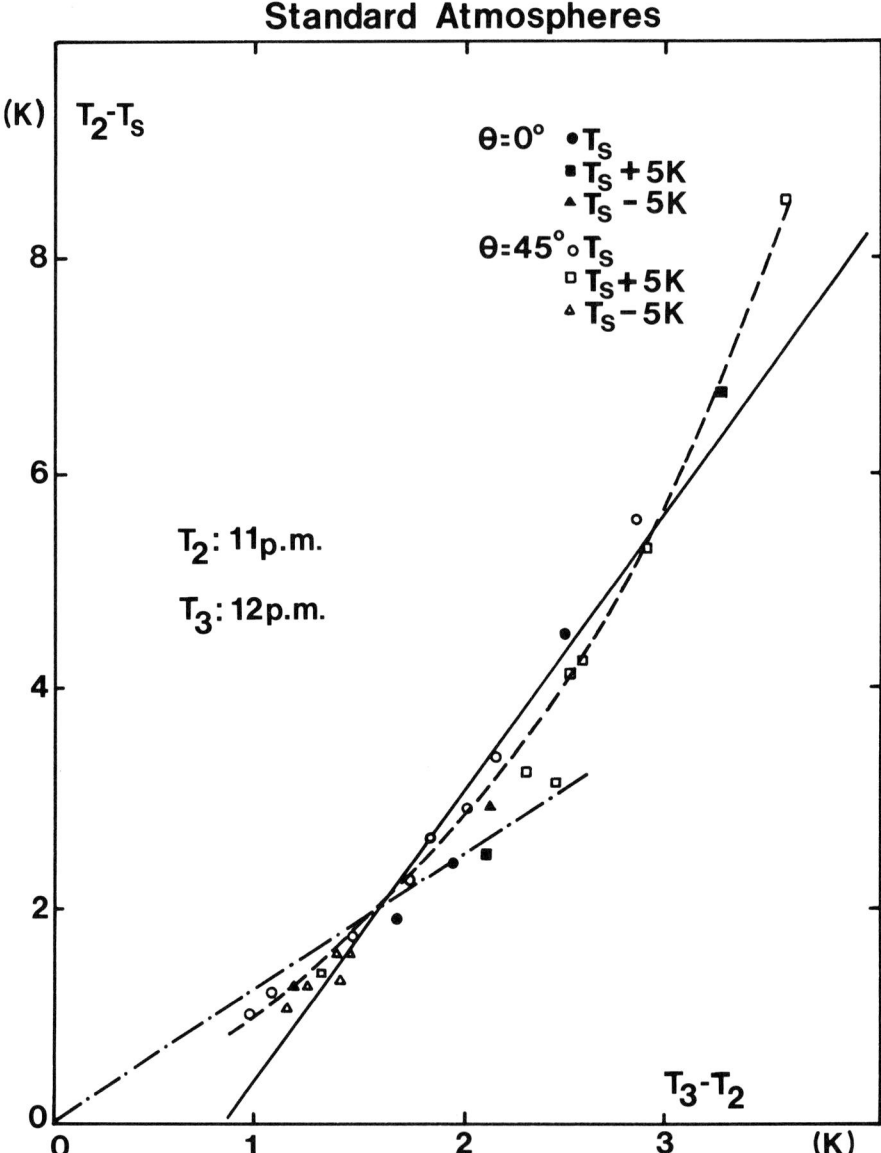

Figure 8: Calculated variation of (T_2 - T_S) as a function of (T_3 - T_S) for different standard atmospheres and for various angles of observation. The straight continuous line is the regression formula given in the text (formula II) (from Th. Phulpin 1978). The line -.- is the prediction of formula (9-b) with a = 1.7.

This formula appears to be correct for atmospheres which lead to a difference in the observed temperatures in the two channels less than 4 or 5 °C. This simple formula allows a very simple algorithm to produce very rapid absolute surface temperature maps. To improve this linear analysis, second order effects which can be accounted for by a measure in a third channel, have to be considered. This is possible with the TIROS N data.

ACKNOWLEDGEMENT

It is a pleasure to acknowledge A. Chedin, D. Imbault, and P. Y. Deschamps for very interesing discussions which allow me to clarify some points presented in this paper; additionally, L. J. Rouse allowed me to use some unpublished results.

REFERENCES

Becker, F., D. Imbault and L. Pontier. 1977. Sea surface temperature and atmospheric transmission: measurements by differential radiometry. Proceedings of the first general assembly by EARSEL, Strasbourg: 93.

Becker, F. and L. Pontier. 1978. Mise en oeuvre d'un radiometre embarquable et mesure aerienne de la temperature superficielle de la mer. Rapport DRME.

Becker, F. D. Blumenroeder, C. Dechambenoy, E. Hechinger, A. Hourani, A. Pellegrin, B. Ramey and J. Trautmann. 1979. Measurement and mapping of the absolute surface temperature of water surfaces by remote sensing. Proceedings of the 13th International Symposium on Remote Sensing of Environment, Ann Arbor.

Chedin, A., N. Scott and N. Husson. 1977. Operation, gestion et etude des informations spectroscopiques atmospheriques. (GEISA) LMD.

DeFelice, P. and L. Pontier. 1977. Determination a distance de la temperature de surface de la mer a l'aide d'un radiometre infrarouge a balayage aeroporte, correction de l'erreur atmospherique. Rapport LMD.

Dechambenoy, C., Le Floch, L. Pontier, M. Raillard and F. Sirou. 1977. Quelques resultats de mesures par navire et avion de la structure superficielle de la mer au large de la Bretagne. Rapport interne.

Deschamps, P. Y., and T. Phulpin. 1978. Correction atmospherique des donnees obtenues par teledetection dans l'infrarouge. Publi Scient. CNEXO. Actes colloque N° 5: 123.

McClatchey, R. A. et al. 1975. Optical properties of the atmos-
     phere. AFCRL 72-0497.
McMillin. 1975. Estimation of sea surface temperatures from two
     infra-red window measurements with different absorption.
     Journ. of Geoph. Res. 80: 5113.
Monge, J. L., L. Pontier and F. Sirou. 1976. Le radiometre ARIES.
     Rapport interne LMD.
Phulpin, T. 1978. Utilisation de la radiometrie infrarouge multi-
     spectrale pour la determination de la temperature de surface.
     These de 3e Cycle, Universite de Lille.
Prabhakara, C., G. Dahi and V. G. Kunde. 1974. Estimation of sea
     surface temperature from remote sensing in the 11-13 window
     region. Journ. of Geoph. Res. 79: 5039.
Scott, N. 1974. A direct Method of Computation of the Transmission
     function of an inhomogeneous gaseous medium. I. Description
     of the method. Journ. of quantitative Spectrosc. and Radiat.
     Transf. 14: 691.

Aerosol Remote Sensing

Adarsh Deepak

ABSTRACT

This paper deals with retrieval of aerosol size distribution
from multispectral measurements of both scattered sky radiance and
direct solar extinction by aerosols. In general, the problem of
making such retrievals from scattered radiance measurements is not
only prohibitively expensive, but practically intractable, unless
some simplifying approximations are made, and even then it remains
quite a formidable task. The difficulty lies not just in the fact
that the scattering by aerosol particles is a complicated process,
which depends not on a single parameter but on at least four intri-
cately mixed parameters, -- namely, size, real and imaginary parts
of complex refractive index, and shape -- but also on the fact that
one must take into account multiple scattering. Essential to
performing such retrievals are two sets of fast algorithms -- one,

for radiative transfer in the atmosphere; and the other, for
retrieval of aerosol physical characteristics. In the remote
sensing technique based on scattered radiation measurements, our
fast multiple scattering approximation to almucantar radiance
distribution in the region close to the sun, namely, the solar
aureole, is described. In this approximation, it is assumed that
for a relatively clear day, the sky radiance is due to single
scattering by molecules and aerosols plus multiple scattering
events alone, the contribution from aerosol multiple scattering
events being assumed negligible. The accuracy of this approximation is discussed in connection with inversion of scattered radiation data. In addition, some preliminary results of aerosol size
distributions retrieved from multispectral extinction measurements
of solar radiation by the nonlinear least squares method are also
discussed.

INTRODUCTION

Remote sensing techniques are, perhaps, the most practical and
economical way of diagnosing and monitoring atmospheric aerosols on
a long-term basis. Aerosol properties of interest include their
shape, size distribution, number concentration, chemical composition, complex refractive index, and spatial distribution. Of
interest, also, is the motion of aerosol particles. Several techniques, both in situ and remote sensing, of varying accuracy have
been developed to measure the aforementioned aerosol properties,
and we will discuss some of the standard and proposed remote sensing techniques in this paper.

Remote sensing is defined as the field which consists of
techniques that obtain reliable information about properties of
objects from a distance. This may be accomplished by measuring (1)
electromagnetic radiation emitted, reflected, scattered, or
attenuated by the object; (2) other force fields, such as gravity
or magnetic fields, created or modified by them; (3) mechanical
(acoustic, seismic) vibrations or waves emanating, reflected, or
scattered from, transmitted through them.

In this paper, discussions will pertain to active and passive remote sensing techniques that are based on electromagnetic radiation. (In the former case, sources of radiation are artificial, e.g., lasers; in the latter, they are natural, e.g., sun.) Nearly all such techniques are based on the measurement of extinction of direct radiation by aerosols, or those of intensity, polarization, or frequency shift of backscatter or multiangle scattered radiation at one or more wavelengths.

Here the discussions will be mainly confined to the physical concepts of remote sensing with a brief description of the experimental and retrieval (mathematical) aspects of remote measurement of aerosol characteristics.

Following is a list of some aerosol remote sounding techniques that are currently being employed or are under development.

Passive Techniques
       Solar Aureole (ground)
       Sunlit Limb Radiances Profile (satellite)
       Diffuse to Direct Ratio (ground)
       Polarimetry (ground/satellite)
       Solar Radiometry (ground)
       Satellite Solar Occultation Radiometry (satellite, such
       as SAM, SAM II, SAGE)
Active Techniques
       Transmissometry
       Lidar
       Doppler Lidar

These experimental techniques measure one or more of the following quantities, vis.,

   Volume Scattering Coefficient, $\beta(\text{km}^{-1})$
   Volume Scattering Function, F_p $[\text{km}^{-1}\text{sr}^{-1}]$, or Volume
      Scattering Phase Function, P $[\text{sr}^{-1}]$
   Degree of Polarization, D_p
   Doppler Frequency Shift, ν_D (sec^{-1})

and employ Radiative Transfer and Inversion Methods to retrieve the following quantities:

   Size Distribution, $n(r)$ $(\text{cm}^{-1}\ \mu_m^{-1})$
   Concentration, $N(0)$ (cm^{-3})
   Complex Refractive Index, $\tilde{m} = m' - im''$
   Velocity, $V(\text{m/sec})$

The problem of remote determination of aerosol properties is, in general, quite difficult except under reasonable simplifying assumptions. Unlike the temperature retrieval problem where a one-to-one relation can be assumed between temperature and frequency, the aerosol retrievals are complicated by the fact that the amount of scattering or extinction of radiation by aerosol particles depends on their size distribution, concentration, complex refractive index and shape. It may be further complicated by the fact that multiple scattering (MS) is also present. Thus, if we try to determine the size distribution, obviously some assumptions have to be made about the other quantities and strategies of making the measurements developed such that the effects of the other quantities is suppressed. Here one takes recourse to the physics of remote sensing, and, to some extent, the retrieval methods. An elegant interplay of the three aspects of remote sensing -- namely, the physics, measurement, and retrieval -- are displayed in the solar method, which has been described in Deepak (1977); Deepak, et al. (1980a); and Deepak, et al. (1980b) will be discussed next.

SOLAR AUREOLE TECHNIQUE

Since 1970, solar aureole method has been used successfully to determine the altitude-integrated (or columnar) size distribution of atmospheric aerosols (Deepak, 1977; Green et al., 1971; Twitty, 1975; Box and Deepak, 1978; Box and Deepak, 1979). Solar aureole is a region of enhanced brightness surrounding the sun's disk within about 20° from it, due to the predominant forward scattering by aerosols. For instance, the contribution to the solar aureole, due to aerosols is 10 to 10^3 times that due to molecules. It is to graphic solar aureole measurement technique in 1970 (Green et al., 1971; Deepak, 1977). The advantage of solar aureole radiance is that it is highly sensitive to the particle size distribution n(r) [cm^{-3} μm^{-1}], but is relatively less sensitive to the effects of particle shape, refractive index, polarization and multiple scattering. Molecular absorption effects can be neglected by working in suitably selected spectral regions. In addition, we make the following reasonable simplifying assumptions:

1. Particles are spherical, so that Mie theory results can be used.

2. MS contribution is significant only for molecules; the aerosol multiple scattering events are not included.

3. One assumes an average refractive index (of about 1.55) for atmospheric aerosols.

4. For zenith angles $\theta > 75°$, spherical earth corrections need to be taken into account; therefore, we choose zenith angles $\theta < 75°$, so that the atmosphere can be treated as plane-parallel.

5. The atmosphere is assumed to be horizontally homogeneous and vertically inhomogeneous.

The retrieval of aersosol size distribution from multispectral solar aureole measurements (PSAMs) made photographically on May 6, 1977, as part of the University of Arizona's Aerosol and Radiation Experiment (UA-ARE), is discussed as follows (Reagan et al., 1978).

Measurements of Almucantar Radiance

The almucantar radiance measurements were obtained by means of a photographic system, for which the experimental steps and the data reduction procedures for making sky radiance measurements are described in detail in Deepak (1977). The digital values of sky radiance at various angular distances from the sun along the almucantar for three wavelengths ($\lambda = 400$, 500, and 600 nm) were then used as input data in the retrieval scheme.

Theory

In much of the earlier work, the retrieval of aerosol size distribution was made tractable by assuming that multiple scattering (MS) is negligibly small compared to single scattering (SS), and could, therefore, be ignored (Deepak, 1977; Green et al., 1971; Twitty, 1975; Box and Deepak, 1978). In the SS approximation, the sky radiance distribution (L_T^{SS}) along the almucantar due to molecules and particulates is given by the relation:

$$L_T^{SS} = L_M + L_P = \Phi_o \sec \theta \, e^{-\tau_T \sec \theta} \{F_{MC}(\psi) + F_{PC}(\psi)\} \tag{1}$$

where $F_{MC}(\psi)$ is the columnar molecular scattering function (sr^{-1}); $P_M = (3/16\pi) (1 + \cos^2 \psi)$, the molecular scattering phase functions (sr^{-1}); $F_{PC}(\lambda,\psi)$, the columnar particulate scattering function (sr^{-1}) is defined as

$$F_{PC}(\psi,\lambda) = \frac{1}{2k^2} \int_{r_1}^{r_2} (i_1 + i_2) \, N_C(r) \, dr \qquad (2)$$

with N_C as the columnar size distribution $(cm^{-2} \, \mu m^{-1})$; i_1 and i_2 are Mie intensity coefficients; r_1 and r_2 are lower and upper limits of radii; Φ_o the incident solar flux; θ the solar zenith angle; subscripts M and P denote molecules and particles, respectively; $\tau_T = \tau_M + \tau_P$, the total optical depth; and ψ the scattering angle given by the relation

$$\cos \psi = \sin^2 \theta + \cos^2 \theta \cos \phi \qquad (3)$$

ϕ is the azimuth angle.

The molecular contribution can be determined by using the molecular density profile for the particular location; so that from Equation (1) one can then experimentally determine $F_{PC}(\psi,\lambda)$ as a function of ψ and λ by replacing L^{SS} by the L_T data as shown

$$F_{PC}^{SS}(\psi,\lambda) = L_T \exp(\tau_T \sec \theta) \cos \theta/\Phi_o - \tau_M P_M(\psi) \qquad (4)$$

This SS approach is well discussed in Green et al. (1971) and Deepak (1977). However, it has been shown by Box and Deepak (1979) that MS contribution to the solar aureole is not insignificant compared to SS, and that the mode radius of the size distribution retrieved on the basis of the SS alone could have errors up to 10 percent for $\tau \sim 0.4$. It was also shown (Box and Deepak, 1979) that one can considerably improve the accuracy of the retrievals by using the MS approximation, which is a modification of the pertur-bation approach suggested by Sekera (1956) and Deirmendjian (1970).

Table 1 shows the difference between the radiance results obtained with the use of our radiative transfer (RT) code and the MS and SS approximations. In the case of our MS approximation, the percentage differences are much smaller than for the SS approxima-tion. Some conclusions can be obtained from Table 2 for the errors in retrieved r_m results obtained with the MS and SS approximations.

TABLE 1: Direct Problem of Computing Almucantar Radiance Using Log Normal Model for n(r) and MS and SS Approximations and RT Code.

θ_0 (deg)	λ	τ_M	τ_P	τ_{O3}	ψ (deg)	% errors in radiance using approximations			
						A = 0		A = 0.2	
						MS approx. %	SS approx. %	MS approx. %	SS approx. %
45	0.4	0.364	0.13984	0.0	0.0	1.60	2.60	1.58	2.87
					3.5	4.10	7.88	3.98	8.93
					13.4	2.65	38.90	1.94	49.30
	0.6	0.069	0.14978	0.05013	0.0	1.16	1.75	1.07	1.87
					3.5	2.33	3.04	2.12	3.30
					13.4	4.11	11.98	2.08	14.80

TABLE 2: Inverse Problem of Retrieving r_m for L_{SS}, $L_{RT}(A=0)$ and $L_{RT}(A=0.25)$ Simulated Measurements Calculated by Using a Modified Gamma Size Distribution Model for $n(r)$, $m = 1.55 - i(0)$ and $r_m = 0.2$ µm.

θ_o (deg)	τ_M	τ_P	Simulated data	Retrieved r_m (µm)			
				SS approx.	% error	MS approx.	% error
60	0.2	0.2	L_{SS}	0.2	0		
			$L_{RT}(A=0)$	0.185	7.60	0.1958	2.15
			$L_{RT}(A=0.25)$	0.1833	9.12	0.1962	1.93

The perturbation approach assumes that the solar aureole radiance is essentially due to SS by molecules and aerosols and MS due to molecules alone, with MS due to aerosols being neglected. This approximation has the advantage that it permits one to retain the simplicity of the SS formulae, and at the same time takes into consideration the effects of MS for nonzero ground albedo A. For the solar aureole region, the MS effects are taken into account by simply replacing $\tau_M P_M$ in Equation (1) by an effective columnar scattering function for the molecular plus aerosol atmosphere, so that the total radiance L_T^{MS} is then given, as in Box and Deepak (1980), by

$$L_T^{MS} = \Phi_o \sec \theta \exp[-\tau_T \sec \theta]$$

$$[(\tau_M + \tau_{MS}) P_M (\psi) + \tau_A P(0^o) + F_{PC}(\psi)] \qquad (5)$$

where τ_{MS} and τ_A are the correction (optical depth) terms to take account of MS due to molecules and ground albedo A, respectively, and defined by the following formulae (Box and Deepak, 1980):

$$A = \frac{A\tau_2}{1 - A\tau_3}$$

$$\tau_{MS}(\equiv \tau_1) = 0.02 \, \tau_{SS} + 1.2 \, \tau_{SS}^2/\mu^{\frac{1}{4}}$$

$$\tau_2 = 1.34 \, \tau_{ss} \, (1 + 0.22(\tau_{SS}/\mu)^2) \qquad (6)$$

$$\tau_3 = 0.9 \ \tau_{SS} - 0.92 \ \tau_{SS}^2 + 0.54 \ \tau_{SS}^3$$

$$\tau_{SS} = \bar{\omega}\tau = \tau_M + \tau_{PS}$$

where τ_{PS} is the particulate optical depth due to scattering in contrast to particulate absorption; ω the SS albedo (averaged); and $\mu = \cos \theta$. The experimental data for the scattering function is obtained by replacing L_T^{MS} in Equation (5) by the L_T data and using the formula:

$$F_{PC}^{MS}(\psi,\lambda) = L_T \ \exp(\tau_T/\mu) \ \mu/\Phi_0$$
$$- \ [(\tau_M + \tau_{MS}) \ P_M(\psi) + \tau_A P_M(0^o)] \tag{7}$$

Since ground albedo A is usually unknown, it must be included among the parameters to be retrieved. Thus one employs the following model for the total columnar scattering function

$$F_{TC}^{MS}(\psi,\lambda) = \frac{1}{2k^2} \int_{r_1}^{r_2} (i_1 + i_2) \ N_C(r) \ dr$$

$$+ \ (\tau_M + \tau_{MS}) \ P_M(\psi) + \tau_A P_M(0^o) \tag{8}$$

This retrieval approach was applied to data obtained during the solar aureole experiment performed in Tucson on May 6, 1977. Almucantar radiance measurements were made for three wavelengths (λ = 400, 500, and 600 nm). From these measurements experimental data were obtained for $F_{PC}(\psi)$ (circles in Figure 1) which were inverted by using a two-term size distribution model, each term being a Haze M distribution

$$N_C(r) = p_1 r \ [\exp(-p_2\sqrt{r}) + p_3 \ \exp(-p_4\sqrt{r})] \tag{9}$$

A nonlinear least squares (NLLS) program described in a separate section was used to invert the spectral data. Initial estimates of the parameters p_i for use in the NLLS program were obtained with the help of a parameterized graphical catalog of phase function plots corresponding to different size distribution models (Deepak and Box, 1979).

Discussion of Results

The size distribution model used to fit the columnar scattering function date for λ = 400, 500, and 600 nm was a sum of two

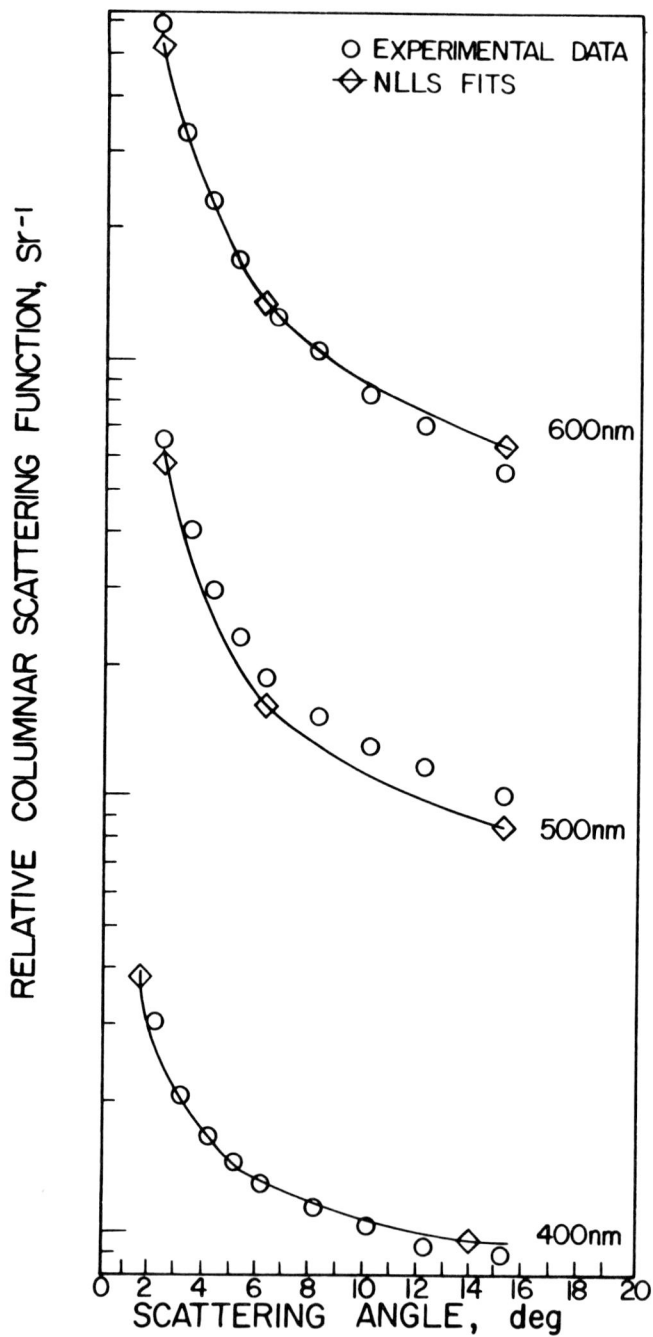

FIGURE 1: Plots of experimental data for $F_{PC}(\psi)$ and NLLS best fits to the data.

Haze M terms. Sensitivity studies were performed to understand the effects on retrieved results due to different factors, such as, aerosol refractive indices, limits r_1 and r_2 of integration, and ignoring multiple scattering. For example, three refractive indices were used: $1.45 - i(0.00)$, $1.50 - i(0.00)$, and $1.50 - i(0.01)$. Essentially no difference was found between the fits obtained for the different refractive indices although the retrieved parameters p_i were slightly different. In addition, the decrease of r_1 from 0.04 μm to 0.01 μm made no difference in the final retrieved $N_C(r)$ results. For scattering angles greater than 12°, it was found that the effects of multiple scattering needed to be taken into account. Retrieved size distribution were relatively more sensitive to r_2 and surface albedo A.

The retrieval size distribution obtained by assuming m 1.55 - i(0) is shown in Figure 2. Preliminary ground truth SD data, obtained by airborne Whitby (◙) and Royco (+) counters, and plotted in Figure 2, were provided by Professor J. Reagan, University of Arizona. Even though the ground truth data is sparse and is for May 7, 1977, the day following the day (May 6) on which solar aureole measurements were made, they are reasonably close to the retrieved columnar size distribution results to give confidence in the latter.

The solar aureole technique is a simple and accurate method for determining the columnar size distribution of tropospheric aerosols. Use of forward scattered solar radiance (i.e., aureole) measured from the ground, as shown here, or from a satellite-borne radiometer scanning up to 10° on either side (horizontal) of setting/rising sun, is, perhaps, the more accurate than the use of direct solar radiation measurements for determining aerosol size distributions in the atmosphere. The deployment of this aureole scanning capacity on satellite radiometers such as SAGE II is, therefore, strongly recommended.

MULTISPECTRAL OPTICAL EXTINCTION TECHNIQUE

Multispectral optical extinction technique, also referred to as solar radiometry or laser beam transmissiometry, essentially

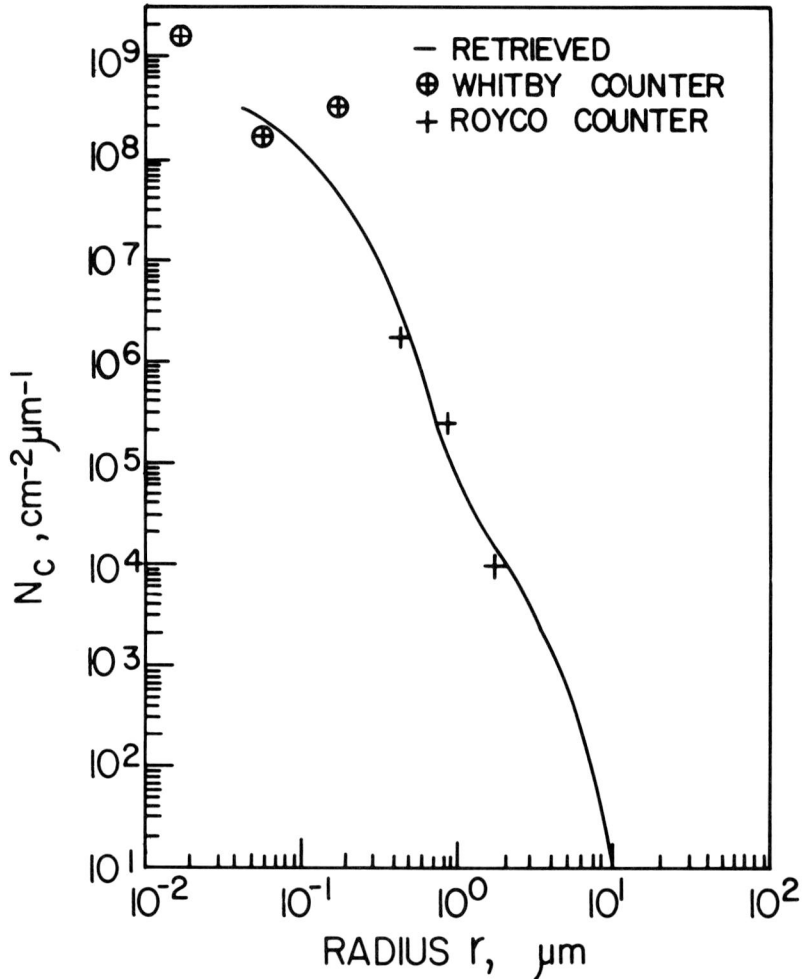

FIGURE 2: Plots of retrieved and measurement data for columnar size distribution $n_C(r)$.

involves the measurements of attenuated irradiance $I(\lambda)$ of solar or laser radiation scattered and absorbed during transit through length L of the medium, and the use of Bouguer-Beer law

$$I(\lambda) = I_o(\lambda) \; e^{-\tau_T(\lambda)} \tag{10}$$

where $I_o(\lambda)$ is the unattenuated irradiance. For solar radiometer measurements, the total optical depth $\tau_T(\lambda)$ can be obtained by the

well-known Langley plot method. With the help of tables and measurements of surface pressure, the molecular and ozone optical
depths $\tau_M(\lambda)$ and $\tau_{O_3}(\lambda)$, respectively, can be computed, so that
the particulate optical depth $\tau_p(\lambda)$ can be obtained from the
relation

$$\tau_p(\lambda) = \tau_T(\lambda) - \{\tau_M(\lambda) + \tau_{O_3}(\lambda)\} \tag{11}$$

τ_p is related to $N_C(r)$ by the relation

$$\tau_p(\lambda) = \int_0^\infty \pi r^2 \, Q_{EXT}(x,m) \, N_C(r) \, dr \tag{12}$$

where Q_{EXT} is the extinction efficiency factor. It is, therefore,
possible to retrieve aerosol size distribution information from
multispectral $\tau_p(\lambda)$ measurements. Several workers have employed
this technique to determine aerosol size distributions by adopting
different retrieval approaches; e.g., modified Twomey techniques
(Twomey, 1963, 1965), well explained by Herman (1977), have been
used by some researchers (Shaw et al., 1973; Chu, 1977), and
approximate methods (Box and Lo, 1976; Box et al., 1980), and table
search methods (Box and Deepak, 1980) have been used by others.
Here, an example of retrieving aerosol columnar size distribution
from optical depth $\tau_p(\lambda)$ measurements by using NLLS method,
described in the next section, will be given.

Measurements of $\tau_p(\lambda)$ made by means of calibrated, seven-
channel solar radiometer during a University of Arizona--Aerosol
and Radiation Experiment, May 6-19, 1977, were provided to us by
Professor John A. Reagan, University of Arizona. Three sets of
$\tau_p(\lambda)$ measurements, obtained by using Equation (11) are shown in
Table 3. The assumed columnar size distribution model was the Haze
H model, viz.,

$$N_C(r) = \tfrac{1}{2} \, ab^3 r^2 e^{-br} \tag{13}$$

for which the parameters p_1 and p_2 retrieved by the NLLS from $\tau_p(\lambda)$
data (Table 3) are shown in Table 4. For details, see Box et al.
(1980).

TABLE 3: Multispectral Measurements for the Particulate Optical Depths.

Wavelength $\lambda(\mu m)$	Particulate optical depth $\tau_p(\lambda)$		
	I	II	III
0.4400	0.0852	0.0603	0.0850
0.5217	0.0756	0.0527	0.0829
0.5556	0.0676	0.0507	0.0687
0.6120	0.0649	0.0490	0.0728
0.6708	0.0629	0.0459	0.0658
0.7797	0.0613	0.0425	0.0683
0.8717	0.0648	0.0364	0.0577

TABLE 4: Retrieved Size Distribution Parameter a and b by Means of NLLS Program.

Observation Data Sets	b_1 (μm^{-1})	$\pm\Delta b_1$ (μm^{-1})	a	$\pm\Delta a$
I	14.60	1.690	212.00	133.00
II	16.50	0.844	298.00	85.30
III	15.10	1.260	271.00	123.00

NONLINEAR LEAST SQUARES (NLLS) METHOD

The nonlinear least squares (NLLS) code is based on the linearization method for solving the nonlinear problem (Deutsch, 1965; Draper and Smith, 1966). This is explained as follows:

Consider the model

$$g = f(x,p) + \varepsilon \qquad (14)$$

where

g = a vector of observed values,
$f(x,p)$ = vector of predicted values,
p = vector of parameters to be estimated,
x = vector of independent variables, and
ε = random error vector.

To estimate the parameters $\underline{p}$, minimize the sum of squares of the errors

$$S = \{\underline{g} - \underline{f}(\underline{x},\underline{p})\}^T \ \underline{W}^{-1} \ \{\underline{g} - \underline{f}(\underline{x},\underline{p})\} \tag{15}$$

where $\underline{W}$ is a weighting matrix.

The best estimates $\hat{\underline{p}}$ of $\underline{p}$ are given by the solution of

$$\frac{\partial S}{\partial \underline{p}} = 0$$

i.e.,

$$\frac{\partial}{\partial \underline{p}} \ \{\underline{f}(\underline{x},\underline{p})\}^T \qquad \underline{W}^{-1} \ \{\underline{g} - \underline{f}(\underline{x},\underline{p}) = 0 \tag{16}$$

or, letting

$$\underline{H} = \frac{\partial}{\partial \underline{p}} \ \{\underline{f}(\underline{x},\underline{p})\} \tag{17}$$

one obtains,

$$\underline{H}^T \ \underline{W}^{-1} \ \{\underline{g} - \underline{f}(\underline{x},\underline{p})\} = 0 \tag{18}$$

However, $\underline{f}(\underline{x},\underline{p})$ is nonlinear in $\underline{p}$, so in order to linearize Equation (18), carry out a first order Taylor series expansion of $\underline{f}$ about an initial parameter estimate $\hat{\underline{p}}_o$ involving first derivatives only, i.e.,

$$\underline{f}(\underline{x},\hat{\underline{p}}) = \underline{f}(\underline{x},\hat{\underline{p}}_o) + [\ \frac{\partial}{\partial \underline{p}} \ \underline{f}(\underline{x},\hat{\underline{p}})]_{\underline{p}=\hat{\underline{p}}_o} \ (\underline{p} - \hat{\underline{p}}_o) \tag{19}$$

substituting Equation (19) in Equation (18) gives

$$-\underline{H}^T \ \underline{W}^{-1} \ \{\underline{g} - \underline{f}(\underline{x},\hat{\underline{p}}_o)\} + \underline{H}^T \ \underline{W}^{-1} \ \underline{H}(\hat{\underline{p}} - \hat{\underline{p}}_o) = 0 \tag{20}$$

Equation (20 is linear in $(\hat{\underline{p}} - \hat{\underline{p}}_o)$ and the solution is given by

$$\hat{\underline{p}} = \hat{\underline{p}}_o = (\underline{H}^T \ \underline{W}^{-1} \ \underline{H})^{-1} \ \underline{H}^T \ \underline{W}^{-1} \ \{\underline{g} - \underline{f}(\underline{x},\hat{\underline{p}}_o)\} \tag{21}$$

or

$$\hat{\underline{p}} = \hat{\underline{p}}_o + (\underline{H}^T \ \underline{W}^{-1} \ \underline{H})^{-1} \ \underline{H}^T \ \underline{W}^{-1} \ \{\underline{g} - \underline{f}(\underline{x},\hat{\underline{p}}_o)\} \tag{22}$$

The above method of solving for the parameters $\hat{\underline{p}}$ is called the Gauss Newton method.

In practice, solution of nonlinear least squares system of equations for $\hat{p}$ is an iterative process, with the new estimate $\hat{p}$ (or $\hat{p}_n$) at the end of each (or n^{th}) iteration, becomes the initial estimate $\hat{p}_0$ for the next or $(n + 1)^{th}$ iteration. The process continues until the solution converges to the required degree. The convergence criteria used in this method were: (1) that the variances χ^2_{n-1} and χ^2_n for the $(n - 1)^{th}$ and n^{th} iterations, respectively, and estimated variance χ^2_{n+1} for the $(n + 1)^{th}$ iteration not differ from one another by more than 1 percent; (2) that the change in each parameter between the $(n - 1)^{th}$ and n^{th} iteration be no more than 0.1 percent.

Another feature of the program used is a search for the optimum update at each iteration. A half interval search is conducted to determine the region which contains the minimum χ^2 and then a quadratic is fitted to three points in this region to determine the factor Υ by which the quantity $(\hat{p} - \hat{p}_0)$ should be multiplied before adding it to $\hat{p}_0$ to give the new parameter update.

In preceding sections, the equations being solved are Equations (8) and (12). A schematic diagram of the NLLS inversion scheme is shown in Figure 3.

CONCLUDING REMARKS

From the above discussions, it is quite evident that the solar aureole technique is a simple and relatively inexpensive method for determining the columnar size distribution of aerosols. The use of the NLLS inversion scheme yields retrievals that are accurate when checked against ground truth measurements of $N_C(r)$. These retrievals were performed by taking into account the multiple scattering contributions to the sky radiance within the region of the solar aureole by the use of a MS approximation. The retrievals are sensitive to the upper limit of integration over radius and the surface albedo on which further work is in progress.

The NLLS method can also be used to retrieve $N_C(r)$ parameters from multispectral optical depth measurements. In addition, work is in progress in developing fast retrieval techniques (Box and

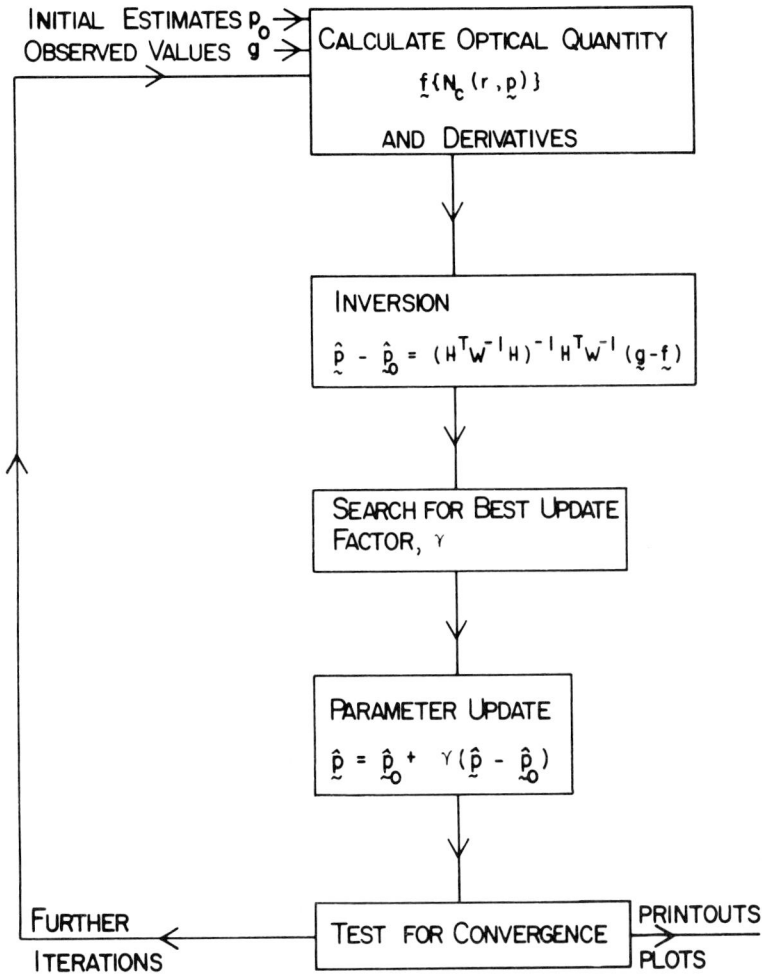

FIGURE 3: Schematic diagram of the NLLS inversion scheme.

Deepak, 1980) which can be used to develop automated inversion systems.

ACKNOWLEDGEMENTS

It is a pleasure to acknowledge the valuable assistance and cooperation of Professor John A. Reagan, University of Arizona, in enabling us to participate in the UA-ARE experiment and providing us with both $\tau_p(\lambda)$ and ground truth, size distribution measurements. The valuable assistance of R. R. Adams and B. Poole, NASA-Langley Research Center, in taking photographs of the solar aureole and data reduction, and Michael and Gail Box, IFAORS, in the analysis of the data, is greatfully acknowledged. This work was supported under NASA Contract NAS1-15198.

REFERENCES

Box, G. P., M. A. Box, and A. Deepak, 1980. On the spectral sensitivity of approximate method for retrieving aerosol size distributions from multispectral solar extinction measurements. IFAORS Report No. 150. (Available from IFAORS, P. O. Box P, Hampton, VA 23666.)

Box, G. P., and A. Deepak, 1980. Fast table search method for retrieval of aerosol size distribution from multispectral optical depth measurements. IFAORS Report No. 148. (Available from IFAORS, P. O. Box P, Hampton, VA 23666; also, to be published as NASA CR in 1981.)

Box, M. A., and A. Deepak. 1978. Single and multiple scattering contributions to circumsolar radiation. Appl. Opt., 17: 3794-3797.

Box, M. A., and A. Deepak. 1979. Retrieval of aerosol size distributions by inversion of simulated aureole data in the presence of multiple scattering. Appl. Opt., 18: 1376-1382.

Box, M. A., and A. Deepak. 1980. An approximation to multiple scattering in earth's atmosphere: almucantar radiance formulation. IFAORS Report No. 134. (Available from IFAORS, P. O. Box P, Hampton, VA 23666.)

Box, M. A., and S. Y. Lo. 1976. Approximate determination of aerosol size distributions. J. Appl. Meteorol., 15: 1068-1076.

Chu, W. P. 1977. The inversion of stratospheric aerosol and ozone vertical profiles from spacecraft solar extinction measurements. In: Inversion Methods in Atmospheric Remote Sounding. pp. 505-527. Ed. by A. Deepak. New York: Academic Press.

Deepak, A. 1977. Inversion of solar aureole measurements for determining aerosol characteristics. In: Inversion Methods in Atmospheric Remote Sounding, pp. 265-295. Ed. By A. Deepak. New York: Academic Press.

Deepak, A., and G. P. Box. 1979. Analytical modeling by aerosol size distribution. NASA Contract Report 159170.

Deepak, A., M. A. Box, and G. P. Box. 1980. Retrieval of aerosol size distributions from scattering and extinction measurements in the presence of multiple scattering. In: Remote Sensing of Oceans and Atmospheres, pp. 95-114. Ed. by. A. Deepak. New York: Adaemic Press.

Deepak, A., M. A. Box, G. P. Box, and R. R. Adams. 1980. Experimental validation of the solar aureole technique for determining aerosol size distributions. IFAORS Report No. 132. (Available from IFAORS, P. O. Box P, Hampton, VA 23666.)

Deirmendjian, D. 1970. Use of scattering techniques in cloud micro-
     physics research: I. The aureole method. Rand Corporation
     Report R-590-PR.
Deutsch, R. 1965. Estimation Theory. Englewood Cliffs, New
     Jersey: Prentice-Hall, Inc.
Draper, N., and H. Smith. 1966. Applied analysis. New York: John
     Wiley and Sons, Inc.
Green, A. E. S., A. Deepak, and B. J. Lipofsky. 1971. Interpreta-
     tion of the sun's aureole based on atmospheric aerosol models.
     App. Opt., 10: 1263.
Herman, B. M. 1977. Application of modified Twomey techniques to
     invert lidar angular scatter and solar extinction data for
     determining aerosol size distributions. In: Inversion
     Methods in Atmospheric Remote Sounding, pp. 469-503. Ed. by
     A. Deepak. New York: Academic Press.
King, M. D., D. M. Byren, B. M. Herman, and J. A. Reagan. 1978.
     Aerosol size distributions obtained by inversion of spectral
     optical depth measurements. J. Atmos. Sci., 35: 2153-2167.
Reagan, J. A., B. M. Herman, D. M. Byren, and M. D. King, 1978
     Third conference on atmospheric radiation, June 28-30, 1978,
     Davis, California. p. 241. Boston, Massachusetts: American
     Meteorological Society.
Sekera, Z. 1956. Recent developments in the study of polarization
     of sky light. Adv. Geophys., 3: 43.
Shaw, G. E., J. A. Reagan, and B. M. Herman. 1973. Investigations
     of atmospheric extinction using direct solar radiation mea-
     surements made with a multiple wavelength radiometer. J. App.
     Meteoral., 12: 374.
Twitty, J. T. 1975. The inversion of aureole measurements to
     derive aerosol size distributions. J. Atmos. Sci., 32: 584.
Twomey, S. 1963. On the numerical solution of Fredholm integral
     equations of the first kind by the inversion of the linear
     system produced by quadrature. J. Assoc. Comput. Mach., 10:
     97.
Twomey, S. 1965. The application of numerical filtering to the
     solution of integral equations encountered in indirect sensing
     measurements. J. Franklin Inst., 279: 95.

Signals and Information

P.B. Fellgett

ABSTRACT

Safe and optimal procedures for the interpretation of observations are those in which we know our ignorance and use this knowledge consistently. Information Theory and the equations of
Inverse Probability enable us to do this.

Among the most significant achievements of physics are to be
found the formulation of measures of quantities obeying the great
conservation laws of physics; the laws of conservation of momentum,
matter and energy, and the laws of thermodynamics. In particular,
the formulation of the concept of quantity of energy led to an
understanding of the laws governing the conversion of energy from
one form to another, and its transmission from one place to
another. These understandings form an essential basis of Energy
Engineering.

195

Similarly, formulation of a definition of quantity of information has led to understanding of the laws of conversion and transmission of information, and hence to Information Engineering. The
modern formulation of this measure of quantity of information is
due to Shannon (1948); its development has a long history going
back at least to Nyquist (1924) and Hartley (1928).

It is often the case in science that an advance has depended
on a "self denying ordinance"; the acceptance of the Uncertainty
Principle being a classic example. Indeed it may be thought that
this self denial is inherent in the basic discipline of science
restricting itself to regarding as meaningful only statements about
matters which are in principle capable of experimental refutation.
Similarly, the formulation of a self-consistent definition of
quantity of information, obeying useful laws, has depended on
concentrating on quantity and not attempting to include consideration of the quality or interest of the information. However, it is
possible, as we shall see, to say what information is about, and
hence to what it is relevant.

The Shannon definition of quantity of information obeys a
conservative law of the second kind; this is to say, information
can be destroyed by error or guesswork, but cannot be increased
(provided, that is, we take a large enough boundary to the information system under consideration).

Information gain is defined as $H_o - H_1$, where H_o is the initial entropy and H_1 the entropy after the message has been received
or the observation made. Entropy may be regarded as a measure of
ignorance, and discrete entropy is defined in the form

$$H = - \Sigma_r \; p_r \; \log p_r \qquad (1)$$

in which p_r is the probability of the rth event, and the choice of
the base of logarithm defines the scale of measurement; usually
$\log_2$ is taken, and the information is then measured in bits. All
that really matters for present purposes is that information is a
function of probabilities.

Because of this fact, we can define an optimum process as one
which conserves the quantity of information relevant to the pro-

blem. Reduction of data always involves making explicit what is implicit in the data concerning the quantities under investigation. If this is done by calculating the relevant probabilities or probability distributions, conditional on the data, relevant information is conserved, since these probabilities carry all the information about the quantities in question. The information gain is the difference between the entropy of the prior probabilities, before the observations were available, and the entropy of the corresponding probabilities after the observations have been taken into account.

A good way of conserving information in the reduction of observations is often to use the Bayes equations of inverse probability. Denote by $p(x)$, $p(y)$ the respective unconditional probabilities of x and y, by $p_y(x)$ the probability of x given y, by $p_x(y)$ the probability of y given x, and by $p(x,y)$ the joint probability of x and y.

Then
$$p(x,y) = p(x) \ p_x(y) = p(y) \ P_y(x)$$

and therefore
$$p_y(x) = \frac{p(x) \ p_x(y)}{p(y)} \tag{2}$$

If x denotes an hypothesis and y a set of observations (x and y may of course be multi-dimensional), it is convenient to renormalize this equation to
$$p_1(x) = k \ p_o(x) \ p_x(y) \tag{3}$$

in which $p_1(x)$ is the probability of x given the observation, $p_o(x)$ is the prior probability of x before the observations, k is a normalizing constant to bring the integral of $p_1(x)$ to unity, and $p_x(y)$ is the likelihood of obtaining the observation y assuming the hypothesis x. This equation accords well with common sense. It implies that our previous expectation of x should be modified, as a result of receiving the observations, by a factor of how likely it would be to obtain these observations if the hypothesis x were true. Here x can represent equally well the values of a set of continuous parameters, or a discrete dichotomy of assumption, for example whether or not a cluster of grains in an astronomical photograph represents a star image.

The use of the Bayes equations is particularly well matched to the many problems of interpretation requiring inversion of a known direct formula. Many such problems can be formulated as the inversion if a convolution; that is to say, roughly speaking, we have blurred information and wish to have sharp information. Use of the Bayes equations in this way almost always gives us a way of proceeding when confronted by a set of observations. The resultant procedure is optimal whenever it can be carried out, and it is unlikely to deceive if done correctly.

These advantages are very powerful, but there is the counter-balancing disadvantage that exhaustive enumeration of the hypotheses is required. Often the hypotheses represent parameter values, and the work becomes very heavy in problems involving many parameters, where we have to try to find the region of maximum likelihood in a multi-dimensional parameter space. This requirement leads only too easily to "chasing factorials".

Hence, the importance of processes not involving comparison of data with each individual model, but with the generalized properties of a class of models. In effect, we make yes/no decisions about class membership as a means of restricting attention, at the earliest possible stage, to a limited range of possibilities. This is the process of discrimination or pattern recognition. It necessarily involves taking non-zero probabilities and treating them as if they were zero or unity. This is known in information theory as "guesswork", and it is a general principle that it always destroys some information. This is the price that has to be paid in return for the ability to treat cases that cannot practically be explored exhaustively.

In summary, it is important to know that optimal procedures exist, and to test any proposed procedures against the criterion of conserving relevant information. Where it is impracticable to carry out (or even in some cases to formulate) strictly optimal procedures, results having a high probability of correctness can be obtained by using the discriminatory power of pattern recognition.

Finally, it is worth recalling that there are no generally "best" procedures for such operations as smoothing, deconvolving, etc. What is best depends on the circumstances, and in particular what is known and what unknown. This underlines the importance of knowing our state of ignorance, and using this knowledge of ignorance in the solutions to the problem. Insofar as information theory quantifies ignorance as entropy, the maximum entropy set of assumptions is always, in this sense, a safe one. An important and well-known example is that assumption of a Gaussian probability distribution is the entropy-maximizing assumption when we are ignorant of everything except the variance or covariance. This approach differs from that of classical statisticians who require the assumption of a particular probability distribution to be justified in terms of a model.

REFERENCES

Hartley, R. V. L. 1928. Bell System Technical Journal 7: 535.
Nyquist, H. 1924. Bell System Technical Journal 3: 324.
Shannon, C. E. 1948. Bell System Technical Journal 27: 379, 623.

Time Series and Its Application to Remote Sensing

Edward J. Wegman

ABSTRACT

Modern time series analysis involves two closely related but rather distinct studies. The first which might be called parametric modelling involves the construction of functional relationships between various observations in a time series. The other involves spectral decomposition for the purpose of frequency analysis of signals. This paper develops a broad overview of these two aspects and suggests a possible application to the analysis of non-imaging sensors such as a radar altimeter.

INTRODUCTION

Modern time series analysis is a vast field upon which many authors have written. It would be foolish to believe one could give any more than a superficial survey in a paper such as this. Therefore, at the outset I would like to recommend texts by Box and Jenkins (1976), Koopmans (1974) and Brillinger (1975). These books provide a detailed view of a terrain for which we now provide only a general roadmap.

Modern time series involves two major aspects: 1. parametric modelling and 2. spectral analysis. The first involves an attempt to interrelate the various numerical values of a time series in much the same way that the dependent and independent variables are related in a regression model. The latter area involves the decomposition of a time series into its wave-like components. We discuss parametric modelling in section 2 and spectral analysis in section 3. We discuss some aspects of filtering theory in section 4 and propose a scheme for analysis of altimeter data in section 5.

PARAMETRIC MODELLING

We begin our considerations with a time series
$$Z_0, \; Z_1, \; Z_2, \; \ldots, \; Z_n.$$
A (discrete or digital) time series differs from the random sample of ordinary statistics in two important ways. First, we assume there are correlations between the various Z_i's so that it makes sense to relate parametrically Z_i to the other Z_t's. Second, we assume that the data are evolving through time so that Z_{t-1} is the observation which chronologically comes just before Z_t.

Perhaps the simplest model for such a time series is
$$Z_t = \alpha Z_{t-1} + a_t \tag{2.1}$$
where a_t are zero mean, constant variance random errors. This simple scheme says that the present value Z_t is proportional to the previous value, but contaminated with some random errors, a_t. Notice that if all the random errors were identically 0, then we could apply the formula $Z_t = \alpha Z_{t-1}$ recursively to obtain
$$Z_t = \alpha Z_{t-1} = \alpha^2 Z_{t-2} = \ldots = \alpha^t Z_0.$$
In this case, Z_t is simply an exponential function. Equation 2.1

might be appropriate for the discharge of a capacitor ($\alpha < 1$), decay of a population ($\alpha < 1$), or growth of a population ($\alpha > 1$).

While it is intuitively appealing to believe the population at time t is proportional to the population at time t-1, for more subtle time series this model may not be adequate. Often, a model that supposes that Z_t is proportional not only to its previous value Z_{t-1}, but also to the rate of change in Z_t is a very satisfactory one. Thus we may write

$$Z_t = \alpha Z_{t-1} + \beta \left[\frac{Z_{t-1} - Z_{t-2}}{(t-1) - (t-2)} \right] + a_t$$

or rewriting a little

$$Z_t = \phi_1 Z_{t-1} + \phi_2 Z_{t-2} + a_t$$

where $\phi_1 = \alpha + \beta$ and $\phi_2 = -\beta$. Such a model might be appropriate for time series measuring the depletion of natural resources such as oil, the growth of government budgets or stock market prices. Obviously the process of adding more terms can be continued,

$$Z_t = \phi_1 Z_{t-1} + \phi_2 Z_{t-2} + \ldots + \phi_p Z_{t-p} + a_t.$$

This recursive reformulation is called a pth order autogressive model.

This formulation while quite useful is somewhat cumbersome to work with. We can introduce a shorthand formulation. We let the symbol B represent a backward shift operator. This is

$$BZ_t = Z_{t-1}, \ B^2 Z_t = B(BZ_t) = BZ_{t-1} = Z_{t-2}$$

and similarly $B^n Z_t = Z_{t-n}$. Hence, the autoregressive model can be written

$$Z_t - \phi_1 Z_{t-1} - \ldots - \phi_p Z_{t-p} = a_t$$

or in operator notation

$$(I - \phi_1 B - \ldots - \phi_p B^p) Z_t = a_t$$

where I is called identity operator. The polynomial

$$\Phi(B) = I - \phi_1 B - \ldots - \phi_p B^p$$

is called the auxiliary polynomial and plays an important role as we shall see later. The general operator formulation of an auto-regressive scheme is

$$\Phi(B)Z_t = a_t \qquad (2.2)$$

In general, the error input into an autoregressive model is a simple sequence of zero mean, constant variance random variables, a_t. There is, in general, no reason why the error structure cannot be more complex. We could for example replace a_t with

$$a_t + \theta_1 a_{t-1} + \ldots + \theta_q a_{t-q}.$$

Again using a backward shift operator, we can write

$$(I + \theta_1 B + \ldots + \theta_q B^q)a_t,$$

or writing again in polynomial form

$$\theta(B)a_t.$$

this error structure is called a _moving_ _average_ so that the model becomes

$$\Phi(B)Z_t = \theta(B)a_t \qquad (2.3)$$

This model is currently referred to as an ARMA (p, q) model meaning _auto_regressive _m_oving _a_verage model although older literature refers to it as an autoregressive model with moving average input. ARMA models are discussed at great length in Box and Jenkins (1976). Of course, the key issue is to identify the parameters ϕ_1, ..., ϕ_p and θ_1, ..., θ_q. To do this requires certain additional assumptions.

Definition:

A time series Z_0, ..., Z_n is said to be _wide_ _sense_ _stationary_ or simply stationary if

1. $E(Z_t) = 0$

2. $E(Z_{t+\tau}Z_t) = \gamma_\tau$ is independent of t.

Actually $E(Z_t)$ equal to any constant would do, but we generally choose 0 for convenience. The function, γ_τ, is called the _covariance_ _function_. It is also called the autocovariance since it is the covariance of Z_t with itself. Of course, $\gamma_0 = var(Z_t)$. The ratio $\rho_\tau = \gamma_\tau/\gamma_0$ is the _autocorrelation_ _function_.

For stationary time series that satisfies Equation (2.2), i.e. that satisfies a purely autoregressive model, one can show that the autocovariance satisfies

$$\Phi(B)\gamma_\tau = 0 \qquad (2.4)$$

where the backward shift is taken as acting on τ. Equation (2.4) is a homogeneous difference equation whereas Equation (2.2) is a corresponding nonhomogeneous one. This suggests that the covariance function γ_τ and the time series Z_t have similar structure save that the time series is driven by a random sequence. Indeed there are many structural parallels between the deterministic γ_τ and the random Z_t which is reason enough for studying γ_τ (or equivalently, ρ_τ).

There are additional reasons, however. For an autoregressive process, the coefficients ϕ_1, ..., ϕ_p are related to the autocorrelations by the so-called <u>Yule-Walker</u> equations:

$$
\begin{bmatrix} \phi_1 \\ \cdot \\ \cdot \\ \cdot \\ \cdot \\ \phi_p \end{bmatrix} =
\begin{bmatrix} 1 & \rho_1 & \rho_2 & \cdots & \rho_{p-1} \\ \rho_1 & 1 & \rho_1 & \cdots & \rho_{p-2} \\ \cdot & & & & \cdot \\ \cdot & & & & \cdot \\ \cdot & & & & \\ \rho_{p-1} & \rho_{p-2} & \cdots & & 1 \end{bmatrix}^{-1}
\begin{bmatrix} \rho_1 \\ \cdot \\ \cdot \\ \cdot \\ \rho_p \end{bmatrix} \qquad (2.5)
$$

Thus, in order to estimate the parameters ϕ_1, ..., ϕ_p for an autoregressive process, it is only necessary to have suitable estimates of ρ_1, ..., ρ_p and use the Yule-Walker equation to fit the autoregressive coefficients. The coefficients of the moving average part and more generally of the entire ARMA model have a non-linear relationship to the correlations. Thus there is not a simple relationship, but, it is in general computable.

The covariance structure has further importance for the statistician, because if we are willing to assume the finite dimensional distributions of the time series are Gaussian, then the covariance structure completely determines the probability structure of the time series. Thus knowing the parametric ARMA model is equivalent to knowing the probability structure.

SPECTRAL ANALYSIS

The theory of spectral analysis represents the other major
aspect of time series analysis. Consider first a simple sine wave
as in Figure 1a. It can be characterized by two parameters, the
amplitude A which measures the height or intensity of the sine wave
and the wavelength, λ, which measure the repeat period of the wave,
i.e. the period it takes to complete one cycle. The <u>angular
frequency</u>, ω_0, is 2π times the inverse of λ_0, $2\pi/\lambda_0$. Corresponding
to the sine wave we could draw a new graph, Figure 1b, which dia-
grammatically represents these two parameters. The function represented in Figure 1b is called the <u>spectrum</u>.

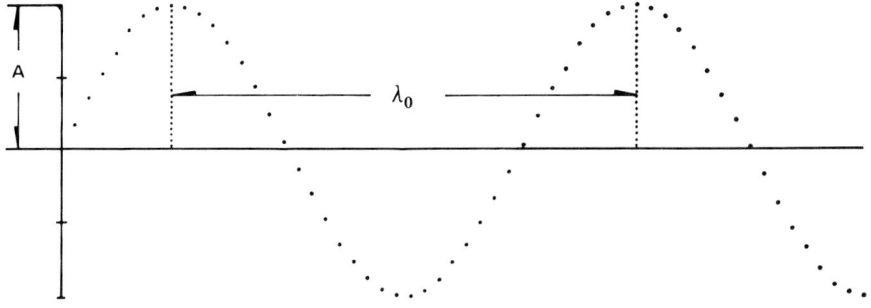

Figure Ia

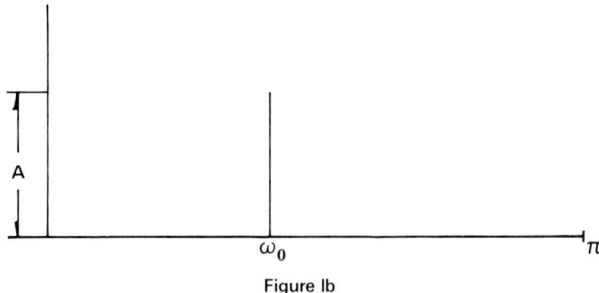

Figure Ib

FIGURE 1: A simple sine wave with wavelength λ_0 is given in 1 (a).
The spectrum of this sinewave with angular frequency $\omega_0 = 2\pi/\lambda_0$ is
given in 1 (b).

Of course, if we were dealing only with simple sine waves, there would be little point in constructing the spectrum. The point is that most wave forms of practical interest are much more complicated than simple sine waves. For example, Figure 2c is the sum of the curves in Figure 2a & 2b. Thus the curve in Figure 2c is the sum of two sine wave components and would have the spectrum represented in Figure 2d. Thus the spectrum gives a very convenient way of representing the sine wave decomposition of any composite wave form. Naturally, there is no reason why we have to be content with the sum of two sine waves. We could, for example, have a spectrum which represents the sum of a finite number of sinusoids, as in Figure 3a or even a continuum of sinusoids as indicated in Figure 3b. In the latter case, we refer to the curve represented as a <u>spectral</u> <u>density</u> and write it, f(ω). This, in fact, is the more usual case. For discrete or digital data, the maximum frequency is bounded by π (which may, of course, be adjusted by adjusting the sampling rate. The number π refers to a sampling interval of 1 time unit.) Figures 1b, 2d, 3a, and 3b are drawn only for the positive x-axis. Generally speaking spectral densities exist on the range (-π, π) with the negative half axis indicating the same periodicity with a 180° phase shift.

For the ARMA models discussed in section 2, the spectral density, f(ω) may be written as

$$f(\omega) = \frac{\sigma_a^2}{2\pi} \left| \frac{\theta(e^{i\omega})}{\Phi(e^{i\omega})} \right|^2 \quad , \quad -\pi \leq \omega \leq \pi, \qquad (3.1)$$

where θ(B) and Φ(B) are respectively the polynomial arising from the moving average and autoregressive parts of the process. In general, the spectral density function (for a discrete time series) is defined by f(ω) such that

$$\gamma_s = \int_{-\pi}^{\pi} e^{is\omega} f(\omega) d\omega \qquad (3.2)$$

Thus like the ARMA models, the spectral density is intimately related to the covariance sequence (it is the Fourier transform of the covariance sequence and vice versa). Notice if we assume all the joint distributions are multivariate normal, then knowledge of the covariance sequence completely specifies the probability struc-

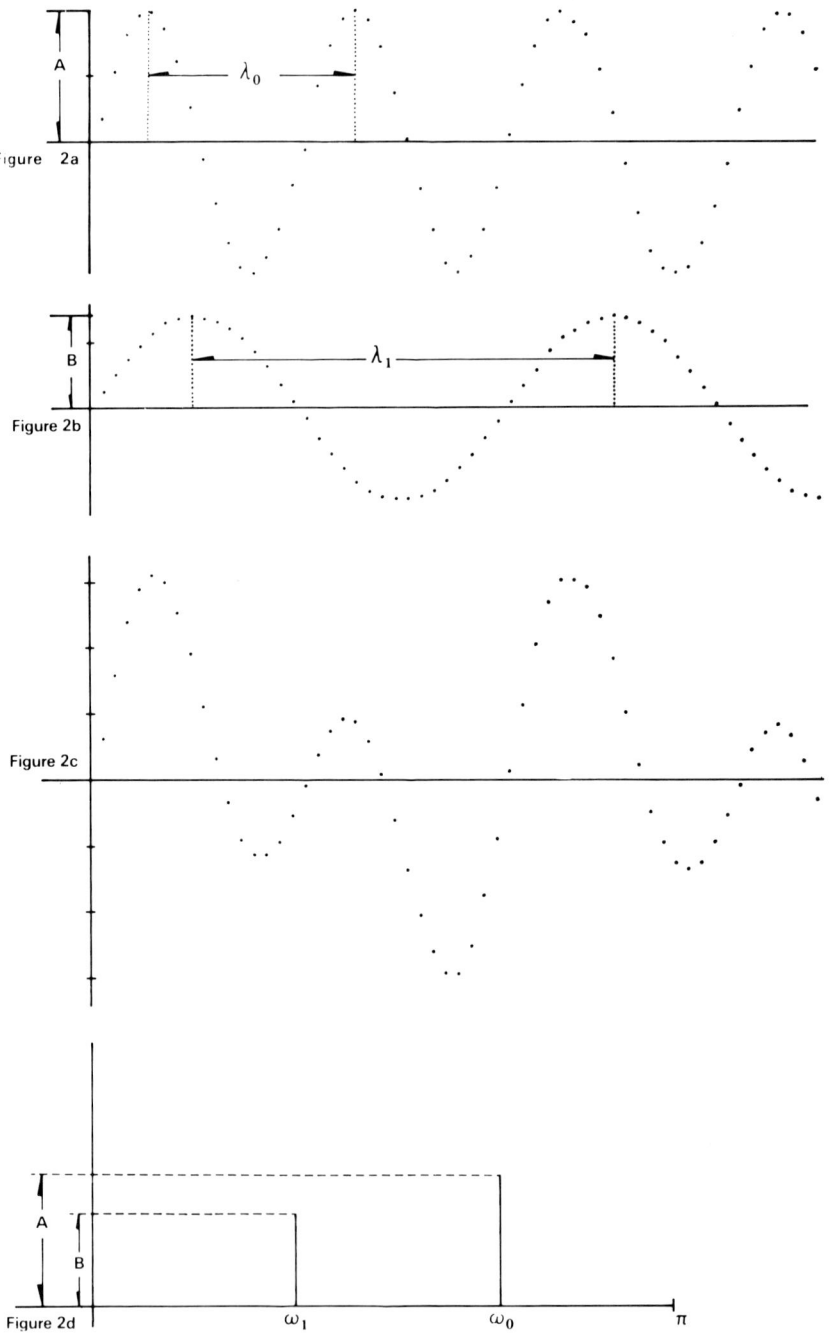

FIGURE 2: The composite waveform given in 2(c) is the sum of the simple sinusoids given in 2(a) and 2(b). The spectrum in 2(d) graphically represents the decomposition of 2(c) into its spectral components.

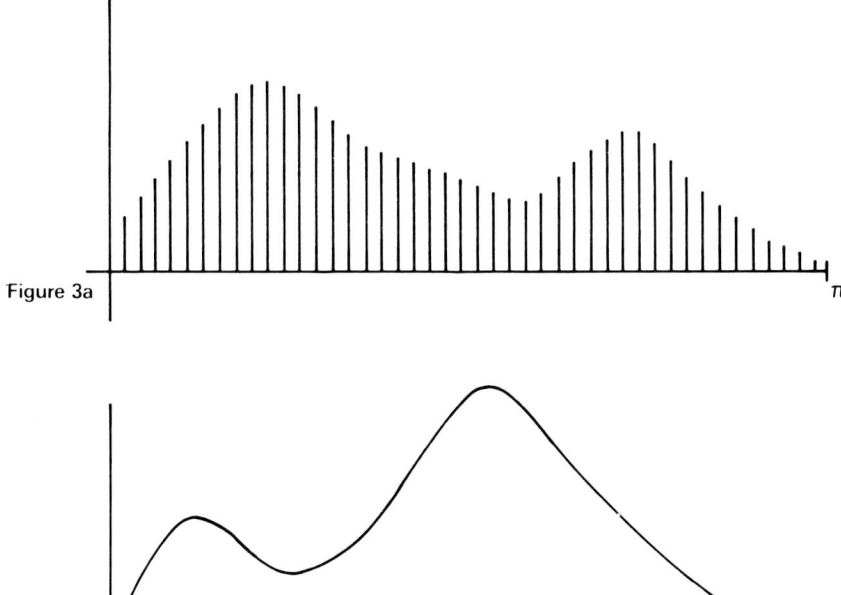

Figure 3a

Figure 3b

FIGURE 3: The spectrum may contain many lines as in figure 3(a) representing a signal with many sinusoidal components. In the limit, it can become a smooth spectral density as in 3(b) which represents a signal having density at every infinitesimal frequency.

ture. But knowing either the ARMA model or the spectral density is equivalent to knowing the covariance. Hence, knowing either is equivalent to specifying the probability structure. From the point of view of statistical inference, this equivalence is very handy.

We note that the inverse relationship to (3.2) is

$$f(\omega) = \frac{1}{2\pi} \sum_{j=-\infty}^{\infty} \gamma_j e^{-i\omega j}, \quad -\pi \le \omega \le \pi \qquad (3.3)$$

If we have two time series X_t and Y_t, we may define the cross-covariance by

$$\gamma_\tau^{xy} = E(X_{t+\tau}Y_t)$$

and by analogy to (3.3), the Cross-spectral density as

$$f^{xy}(\omega) = \frac{1}{2\pi} \sum_{j=-\infty}^{\infty} \gamma_j^{xy} e^{-i\omega j}, \quad -\pi \leq \omega \leq \pi \qquad (3.4)$$

Several facts are of interest regarding spectral densities. If a time series, say Z_t, is just a sequence of independent and identical distributed ramdom variables (IID) which is the usual statistical case, then we can write

$$Z_t = a_t$$

which is an ARMA model with $\Phi(B) = I = \theta(B)$. Thus the spectral density is just

$$f(\omega) = \frac{\sigma_a^2}{2\pi}, \quad -\pi \leq \omega \leq \pi$$

which is constant. Since all frequencies appear with equal intensity we refer to Z_t as <u>white noise</u> by analogy with white light.

Notice from (3.2), that

$$\gamma_0 = \int_{-\pi}^{\pi} f(\omega)d\omega.$$

Thus unlike an ordinary probability density, the spectral density has an area under it equal to γ_0 variance of Z_t. This is analogous, in electrical engineering terms to the <u>power</u> of a wave and is sometimes referred to as the power. The density f is sometimes called the <u>power</u> <u>spectrum</u>.

For the ARMA models, both $\Phi(B)$ and $\theta(B)$ are polynomials, in this case $f(\omega)$ is the ratio of polynomials and Z_t is said to have a <u>rational</u> spectral density. The ratio $H(z) = \theta(z)/\Phi(z)$ is called the <u>transfer</u> <u>function</u>. Electrical engineers call the ARMA model a <u>poles</u> <u>and</u> <u>zeros</u> <u>model</u> since the p roots of Φ are "poles" of H and the q roots of θ are "zeros" of H.

FILTERING

Perhaps the most potentially useful idea in time series is <u>filtering</u>. A filter is nothing more than a functional which transforms one time series into another. If in the ARMA model, we replace a_t with X_t and Z_t with Y_t, we obtain the model,

$$\Phi(B)Y_t = \theta(B)X_t$$

or equivalently

$$Y_t = H(B)X_t \tag{4.1}$$

In general, $H(B)$ will no longer be a simple polynomial but rather an infinite series. Thus (4.1) can be written

$$Y_t = \sum_{j=0}^{\infty} h_j X_{t-j}.$$

For suitably convergent series, this is called a <u>linear</u>, <u>time-invariant</u> <u>filter</u>. The numbers, h_j, are called the <u>filter</u> <u>weights</u> or sometimes the <u>impulse response function</u>. The latter terminology is particularly used in a continuous time version:

$$Y_t = \int_0^{\infty} h(\tau)X_{t-\tau}d\tau.$$

The functions, h_j and $H(z)$ are related by the equation

$$h_j = \int_{-\pi}^{\pi} H(\omega)e^{ij\omega}d\omega, \tag{4.2}$$

i.e. they are Fourier transform pairs. The impulse response function h_j (or $h(\tau)$) can be given an interpretation. It is the response of the system modelled by the filter to an instantaneous, unit pulse. For most purposes, knowledge of the impulse response function and of the transfer function are equivalent.

For a noisy linear system which can be characterized by

$$Y_t = \mu + \sum_{j=0}^{\infty} h_j X_{t-j} + \varepsilon_t,$$

the μ and h_j which minimize $E(\varepsilon_t^2)$ are given by

$$\mu = EY_t - H(0)EX_t$$

and

$$h_j = \int_{-\pi}^{\pi} H(\omega)e^{i\omega j}d\omega.$$

The transfer function may be written in terms of the spectral density

$$H(\omega) = f^{yx}(\omega)/f^{xx}(\omega). \tag{4.3}$$

Thus the impulse response function may be written in terms of the transfer function, which in turn may be written in terms of spectral densities, which in turn may be written in terms of covariances. The estimation of such functions has received some considerable attention in the literature (see Brillinger, 1975; Wegman, 1980, 1981). We will not further pursue the details of the methodology in this article except to point out that there is a chain of relationships between the raw data and such estimators.

SEMI-IMAGING REMOTE SENSORS

We suggest in this section a possible application of time series methodology to the processing of data from such remote sensors as radar altimeters. At least one major problem with such imaging sensors as the synthetic aperature radar (SAR) is the high volume of data telemetry required by these instruments whereas a radar altimeter has modest telemetry requirements and correspondingly modest spatial resolution. Our suggestion is that there is information on spatial distribution buried in the returned signal of a radar altimeter which may be elicited by proper time series processing. Our ideas are conceived in relative ignorance of the engineering aspects of remote sensing instruments and we leave the test of their mettle and their implementation to others.

We consider first a pulse-limited radar (see the article by C. R. Francis elsewhere in this volume) which has a pulse sufficiently short so that not all the target is illuminated simultaneously. The radar pulse expands in a spherical shell at successive instants. The footprint of the instrument will in general be elliptical (see Figure 4). The intersection of the radar pulse with the surface of the earth will first be elliptical (or circular) and then as time continues, annular. The annuli represent points to which the travel time of the pulse from the satellite to the surface of the earth is constant. These we call them isochronal annuli (see Figure 5a). The signal returning to the satellite from one of these isochronal annuli represents the radar reflectivity of this annulus. Thus the entire returned signal represents data about a sequence of annuli and we need only untangle the returned pulse to obtain a form of spatial resolution.

Notice that increasing time in the returned signal corresponds to increasing nominal radius in the annulus. A typical pulse wave form is given in Figure 5b.

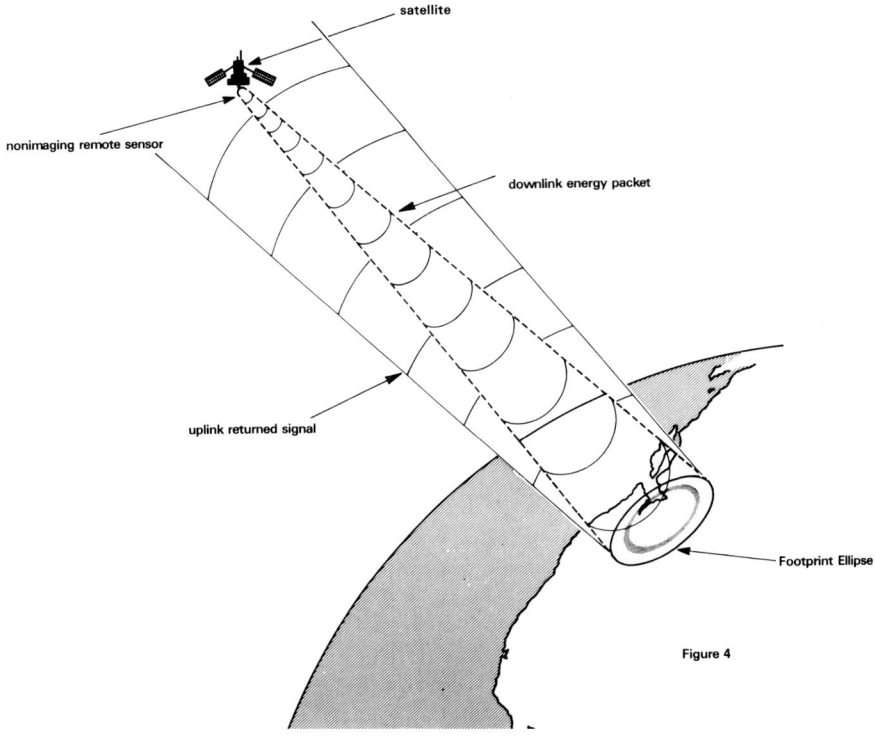

FIGURE 4: Schematic diagram of pulsed radar altimeter system. Note the elliptical footprint together with its expanding elliptical annuli.

Suppose we now abandon our pulse-limited radar in favor of a continuously operating one (this is necessary to provide the stationarity assumption of the time series methodology). Then we will have essential all the annuli returns coming in simultaneously. Fortunately, we can model this in a rather simple way by

$$Y_t = \sum_{\tau=\tau_1}^{\tau_2} h_\tau X_{t-\tau} + \varepsilon_t \qquad (5.1)$$

where τ represents the travel time to an annulus and back to the satellite, h_τ represents the intensity of radar reflection for the

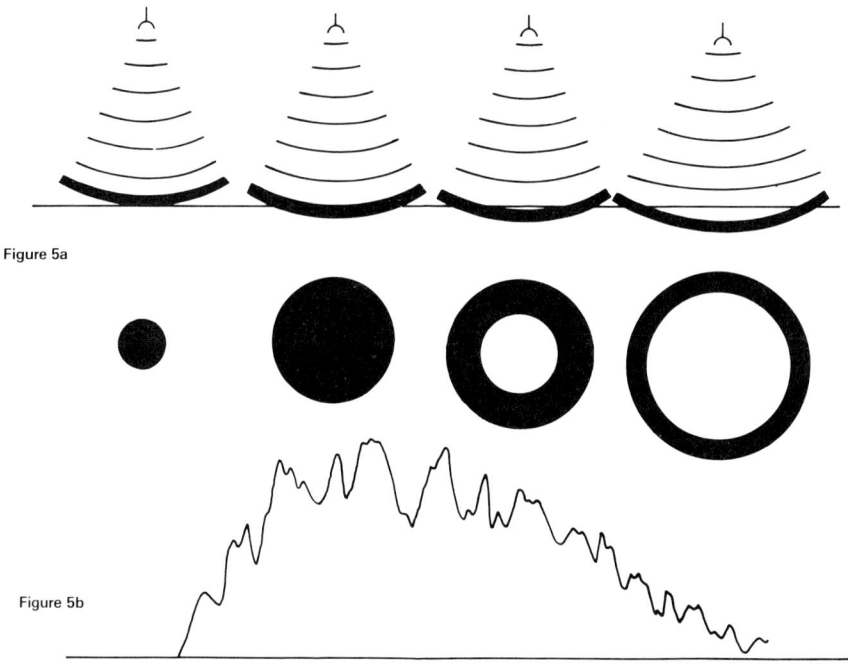

Figure 5a

Figure 5b

FIGURE 5: As a radar pulse intersects the earth's surface an enlarging disc and later an enlarging annulus is formed. As the disc increases in 5(a), the reflected signal increases in 5(b). As the annulus continues to expand to the limits of the footprint in 5(a), the signal levels off and gradually begins to decay in 5(b).

annulus whose travel time is τ, X_t is the emitted signal, Y_t is the returned signal, ε_t is the noise, τ_1 is the minimum total travel time and τ_2 is the maximum total travel time.

The function, h_τ, which represents the reflectivity of a particular annuli, is the impulse response function. As pointed out in section 4 we can relate h_τ to the covariances which can be estimated by knowing the signals Y_t and X_t. The function, h_τ, carries a physical interpretation and depending on how finely we can differentiate the annuli, we can increase our spatial resolution.

We offer the following comparison only as a theoretical computation which could change depending on hardware implementation. Consider a nonimaging sensor with circular footprint of 10km diameter. The area of this footprint is about 78.5km^2 and with only the usual processing we could not expect to resolve anything better than this. By contrast an imaging sensor covering a total of 78.5 km^2 with a 128 x 128 pixel image could expect to resolve areas down to .00479 km^2, that is, we can expect a resolution improvement by a factor of 2^{14}. Suppose we process by the method put forward here and that the isochronal annuli can be resolved to a thickness of .1 km. Then the area of the annuli varies from .157 km^2 to 1.56 km^2, which represents an improvement over the nonimaging sensor by a factor of between 50 to 500. Finally, we note that if one annulus shows a particularly strong return, we can have a second look at the surface after the satellite has moved. This would provide a second annulus, and the source of the strong response would then be narrowed to the intersection of the two annuli. The intersection could be very small indeed, say on the order of .02 km^2 which is comparable to the resolution of the imaging sensor.

ACKNOWLEDGEMENT

I am grateful to Mr. Ferd Diemer who encouraged me to put forth my "half baked" idea and to Dr. C. R. Francis whose presentation and diagrams helped bake my ideas. Finally, I am grateful to Mrs. Nancy Hall at the Naval Surface Weapons Center, White Oak, Md. who typed this report.

REFERENCES

Box, G. E. P. and Jenkins, G. 1976. Time Series Analysis, Forecasting, and Control, Holden-Day, San Francisco.

Brillinger, D. R. 1975. Time Series: Data Analysis and Theory, Holt, Rinehart and Winston, New York.

Thomas, D. P. and Francis, C. R. 1981. "Microwave, HF, IR and Visible Spectrum Sensors," this proceedings.

Koopmans, L. H. 1974. The Spectral Analysis of Time Series, Academic Press, New York.

Wegman, E. J. 1981. "Vector splines and the estimation of filter functions," To appear Technometrics, February, 1981.

Wegman, E. J. 1980. "Optimal estimation of time series functions," To appear IEEE Transactions on Acoustics, Speech and Signal Processing, December, 1980.

Models and Algorithms in Automatic Analysis of Remotely Sensed Imagery

Bruno Apolloni

ABSTRACT

A critical survey of the analysis of remote sensed images is presented. The paper focuses on models of ground regions, and the kinds of images for which the various methodologies are suitable. The most promising trends on optimal collections of theoretical tools are stipulated and the limits of completely automatic analysis are also pointed out.

INTRODUCTION

For about ten years remote sensed imagery has been considered a useful tool in the large scale analysis of earth surfaces and many institutes have been engaged in looking for a way to utilize remote sensed imagery supplied cheaply by the satellites.

Many studies are being made, many companies are purchasing and many others are selling works and devices for telesurveying projects; interesting results are coming forth from various fields, such as geology, ocean science, hydrology, land cover analysis, etc. However, it is not as yet a completely delineated discipline.

One important topic that is still under discussion is the alternative between human eye interpretation and automatic analysis of the image. At the end of the pretreatment phase of the data, devoted to cleaning and enhancing the image so as to make it as "meaningful" as possible, one needs to extract its information content.

Human interpretation, supported by interactive computer utilities, is without doubt the more efficient way to proceed in the analysis of an image of no great size. This is true because the power of synthesis of the human eye takes into account a great number of particulars at the same time, and its flexibility updates and arranges the model of the image and the tasks and strategies of the analysis in an interactive way as a function of the partial information achieved during the analysis itself. On the other hand, completely automatic analysis of the image pursues the objective of surveying regions of a considerable size in a reasonable time, allowing for the acquisition of global information on earth surfaces which is one of the most appealing aspects of telesurveying projects. In this sense, the interest of researchers has been focused on land use automatic analysis, but until now only few applications have shown much efficiency in the recognition of land classes. More precisely, a great number of methodologies have been studied, implemented, and improved. Generally all the corresponding papers, whatever the methodology, end with some pictures and tables which show a high ratio of correctly classified pixels (80 to 90%), but these results correspond to the special regions considered by the authors (Holmes et al., 1976). So the questions that the user asks himself in employing one of the methods are the following: which one of the methods has to be employed, and why the correctly classified ratios which he finds are so much less than the ones shown in the papers. The answer comes from the tautological consideration that one can find only in the image for which one is searching, and therefore the choice of the methodology and the discrepancies between the results depend on the degree of compliance among the special characteristics of the methodology, the mathematical model of the image hypothesized by the user, and the real behaviour of the phenomena which give rise to the imagery.

The aim of the paper is to present a review of the methodologies available for land cover automatic analysis, putting in evidence the underlying models and the kinds of images for which they are suitable. Data processing of remote sensed imagery is generally characterized by a large number of steps, which one tries usually to subdivide into several almost independent sets (Figure 1). The paper deals with the central processor step, which achieves in input the pretreated image and gives as output the classified image. Preceding or subsequent steps will be considered everytime they are strongly influenced by the special characteristics of the central processor. Such topics as pre- and post-processing, data compression, and so forth are adequately covered by exhaustive reviews (Kanal, 1974; Fu, K. S. Ed., 1976; Mitchell et al.; Cappellini et al., 1978; Chen, 1978; Thomas, 1978).

MATHEMATICAL MODELIZATION OF THE IMAGERY

Mathematical models must state analytical relations between the signal received by the remote sensors and the desired map of ground points classified as belonging to some land cover classes. Owing to the randomness of the ground phenomena, the model has to be a probabilistic one, requiring the estimation of some parameters. The choice of the estimators must be optimized with respect to a risk function which depends on the characteristics of the use to be made of the information once extracted. The signal received by the sensors is generally described by the spectral intensity diagram, the time at which it is received, and the position of the sensor with respect to the ground. Some ancillary data are also often available. These data specify the environment of the data transmission such as orbit of the sensor platforms, atmos-

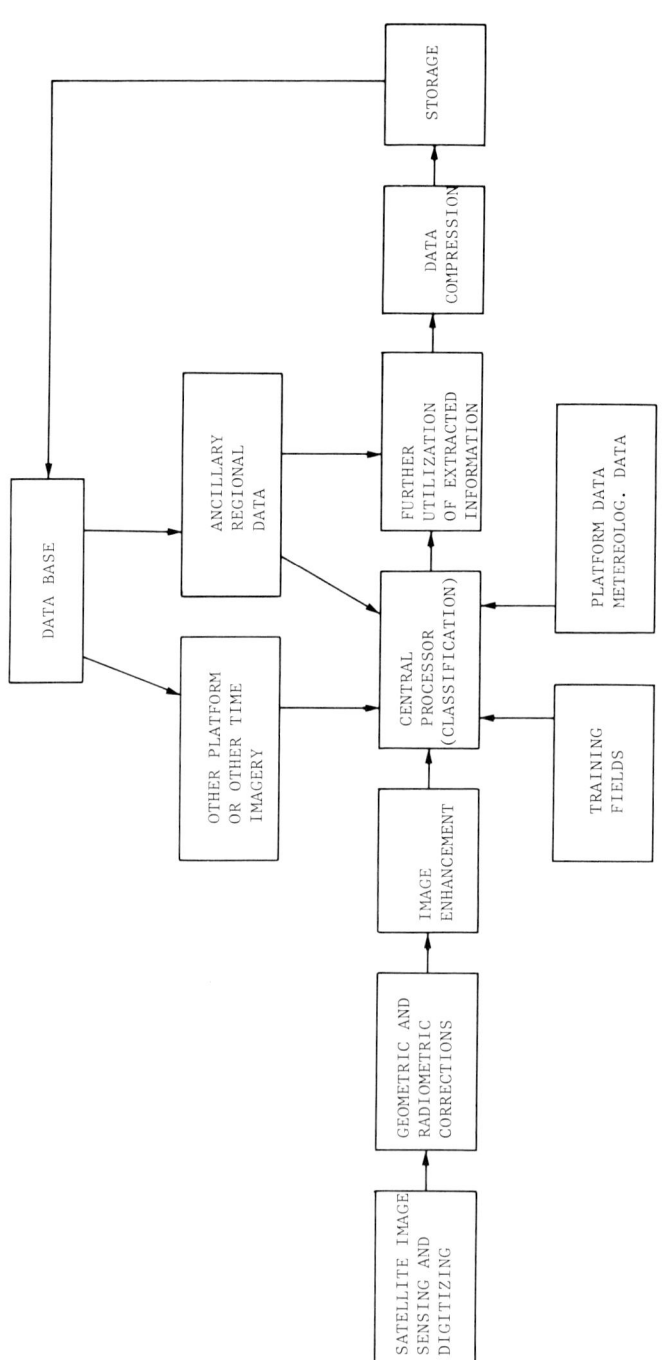

FIGURE 1: Block diagram of processing.

pheric conditions, and so on.

To provide a rational framework for surveying, (Apolloni, et al., 1978) the observation and feature space are defined so that the image analysis may be considered as the proper transformation of the data from the first to the second space, and a subsequent discrimination among the transformed points (Figure 2). The observation space is defined by the following coordinates: spectral wavelength coordinate, physical spatial coordinates of the sensed earth's region, time of sensing operation coordinate, and intensity of the signal (gray level) coordinate. The resolution element of a point in this space is defined with respect to the transmission wavelength band, spatial resolution cell (pixel), time, and radiometric resolution of the sensor (gray level resolution). The data are the observed points of this space and are representative of what is contained in the pixel; to distinguish one content from another, we shall broadly use the term class. A map is a subspace of the observations space where each point is assigned to a class, generally constituted by a spatial coordinate subspace. However, it is also often considered a temporal coordinate, or spectral and temporal coordinate subspace (e.g. to examine the phenological cycles). A feature is any computable function of the observed points. The feature space is described by a selected set of features. One makes the hypothesis that features are strongly dependent on the classes of the points from which they are computed. In other words one is looking to find a set of features which result (as greatly discriminant) with respect to these classes.

It is meaningful to classify the features into levels of increasing complexity. At the first level the feature (referred to as point feature) is computed by the coordinates of only one observation point at a time. Thus, for instance, with LANDSAT data a point feature is the gray level coordinate at the observation point. At the next level, the features are derived from a mono- or multi-dimensional neighborhood. For example, one can consider in the observation space a geographical subspace neighborhood, take gradients or perform other types of transformations, and assume

them as features. At the first two levels, the features can be
considered as local features. At higher levels features are to an
increasing extent "global". They are entities connected with
measurements taken along curves (patterns, surfaces, volumes). At
each level features may be interrelated to different extents, but
higher level features automatically account, to various degrees,
for the interrelations among lower level features.

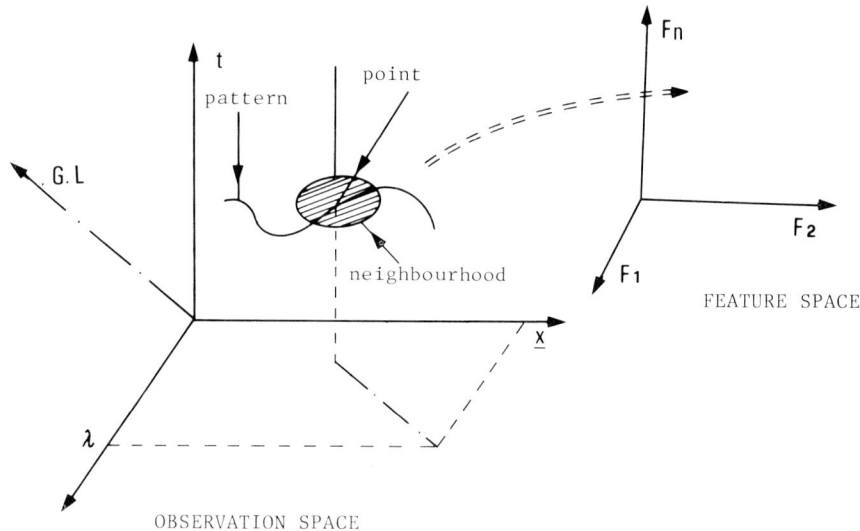

FIGURE 2:   Feature extraction
           $\underline{X}$ = physical spatial coordinates;
           t = time coordinate;
           λ = wavelength coordinate;
       G.L. = gray level coordinate;
           F_i = feature coordinate.

Owing to the complexity of ground radiation phenomena, to
state general relations between features and classes, one prefers
to formulate simplified models in which some parameters have to be
estimated in a statistical way. Essentially one takes the statistical characteristics of the features in each class by means of
training fields; the mathematical model suggests the characteristics to take into account in the training fields, the more discriminant features, and the kinds of relations between the first
and the second ones. Also, the coordinates of the observation

space need some model to be evaluated. With respect to the geographical coordinates, one knows the coordinates of the sensor, and
to obtain those of the sensed point one generally needs some relations involving ancillary data such as trajectory and attitude
angles of the platform, etc. The size of the resolution cell is
not an altogether small problem, especially in the network of radar
sensors (Harger, 1970). A critical topic is constituted by the
gray level model for pixels where there are present moisture
regions belonging to different classes. The gray level of the
resolution cell of a synthetic aperture radar image is strongly
dependent on the pretreatment of the image (Murino, et al., 1979).

The ultimate goal of image analysis is always the extraction
of only the desired information. Therefore, an ideal image
analysis must preserve only the information that is of interest for
subsequent analysis, throwing out what is not desired. Generally
one refers to Bayesian risk functions in order to choose the
classifier of the pixels. Only in the last few years, as it will
be seen later, two steps analysis and global context analysis
approaches introduced some loss functions which expressly depended
on the user's model of application of information taken from
remotely sensed images.

METHODOLOGICAL APPROACHES

The task of each methodology in processing the imagery is to
attribute the points of the feature space to one of the preassigned
classes. The aim of optimal methodology is to take into account
all the informational content of the observations space, using
ancillary information as a priori knowledge or as boundary constraints. In this framework a challenge is to overcome the pre-
vailing trend that considers the information content of some coordinates of observation space as "separate" or "disjoint" and to
develop efficient techniques which consider the contents as belonging to an "informational continuum" sampled, for technological or
practical reasons, in a discrete manner. An optimal criterion of
choice among the methodologies is to consider jointly the accuracy

in extracting the information and the corresponding computational complexity.

The largely used division of methodologies into two categories, the decision theoretic (or statistic) and syntactic, will be adopted here. (Duda and Hart, 1973). The decision theoretic approach deals with methodologies which search for a partition of the feature space for the optimal solution of identification or classification problems. In a broad sense, the syntactic methodologies use a priori available information that can be cast into a set of rules (called morphological or synatactic rules) describing relations among higher level features called primitive structural elements (words, morphs). The syntactic approach tries to recognize allowable higher order structures in terms of primitives and their interrelationships.

In the last year a great amount of work has been devoted to exploit jointly the usefulness of both the syntactic and statistical approach in the analysis of the image, and also in the frame of each approach many kinds of methodology have been developed so as to give rise to an intriguing network of cases. In this paper the key in surveying the methodologies will be the distinction, where possible, between the two approaches, and in the frame of each one, to differentiate on the basis of the complexity of the model. The measure of complexity is taken to be the amount of information and of interrelations that the model is capable of taking into account (Figure 3).

DECISION THEORETIC APPROACH

Point by point classification

Point features. For many years data processing in remote sensing analysis has been mostly restricted to Bayesian pixel methodologies, with the features vector constituted by gray levels in the four spectral bands of the MSS sensor. This methodology derives from a model of earth's regions in which each pixel's emission is randomly distributed around a class mean value, quite independent of the pixels neighboring it along the space or time

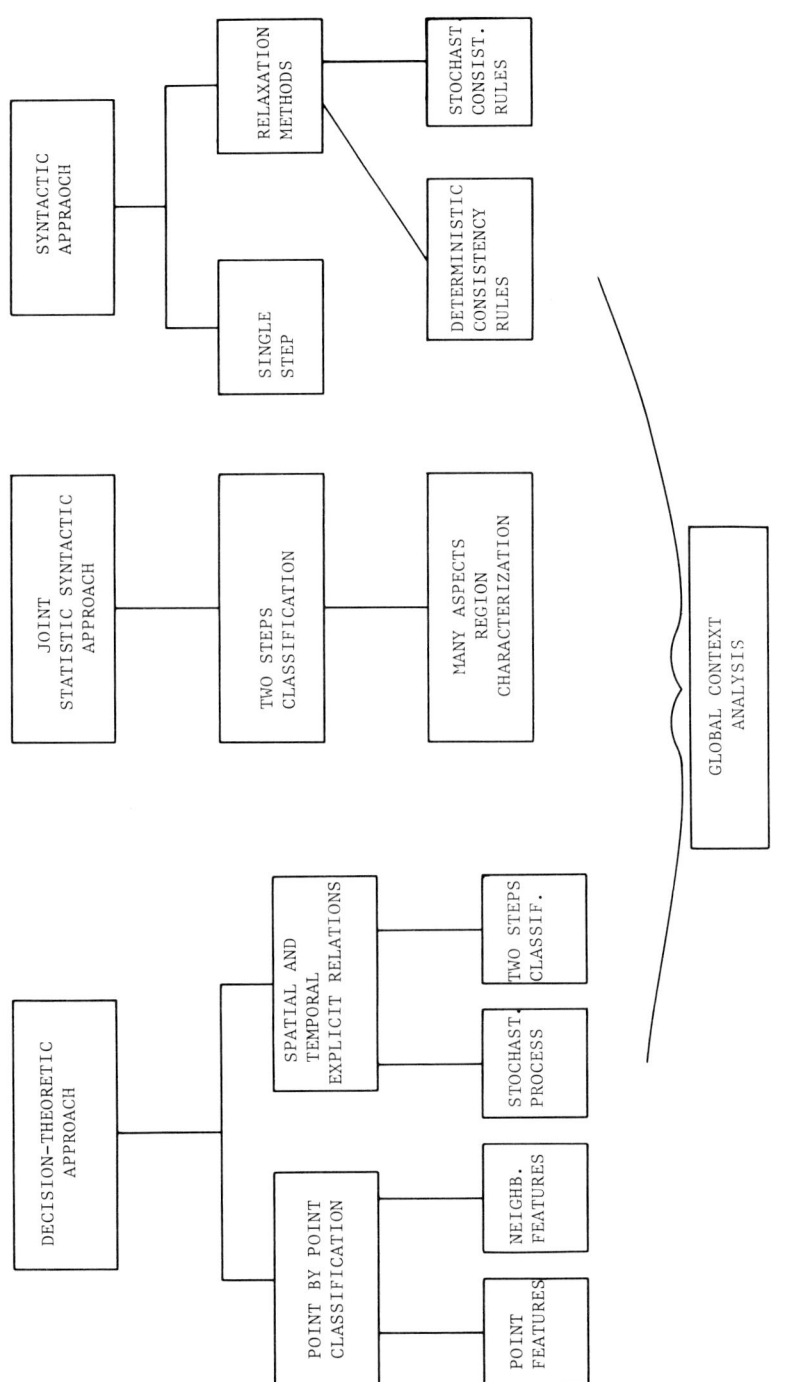

FIGURE 3.   Schematic partitioning of the methodologies.

coordinates.   In other words, the only correlation considered among
the axes of the observation space is the one among gray levels and
spectral wavelengths.   Such a model was suggested by three kinds of
factors:

-    the difficulty of physical models of radiation of soil
     made it desirable to leave out any treatable, but too
     simplified, approximation.

-    the first results obtained from the analysis of simple
     regions allowed one to suppose that their accuracy, even
     if not excellent, would have been improved by proper
     tuning of the algorithms.

-    the comparison of the computer time required by this
     method and the one required by any syntactic method
     suggested that the computer time was less in direct
     relation to the simplicity of the model.

This first model of the earth region gives rise, in the framework
of supervised methods, to discriminant analysis with linear or
quadratic decision functions, and in the framework of unsupervised
methods to cluster analysis.   The typical image for which these
analyses give good results is constituted, as sketched in Figure 4
along a spatial coordinate, by regions whose pixels have gray
levels distributed around a constant mean, and the differences
among the means of different classes are so large that the distribution of gray levels overlap by only small amounts.   When using
unsupervised methods, one needs the regions to be sufficiently
large; otherwise, it is possible that points of small regions are
considered as far points of statistical populations corresponding
to large regions, and therefore would be phagocyted by them.   The
computational time increases with the number of classes in the
image and often is higher than in other methods, because for each
pixel the same number of operations must be performed without
bypass rule.   Improvements in these methods have been realized in
two ways: deeper statistical considerations, and decision tree
schemes.

With respect to the supervised methods case, the main body of
problems continues to be those relative to the use of training

fields, as evidenced by the numerous studies on the best use of these fields and on spectral signature extension (Havens et al., 1977; Henderson, 1976). Using unsupervised methods, the researchers' attention has been focused on the following: 1) globally sensitive methods, wherein the clusters are represented by pivots or kernels. The main topic in these studies is dynamical clustering methodologies. The motivation for this interest lies in the need to take into account the global characteristics of the clusters by means of point by point algorithms (Frentres and Frye, 1975; Haralick, 1979; Wong and Lui, 1977). 2) locally sensitive methodologies, which make use of local feature patterns as reflected for instance in the probability density function (peak selection, mode selection). To determine the thresholds needed to partition the feature space, both automatical and manual procedures have been proposed (Kittler, 1976; Mizosuchi and Shimuka, 1976; Chikara and Register, 1977). The latter ones, based on the analysis of histograms of marginal distributions, are more widely diffused. One way to take more details from observations space is to consider a great number of features for each pixel (principal components, hue, saturation, higher order features, etc.), but accuracy increases only slightly as more and more features are used. The decision tree approach allows for a more efficient use of these features.

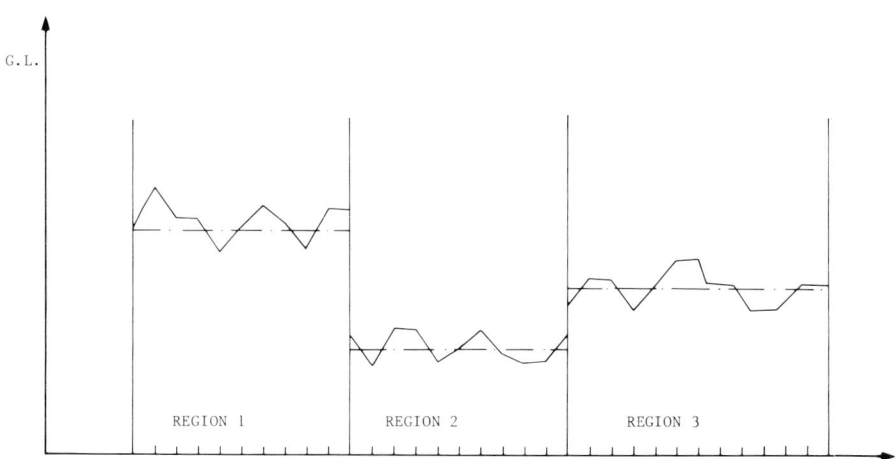

FIGURE 4: Model of regions in agreement with point features. G. L. = Gray Level; X = one among the physical spatial coordinates.

Decision tree scheme is typical of multistage methods. Sequential decision functions are used in a tree structure, until one
arrives at the final choice. At each node features and decision
rules can be selected in an adaptive manner to enhance their dis-
criminitive power (Figure 5). The design of the decision tree,
entailing the choice at each node of the features that least overlap and of the corresponding more discriminant decision rules, can
be committed to manual or automatic optimizing procedures. The
total number of computations required may be significantly less
when all features are used in a single stage decision method; and,
since at each node only a set of available features is employed,
the overall number of training fields can often be significantly
reduced. Several types of multistage decision schemes are known
(Apolloni and Murino, 1979; Fu, 1976; Kulkarni and Kanal, 1976; Li
and Fu, 1976). A recent assessment of the design and of the potential of decision tree classifiers is given in Swain and Hauska,
(1977).

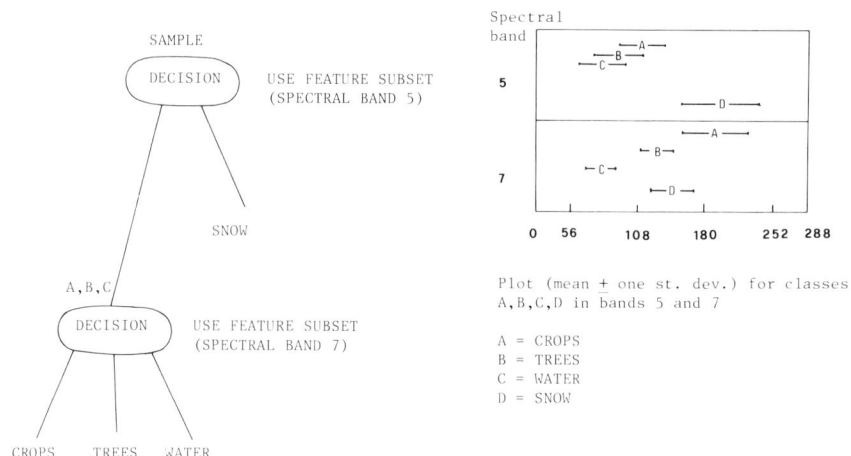

FIGURE 5: Decision tree example.

Neighbourhood features. In the framework of a point by point
classification scheme, statistical relations among the observation
space points can be taken into account only in the features. To do

so one needs to use neighbourhood features, and those usually considered as texture features. Texture features are statistical measures of periodical recurrences of gray levels of neighbouring pixels. Such features are functions of second or higher order moments of the gray levels and account for contrast in the image, not for mean intensity. Due to the excellent results obtained in some cases, and the capability of handling the structure of the image with reasonable computational complexity, a large number of studies has been made on this subject. Some 80 texture measures have been designed and studied; some of these turn out to be highly correlated. The features are usually computed for fixed size neighbourhoods and their values are usually referred to the neighbourhood itself. At times, however, as in the so called nine-points features, the texture statistics are referred to the central pixel. A rather exhaustive comparative study of texture features for terrain classification is given in Dyer and Rosenfeld, (1976), Haralick et al., (1973), Sheinm, (1975), Wezka et al., (1876) and Zavalishin et al., (1975). The classification algorithm is always a point by point algorithm in which the point is constituted by the whole set of neighbouring pixels, when the feature refers to them and therefore, the methods are not efficient when classifying the pixels near the edge of regions with different textures. However, the methods are appropriate in the imagery where there is gray level behaviour as in Figure 6. The computer time is less than using point features if one classifies a set of pixels at a time.

Spatial and temporal models

Two kinds of methodologies have been developed in connection with models which are expressly based on the relations among the observed points: 1) joint probability analysis 2) two-steps analysis. In what follows we will refer almost indifferently to spatial and temporal coordinates of the observation space. It appears obvious in principle to extend such considerations to a spatial-temporal domain, but it does not seem that this extension proves to be useful in most cases.

Joint probability analysis

In a merely statistical point of view the value of the feature

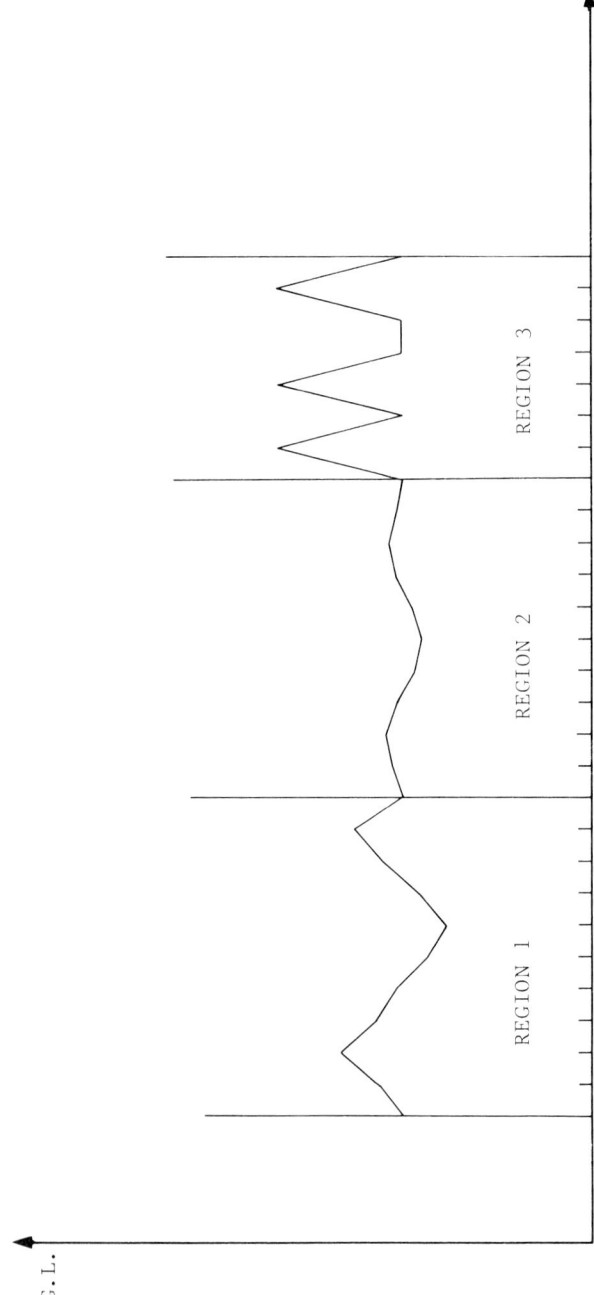

FIGURE 6: Model of regions in agreement with texture features. G.
L. = Gray Level; X = one among the physical spatial coordinates.

can be considered as a multidimensional stochastic process in spatial (and temporal) coordinates. A classification method which is capable of considering the entire process can minimize the misclassification risk evaluated with respect to the whole geographical region. The first implemented models employing point features are based on the following simplifying hypotheses: a) The correlation among the features differs from zero only in a small neighbourhood of pixel considered. Therefore, one needs to optimize, each time, the classification only with respect to such neighbourhood. b) The features are class conditional independent. Therefore, the conditional probability of features corresponding to a pixel, given the class to which it can be attributed, is independent of the values that the features assume in the other pixels of the neighbourhood, i.e. given the classes of two pixels, the corresponding features are stochastically independent of the classes of the pixels of the neighbourhood. In such a way it is possible to express the joint probability that two pixels belong each to a class, as the conditional probability of the feature of the first one, given its class, times the conditional probability of the feature of the second one, given its class, times the joint probability of the two classes. The conditional probability of one class, given the other, is called transitional probability. The sequence of simplifying assumptions is described, in mathematical terms, in Figure 7. This approach leads to classification rules slightly more complex than those pixel by pixel, while the classification accuracy is generally higher. The applications so far published (Swain, 1978) introduce, in the above sense, a correlation between multitemporal images. Other models are being developed which introduce spatial correlations.

The use of models of stochastic processes gives rise also to interesting results in the enhancement of the image (Haralick and Watson, 1979) and in extraction of textural features (Galalowitz, 1978).

Two steps analysis. Image analysis can be developed into two steps; segmentation of the image into homogeneous regions, and

$$\text{SET OF CLASSES:} \quad \{ c^1, c^2, \ldots\ldots, c^M \}$$

$$\text{ADJACENT (IN TIME OR IN SPACE) PIXELS: i, j}$$
$$\text{HAVING FEATURES} \qquad\qquad : X_i, X_j$$

JOINT DENSITY

$$f(X_i, X_j, c_i^k, c_j^h) = f(X_i, X_j \mid c_i^k, c_j^h) \; f(c_i^k \mid c_j^h) \; f(c_j^h)$$

$$-\text{CLASS INDEPENDENCE:} \quad f(X_i, X_j \mid c_i^k, c_j^h) = f(X_i \mid c_i^k, c_j^h) \; f(X_j^h \mid c_i^k, c_j^h)$$

$$-\text{ADDITIONAL HYPOTHESES:} \; f(X_i \mid c_i^k, c_j^h) = f(X_i \mid c_i^k)$$
$$f(X_j \mid c_i^k, c_j^h) = f(X_j \mid c_j^h)$$

$$f(X_i, X_j, c_i^k, c_j^h) = f(X_i \mid c_i^k) \; f(X_j \mid c_j^h) \; f(c_i^k \mid c_j^h) \; f(c_j^h)$$

FIGURE 7: Joint distribution density of adjoining pixels.

classification of them into different classes (Figure 8).   These
steps are based on two different submodels (Haralick and Watson,
1979; Nagao et al., 1978).   The first one deals with the characteristics of the regions, i.e. uniformity of shading of brightness,
textural properties, etc.   The second one deals with the specific
characteristics of the different classes belonging to the image.

The second step, when performed within a decision theoretic
approach, is reported as per-field classification, and can be
briefly described as follows: Given a set of points $(X_1,$
$X_2,\ldots\ldots,X_n) = (\underline{X})$ all of the same field, having a distribution
$G(X)$, the classifier decides to which F_j (X) they belong, where
$(F(X))$ is the set of distributions associated with the set of
classes (C_j).   The decision is made by means of a distance $d(C,F_j)$
which measures the separation between these distributions;
generally, either maximum likelihood or Battacharyra distances are
used.

For the segmentation step, several methods have been proposed
to form homogeneous fields.   They are known as partitioning
methods, and can be grouped into two broad classes: edge-detection
and object-detection methods.   Methods of the first class are

HOMOGENEITY TESTS

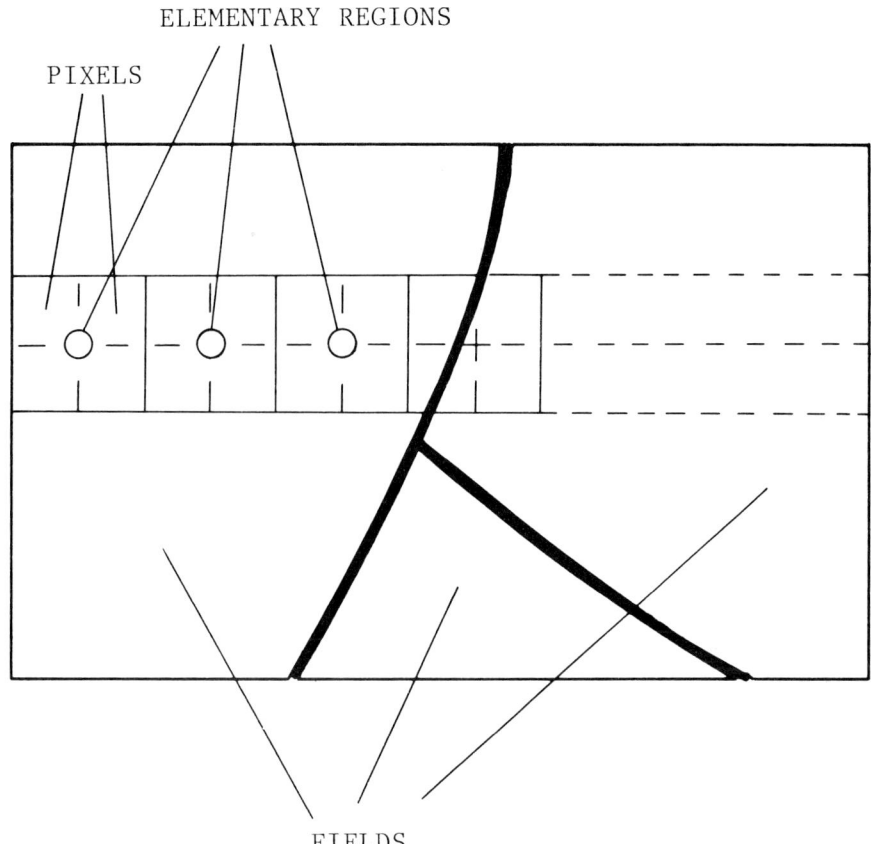

FIGURE 8: Two steps analysis.

essentially syntactic and will be described later on. At any rate
they appear not to be very efficient when the regions are not
already labeled. The second methods consist of finding homogeneous
regions in the image, either by merging similar neighbourhoods
(conjunctive methods) or by starting with a very simple partition
criterion of homogeneity (disjunctive methods).   Disjunctive
methods are not very efficient.   They have the same accuracy but
require lengthier calculations (Robertson, 1973).   Conjunctive

segmentation is very promising and many researchers are working on it. The basic idea is that of increasing the dimension of an initial set of pixels by annexing to it the adjacent sets (of predetermined dimensions, generally greater than one) which are homogeneous with the initial one. The rules to test the homogeneity are the critical ones in that they answer the following question: for which kind of object is one looking? Therefore, they come from a simulation of eye synthesis actions, or from considerations on the subsequent use of the image, and so on. Among the works are the following: 1) some (Gupta and Wintz, 1975; Ketting and Landgrebe, 1976) evaluate only the shift of the first or second moments of the features into the adjacent sets, with respect to the global values into the growing regions, 2) others (Glen and Pavlidis, 1978, Fukada, 1978, Kavsand, 1978; Nagao et al., 1978) take some considerations of the behaviour of the features among the regions, which accounts for brightness, smoothness, structure of the image, and so forth and on other characteristics of the image such as shape or elongation of which one will speak later in the frame of joint syntactic-statistic methodologies.

SYNTACTIC APPROACH

The application of the syntactic approach to remote sensing is comparatively recent (Rosenfeld, Ed., 1976; Yu and Fu, 1978). Syntactic methodolgies have for many years been mostly employed in connection with very simple models of images. Essentially, the analysis of these images is performed in a combinatorial scheme by examining, at each level of aggregation of the growing segments, in an enumerative way, the possible conformities of the sets of segments with the grammar and syntactic rules which define the higher level segments.

This procedure is severely influenced by the following characteristics of remotely sensed imagery: great volume of data, complexity of the imagery, and random nature of the data. The first of these strongly limits the allowable computational complexity. The second causes many difficulties and makes it almost impossible to establish syntactic rules for large neighbourhoods of points. The third one imposes the requirements that the morph extraction and grammar rules not be sensitive to noise and must be able to resolve ambiguities.

In this context, the design of methods which correctly detect a pattern, on the basis of one local measurement, i.e. by a single step of reading, is very difficult if not altogether impossible. Therefore, relaxation procedures have been devised where the logical relations formalized by the grammar syntactic rules constitute a feed-back to segment extraction to contrast the bias given by ambiguities and the noise of the data (Zucker, 1976). The global pattern extraction problem is decomposed by the relaxation methods, into a network of local extraction processes, whose intercommunications are governed by local consistency relations. This gives rise to a loop where the consistency relations between a segment and its neighbors update the label or the structure of the first and so on until a stable segmentation is achieved. From a general point of view, consistency relations may be deterministic or probabilistic. In satellite image analysis, only the second of these appears to be suitable. Relations are thus expressed in terms of the probabilities that a segment is connected with others and the whole pattern recognition is accomplished by maximizing the probability of the pattern.

To date, from a general point of view, applications of syntactic methods to remote sensing are rather limited, in number and scope (Bajesy and Tavakoli, 1976; Fu, 1976; Keng and Fu, 1976; Lu and Fu, 1978; Zucker et al., 1977; Zucker, 1978). However, they appear to be efficient for forms resulting from simple primitives. Typical instances of applications are the identification of rivers and highways. A special topic is represented by the edge detection upon which much work has been done in past years (Mascarenhas and Prado, 1978; Schachter et al., 1977; Triendl, 1978; Wechsler amd Kidode, 1977). For more than segmentation tasks, such methods can be employed for the enhancement of borders of the regions extracted by means of the methods shown in the preceeding section. In this sense, some papers (Schachter et al., 1977; Triendl, 1978) deal

directly with previously segmented regions among which they identify the edges with an accuracy often greater than one pixel.

JOINT STATISTIC-SYNTACTIC APPROACH

In recent years the analyses of the imagery are distinguished
on the basis of the underlying image model rather than on the
approach exploited. Two step analysis is used to best state what
one will find by means of double level (region and class) models
(Haralick and Watson, 1979), and it is possible to find this by
statistical and syntactical rules (Rastigin and Erenstein, 1975).
In other words, segmentation can be considered as a grouping of
observed units of the image, performed on the basis of any model of
region, and accounts also for cleaning and enhancement operations
which usually are done independently, see the block diagram of
Figure 1, in a pretreatment step. So, these segments can be
characterized with respect both to segmentation and to classification, by size, orientation, brightness, mean hue, and other vector
valued features. This approach appears very promising but requires
more theoretical and experimental work to identify the most discriminant among these "mixed" features and to define classification
criteria which attribute relative weights to statistical and syntactic tests.

As an example of a joint approach in segmentation we quote
(Apolloni and Murino, 1978) the axiomatic definition of region and
edges of an image which is defined as constituted only by these two
elements. "The axioms which define the region and the edge are as
follows:

1)   An edge is the contour line of a region and is charac-
     terized by discontinuity in the distance in the feature
     space between adjacent pixels. It belongs therefore to
     the set of those pixels whose distance from adjacent
     pixels is greater than a preassigned threshold.........

2)   A region is a particular set of adjacent pixels homo-
     geneous among themselves.

The homogeneity criterion is the following:

3) Two adjacent pixels are considered homogeneous if the absolute value of the distance in feature space is less than the previously mentioned threshold.

The topological characteristics of the region are the following:

4) A pixel belongs to a region if it is homogeneous with an adjacent pixel of the growing region.

5) m pixels along one direction may belong to a region if they are homogeneous and $m \geq \bar{m}$, where $\bar{m}$ is assigned.

6) A set of homogeneous pixels does not constitute a region if there is not a given square window of given size W_1, which includes a given number K_1 of these homogeneous pixels. This axiom is in essence a cleaning algorithm."

Such definitions call for two successive filtering steps:

1) a low-pass filter, wherein upperbound frequency decreases when the amplitude increases, and retaining, at same time, high frequencies at the edges of the regions,

2) a further low-pass filter on the labeled regions so that very small regions are phagocyted by larger regions in which they are included. Thus, the corresponding segmentation method is suitable for shading regions separated by jumps of the features (Figure 9); the second filter accounts for consistency constraints on the regions.

As an example of a joint approach in classification, one reports from Nagao et al., 1978, the syntactic rule for classifying the boundary of crop fields in aerial photographs, which may be employed jointly with some statistical rules to classify special kinds of crop fields. "Boundaries of large homogeneous regions are examined, and those which have straight boundaries are recognized as crop fields........The straightness of a region boundary is calculated as follows:

1) For each point p_i (i=I, ..., N) on the boundary, calculate the angle α_i between two straight lines connecting p_i with p_{i-n} and p_{i+n}.

2) Let N' denote the number of points where $\alpha_i \leq 22.5°$

3)   If $N'/N \geq 0.6$, then this region is regarded as having a
     straight boundary."

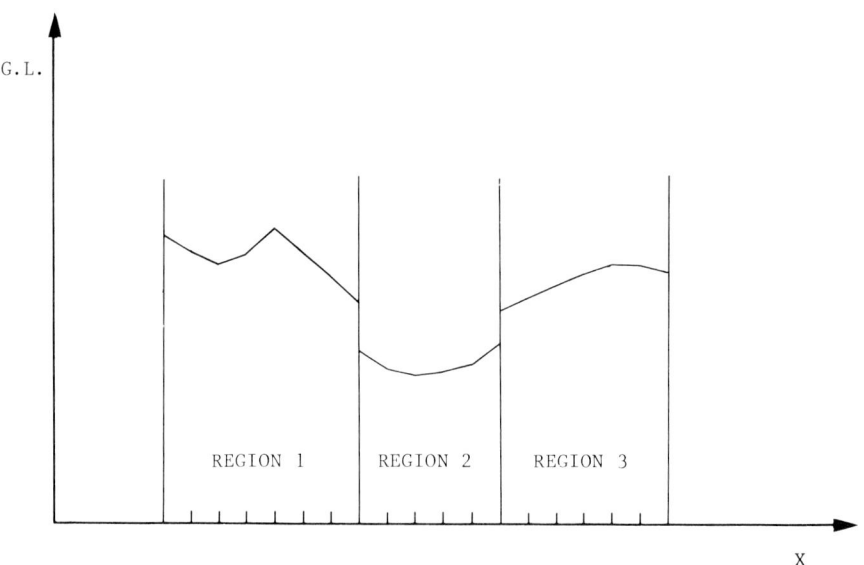

FIGURE 9: Model of regions in agreement with a syntactic-statistic
approach. G. L. = Gray Level; X = one among the physical spatial
coordinates.

GLOBAL CONTEXT ANALYSIS

Global context analysis may be defined as a final syntactic
analysis of the already classified regions. In essence one wants
to test if the extracted information complies with some consistency
rules regarding the meaning of the entire image. Such analysis may
give rise to a feed-back procedure, to a sequential classification
of the regions, or to a further partitioning of the regions into
subclasses.

At the moment, few examples are present in literature. Some
interesting applications are the removal of the shadows caused by
isolated clouds, or the identification of bridges on rivers (Bajesy
and Tavakoli, 1976). As an example we quote from Nagao et al.,
1978, the three rules of consistency for residential areas by
aerial photographs.

1)   not all areas contain vegetation

2)   the area includes shadow and shadow-making regions

3)   there exists many small constituent regions which have rectangular shape.

CONCLUDING REMARKS

Some ten years after the first studies in the utilization of satellite MSS data, the general feeling is that, in the processing of the image, the individual theoretical tools, as features, statistical tests, decision schemes, syntactic rules, relaxation methods, etc., are available at a sufficient level of sophistication; now one needs to look for proper collection of them in optimal procedures. It is this direction that the main interest of the researchers now lies.

Deep analysis of the model of the regions, and of the further use of the extracted information allows one to identify what one is searching for. Multi-step analysis and global context analysis are very flexible approaches to exploit the above mentioned theoretical tools, and to take into account the ancillary data. Special regional data bases (Degani, 1979, Haralick, 1979; Pequet, 1979) can insert remotely sensed data into regional information systems, allowing for the realization of a global context to a very large extent, and an optimal compression of the enormous volume of data available on the region, sensed both remotely (Rogers et al., 1977) and in the field. More and more, advanced technological devices make increasingly possible the handling of all these data in a reasonable time.

On the other hand, it is necessary to take clearly into account the limits of completely automatic analysis of the image, and therefore one must accurately identify the application range of such methods to avoid a great expense of human resources, of computational time, and of special devices, in pursuing the same results which can be obtained in a very reasonable way by human eye inspection.

REFERENCES

Apolloni, B., P. Murino 1978. On the identification of a mixture of samples from two multivariate distributions. Acta Astronautica, Vol. 6: 1031-1042.

Apolloni, B., P. Murino, L. G. Napolitano 1979. Recent developments in automatic analysis of remotely sensed imagery. In: Astronautics for Peace and Human Progress. pp. 233-246. Ed. by L. G. Napolitano. London: Pergamon Press. 1979.

Apolloni, B., P. Murino 1980. Region extraction for thematic analysis of remote sensed images. Acta Astronautica, 7.

Bajesy, R., M. Tavakoli 1976. Computer recognition of roads from satellite pictures. IEEE SMC, 6: 628-637.

Cappellini, V., A. G. Constantinides, P. L. Emiliani. Digital filters and their applications. New York: Academic Press. 1978.

Chen, C. H. 1978. Finite sample consideration in statistical pattern recognition. IEEE Conf. PRIP. 1978.

Chen, P. C., T. Pavlidis 1978. Segmentation by texture using a co-occurence matrix and a split and merge algorithm. Proc. 4th Int. JCPR, Kyto, Japan, November 78: 565-569.

Chikara, R. S., D. T. Register 1977. A nonsupervised classification method for partitioning of a large multidimensional data set. IEEE Annual Mtg. Am. Stat. Assoc., Chicago: 428-440.

Degani, A. 1979. Basic concepts in the design of an information system for regional automated map analysis and planning. In: Proc. ASI on MAP Data Processing. Ed. by R. M. Haralick. Sijthoff & Noordhoff International Publishers. 1979.

Duda, R. O., P. E. Hart. Pattern classification and Scene Analysis. 458 pp. New York: J. Wiley. 1973.

Dyer, C. R., A. Rosenfeld 1976. Fourier texture supression of aperture effects. IEEE SMC, 6: 703-705.

Frentres, C. D., R. G. Frye 1975. Wild life management by habitat units: plan of action. NASA-TMX 58168, 1-A: 245-262.

Fu, K. S. 1976. Pattern recognition in remote sensing of the earth's resources. IEEE GE, 14: 10-18.

Fu, K. S. (Ed.) 1976. Digital Pattern recognition. 213 pp. Berlin: Springer Verlag. 1976.

Fukada, Y. 1978. Spatial clustering procedures for region analysis. Proc. 4th Int. JCPR, Kyoto, Japan, November 78: 329-331.

Gagalowitz, A. 1978. Analysis of texture using a stochastic model. Proc. 4th Int. JCPR, Kyoto, Japan, November 78: 541-544.

Gupta, J. N., P. A. Wintz 1975. Boundary finding algorithm and its applications. IEEE CS, 22: 351-362.

Haralick, R. M., K. Shanmuham, I. Dinstein 1973. Textural features for image classification. IEEE SMC, 3: 610-621.

Haralick, R. M. 1979. Data structures for map data processing. In: Proc. ASI on Map Data Processing. Ed. R. M. Haralick. Sijthoff and Noordhoff International Publishers. 1979.

Haralick, R. M., L. Watson 1979. A facet model for image data. IEEE Conf. PRIP, Chicago, August 79.

Harger, R. O. Synthetic-aperture-radar systems: theory and design. 237 pp. New York: Academic Press. 1970.

Havens, K. A., T. C, Minter, S. G. Thadani 1977. Estimation of the probability of error without ground truth and known a priori probabilities. IEEE GE, 15: 147-152.

Henderson, R. G. 1976. Signature extension using the MASC algorithm. IEEE GE, 14: 34-37.

Holmes, Q. A., D. Goodenough, J. D. Erikson 1976. Remote sensing data processing: Two years ago, today, and two years from today. 3rd Int. JCPR, San Diego. Calif: 125-135.

Kanal, L. 1974. Patterns in Pattern Recognition: 1968-1974. IEEE IT, 20: 97-121.

Kavsand, T. 1978. Scene segmentation and segment clustering experiments. Proc. 4th Int. JCPR Kyoto, Japan, November 78: 426-429.

Keng, J., K. S. Fu 1976. A syntax-direct method for land-use classification of LANDSAT images, Proc. Symp. on Current Mathematical Problems in Image Science. Monterey, Calif.

Ketting, R. L., D. A. Landgrebe 1976. Classification of multi-spectral image data by extraction and classification of homogeneous objects. IEEE GE 14: 19-26.

Kittler, J. A. 1976. A locally sensitive method for clsuter analysis. Pattern Recognition, 8: 23-33.

Kulkarni, A. V., L. N. Kanal 1976. An optimization approach to hierarchial classification design. Proc. 3rd Int. JCPR, San Diego, Calif.

Li, R. Y., K. S. Fu 1976. Tree system approach to LANDSAT data interpretation. Proc. Symp. on Machine Processing of Remotely Sensed Data, Purdue University, W. Lafayette, Indiana.

Lu, S. Y., K. S. Fu 1976. A syntactic approach to texture analysis. Computer Graphics and Image Processing, 7.

Mascarenhas, N. D. A., L. O. C. Prado 1978. Edge detection in images: a hypothesis testing approach. Proc. 4th Int. JCPR, Kyoto, Japan, November 78: 707-709.

Mitchell, O. R., E. J. Delp, P. L. Chen 1977. Filtering to remove cloud cover in satellite imagery. IEEE GE, 15: 137-141.

Mizoguchi, R., M. Shimuka 1976. Non-parametric learning without a teacher based on mode estimation. IEEE C, 25: 1109-1117.

Murino, P., B. Apolloni, L. G. Napolitano 1979. Processing and using S. A. R. data. In: Space Developments for the Future of Mankind. Ed. by L. G. Napolitano. London: Pergamon Press. 1979.

Nagao, M., K. Matsuyama, Y. Ikeda 1978. Region extraction and shape analysis of aerial photographs. Computer Graphics and Image Processing, 9: 394-407.

Pequet, D. J. 1979. Problems encountered in modifying Raster-Formatted Map Data. In: Conf. ASI on Map Data Processing. Ed. by R. M. Haralick. Sijthoff and Noordhoff International Publishers. 1979.

Rastrigin, L. A., R. Kh. Erenstein 1975. Decision processes by collection of decision rules in pattern recognition problems. Avtomat. I. Telemekhanika, vol. 36, n. 9, part. 2: 133-144.

Robertson, T. V. 1973. Extraction and classification of objects in multispectral images. Proc. Symp. on Machine Processing of Remotely Sensed Data, Purdue University, W. Lafayette, Ind. : 28-35.

Rogers, R. H., L. E. Reed, W. R. Enslin, K. E. Keinfenheim, T. N.
    Haga, R. Karwowski 1977. Land use inventory through merging
    of LANDSAT (satellite), aerial photography and map sources.
    Full Mtg. Soc. of Photogrammetry and Am. Congr. on Surveying
    and Mapping, Little Rock, Arkansas.
Rosenfeld, A. (ed.) 1976. Digital Picture Analysis. 384 pp.,
    Berlin: Springer Verlag. 1976.
Schachter, B. J., A. Lev, S. W. Zuker, A. Rosenfeld 1977. An
    application of relaxation methods to edge reinforcement. IEEE
    SMC, 11: 814-816.
Sheinin, R. L. 1975. A linguistic approach to processing of com-
    plex textual images. Avtomat i telemekhanika, vol. 36, n. 3,
    part 2: 112-122.
Swain, P. H., H. Hauska 1977. The decision tree classifier:
    design and potential. IEEE GE, 15: 142-147.
Swain, P. H. 1978. Bayesian classification in a time-varying
    environment. IEEE SMC, 8: 879-883.
Thomas, J. O. 1978. Digital image processing with special
    reference to data compression in remote sensing. In: Proc.
    ASI on Digital Image Processing and Analysis. Ed. by J. C.
    Simon. Sijthoff and Noordhoff International Publishers.
    1978.
Triendl, E. E. 1978. How to get edge into the map. Proc. 4th Int.
    JCPR, Kyoto, Japan, November 1978: 846-850.
Wechsler, H., M. Kidode 1977. A new edge detection technique and
    implementation. IEEE SMC, 12: 827-836.
Weszka, J. S., C. R. Dyer, A. Rosenfeld 1976. A comparative study
    of texture measures for terrain classification. IEEE SMC, 6:
    269-285.
Wong, A. K. C., T. S. Lui 1977. A decision-direct clustering
    algorithm for discrete data. IEEE C, 26: 75-82.
Yu, T. S., K. S. Fu 1978. Statistical pattern recognition using
    contextual information. Rept. TR-EE-78-17, Purdue University,
    W. Lafayette, Ind.
Zavalishin, N. V., I. B. Muchnik, R. L. Sheinin 1975. Automatic
    classification of textural images. Avtomat. i Telemekhanika,
    vol. 36, n. 2, part. 2: 95-104.
Zucker, S. W. 1976. Relaxation labelling and the reduction of
    local ambiguities. Proc. 3rd Int. JCPR, San Diego, Calif.:
    852-861.
Zucker, S. W., R. Hummel, A. Rosenfeld 1977. An application of the
    relaxation labelling to line and curve enhancement. IEEE C,
    26: 922-929.
Zucker, S. W. 1978. Local structure consistency and continuous
    relaxation. In: Proc. ASI on Digital Image Processing and
    Analysis. Ed. by J. C. Simon. Sijthoff and Noordhoff
    International Publishers. 1978.

Initial Plans for Remote Sensing of the North Sea Using Coastal Zone Colour Scanner (CZCS)

A.P. Cracknell

ABSTRACT

In this paper I outline the general plan of the work which we have recently begun on the remote sensing of the North Sea. Our primary interest is in the use of the Coastal Zone Colour Scanner (CZCS), which is currently being flown on NIMBUS-7, for the study of suspended sediment concentration, chlorophyll concentration, sea-surface temperatures, and oil and sewage pollution. We are also interested in the use of multi-spectral scanning (MSS) data from other satellites and have already obtained some interesting results using LANDSAT data. Some examples of these results concerning pollution, fronts, and the mapping of sandbanks are presented.

INTRODUCTION

The following is a description of a project on which we only started about a year ago and in which we have obtained some preliminary results. However, we are still far from the final and successful completion of the project. The aim of this project is to investigate the possibility of using data received from multi-spectral scanners, mounted in satellites, to study the North Sea. Correlation between the satellite data and "sea truth" measurements is being attempted using that part of the North Sea which is situated close to Dundee, Scotland.

Multi-spectral scanning, using satellites, was developed originally for military purposes. On the civilian side great progress has been made in the surveillance of land areas and many practical applications have now been demonstrated to be feasible. Newcomers to the field of Earth-observation (remote sensing) from satellites often find it difficult to see the wood for the trees in the vast number of satellite programs, with a wide variety of names of satellites, programs, missions, etc. A useful summary of Earth-observation satellite programs is given by Hammond (1977). In practice, the most widely used data at present are those received from the LANDSAT-2 satellite. The capital cost of the LANDSAT program has been borne by the U.S. taxpayer, via NASA. The multi-spectral scanner (MSS) data from this satellite, digitised in the form of computer compatible tapes (CCTs) or processed to produce "photographic" images, are made available to interested users. This is not done free of charge but the charge involved is more closely related to the direct costs involved in receiving the data and preparing the "product" for the user, rather than including any contribution to the enormous costs of the program of satellite construction, launching and control. The user (Government Department, University, Industry, etc.) then has to arrange and pay for the processing of CCTs or the analysis of photographic products for his own purposes. The applications which have been found for MSS data from satellites for the civilian surveillance of land areas are many and varied. They include agricultural uses (crop

patterns, irrigation networks, yield estimates, disease detection),
land-use surveys (commercial and industrial versus housing and
recreational use of urban land), cartography (mapping of large or
inaccessible territories), water resources surveys, mineralogical
and geological prospecting, dust storms, etc. (see, for example,
many of the papers in Plevin et al., 1978).

Work on the observation of the surface of the sea by multi-
spectral scanning from satellites is less advanced than that of the
land areas. One reason for this is the fact that the technical
problems involved in the interpretation of the data are more difficult. To distinguish between an agricultural area and an urban
area from LANDSAT data is now (relatively) straightforward, while
to distinguish, say, corn from other crops is feasible. The variations in the reflected spectrum from the sea are generally much
smaller and much less sharp than from the land. Nevertheless the
feasibility of the use of MSS data has been demonstrated in connection with the observation of acid waste and sewage sludge off the
East Coast of the United States (Johnson et al., 1977) and of
variations of blue-green algal concentrations in the Baltic Sea
(Ulbricht and Schmidt, 1978). Thus the general situation with
regard to marine applications is that MSS data are fairly readily
available (though generally not entirely free of charge) but that
so far only rather limited attempts have been made to correlate the
data with sea-surface, or very shallow sub-surface, conditions.

The type of information that we would hope to be able to
obtain by marine remote sensing, using multi-spectral scanning,
would include information about foreign "bodies" on the surface of
the sea (oil slicks, floating wreckage, floating or slightly submerged vegetation, sewage, or industrial-waste pollutants) and
about the conditions of the "pure" surface itself (temperature,
wave formations, surface currents, etc.). In the long term we
envisage the development of a satellite-based continuous real-time
surveillance system to detect pollution of the sea by the various
agents indicated above. However, we recognize that this is a very
long term aim and it is likely to be many years before we can
expect to see such a system in operation. Moreover, the problem of

cloud cover would probably require a switch from optical MSS techniques to microwave synthetic aperture radar (SAR) techniques to
provide an all-weather capability. With reference to oil slicks,
we do not refer to "blow outs" or large spillages (Torrey Canyon,
Amoco Cadiz, etc.) where there are plenty of people around to
observe that the spillage has occurred. Rather we envisage being
able to check up remotely on the unscrupulous or unfortunate master
of a ship who, on the open sea miles away from anyone else, quietly
empties his bilges or spills some oil or some other noxious
chemical and who might otherwise escape detection. Looking for
wreckage from regular satellite scans, on a real-time basis and if
the resolution can be made good enough, could supplement or even
possibly replace having to send aircraft or helicopters to search
for wreckage following reports of a ship or aircraft going missing.
Many other applications of such a possible operational system, such
as fishery protection, siltation monitoring, navigational aids,
etc., are discussed in a Canadian Government Task Force Report
(Morley, 1977) - although some of their problems, such as the
monitoring of icebergs, do not concern us in the North Sea.

The multi-spectral scanner with which we are most involved is
the Coastal Zone Colour Scanner (CZCS), which is currently being
flown on NIMBUS-7. The main reason for this is that this scanner
has been designed to be particularly appropriate for studying the
"colour" of the water in coastal regions. Moreover, we have easy
access to these particular data because of the Dundee involvement
in the EURASEP project (European Association for Scientific
Experiments on Pollution) (see below). However, we also make use
of data from other satellites, where appropriate, to complement the
CZCS data. In particular, in the period during which we have had
to wait for CZCS data we have been practising on LANDSAT data. In
the present paper some results of this LANDSAT work will be
described.

GENERAL VIEW OF MARINE REMOTE SENSING ACTIVITIES IN DUNDEE

We in Dundee are particularly well-placed to attempt to correlate satellite and sea-truth data because of the existing involve-

ment of the University's Electrical Engineering and Electronics
Department (P. E. Baylis and R. J. H. Brush) in satellite-tracking
work and of the University's Tay Estuary Research Centre (J. A.
Charlton, A. M. Jones, and J. McManus) in estuarine and marine data
collection on the surface. Over the last few years the electrical
engineers have developed a very successful "home-made" data-reception system which is now functioning and producing pictures from
data obtained from meteorological satellites. These pictures are
now being supplied to many users throughout the United Kingdom and
also overseas. Since the work of this data-reception facility is
described in another paper presented at this conference, by P. E.
Baylis, I shall say no more about it here. For many years the Tay
Estuary Research Centre has conducted research within the Tay
Estuary, in the neighbouring coastal waters, and in coastal waters
elsewhere in this part of the British Isles. This centre has no
academic staff of its own but provides the facilities for research
work for people from a number of different departments including
Biological Sciences, Civil Engineering, and Geology. It was the
capacity of the Tay Estuary Research Centre to collect "sea-truth"
data, or "sea-information" as it is more fashionably described now,
that led to the selection of the part of the North Sea near to
Dundee as one of the "test sites" for the EURASEP project.

What was missing, and to some extent is still missing, in
Dundee was a digital processing and interpretation capability to
correlate satellite data with "ground-truth" or "sea-truth" data.
The project described in this paper represents an attempt to go
some way towards filling this gap. Indeed, it was at the suggestion of Dr. J. A. Charlton, of Dundee University's Civil
Engineering Department and the Tay Estuary Research Centre, who is
chairman of the EURASEP sea-truth working group, that I became
involved in this work. My own background is that of a theoretical
solid-state physicist with considerable experience in the application of large-scale computing facilities and techniques to the
solution of physical problems.

THE COASTAL ZONE COLOUR SCANNER (CZCS)

The Coastal Zone Colour Scanner (CZCS) is one of several

instruments mounted in the NIMBUS-7 satellite which is currently in orbit. The satellite itself bears a remarkable resemblance, which is not entirely accidental, to the LANDSAT series of satellites. The CZCS is a multispectral scanner, similar to those mounted in the LANDSAT satellites, but having six bands or channels. The wavelength ranges of these bands or channels are given in Table 1. Some of the more important parameters of the scanner in operation are indicated in Figure 1. Note that the spatial resolution is about ten times lower than with LANDSAT. However, what is lost in spatial resolution is gained in frequency of coverage.

The NIMBUS-7 satellite on which the Coastal Zone Colour Scanner is being flown was launched about 8 months ago. But, as has already been remarked by at least one previous speaker at this conference, there has been a delay in the establishment of facilities for receiving the data from NIMBUS-7 and producing CCTs and imagery. All that we, in Dundee, have received so far is one computer compatible tape (CCT) which contains data for one pass over the Gulf of Mexico. This tape has been of some value to us for testing some of our software and we have indeed been able to obtain pictures from it, see Figure 2, but it is of limited value for any real studies of the North Sea. While we have been waiting for CZCS imagery we have been making use of some of the LANDSAT imagery which is readily available. This LANDSAT imagery is of considerable value and interest to us and we have obtained a number of interesting results.

THE EURASEP PROJECT

This project is coordinated by the European Economic Community's Joint Research Centre at Ispra in Italy. The objectives are to study marine pollution using multi-spectral scanning data received, in particular, from the CZCS flown in NIMBUS-7. It is acknowledged that it is impossible to interpret multi-spectral scanning data without, at some stage, having access to "sea-truth" or "sea-information" data. To acquire the sea-truth data a number of "test sites" have been established for the EURASEP project.

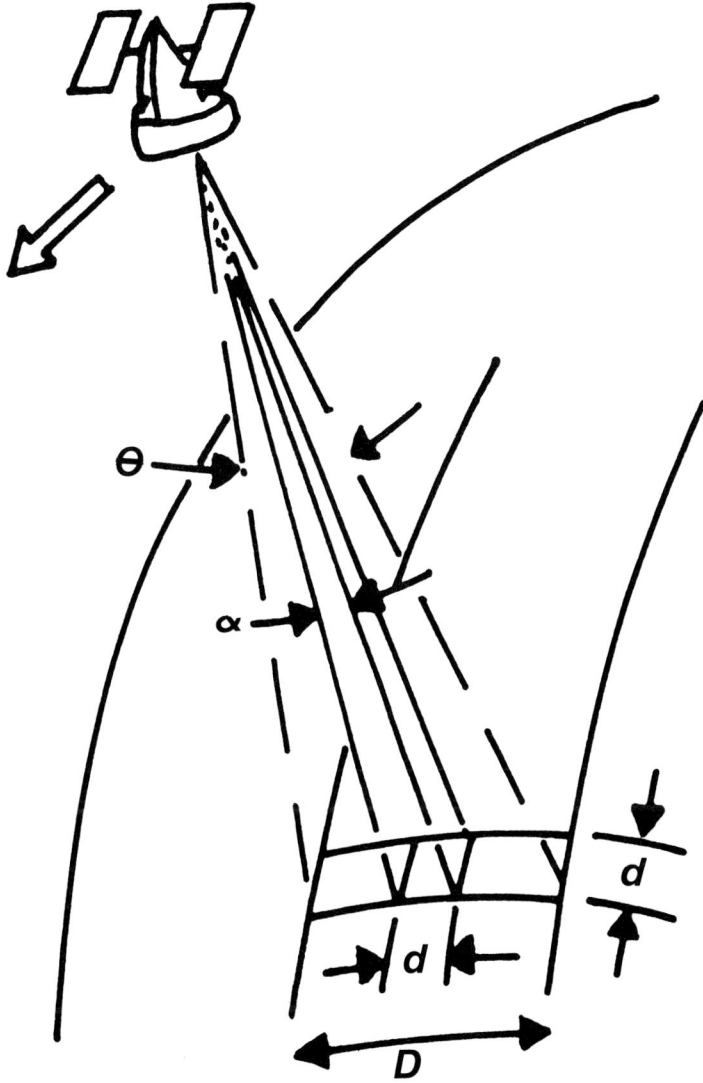

FIGURE 1: CZCS scanning arrangement, θ = 1.373 rad (78.68°), α = 0.865 x 10^{-3} rad (0.0496°), d = 0.825 km, D = 1639 km, orbital altitude = 955 km.

Workers at these test sites collect sea-truth data to coincide with overpasses of NIMBUS-7 which carries the CZCS. By correlating the CZCS data with the sea truth for these sites it is hoped that it will be possible to develop alogrithms which can then be applied to

TABLE 1: Wavelengths and spectral bandwidths of CZCS.

	CHANNEL					
	1	2	3	4	5	6
Centre wavelength λ (μm)	0.443 (blue)	0.520 (green)	0.550 (yellow)	0.670 (red)	0.750 (far red)	11.5 (thermal infrared)
Spectral bandwidth $\Delta\lambda$ (μm)	0.433–0.453	0.510–0.530	0.540–0.560	0.660–0.680	0.700–0.800	10.5–12.5

FIGURE 2a: NIMBUS-7, 1st November 1978, Coastal Zone Colour
Scanner images, channel 1, Gulf of Mexico.

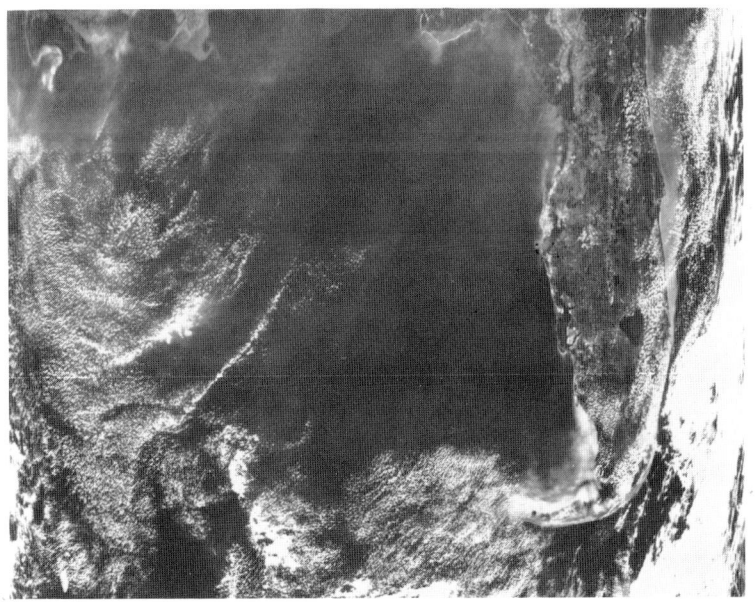

FIGURE 2b: NIMBUS-7, 1st November 1978, Coastal Zone Colour
Scanner images, channel 2, Gulf of Mexico.

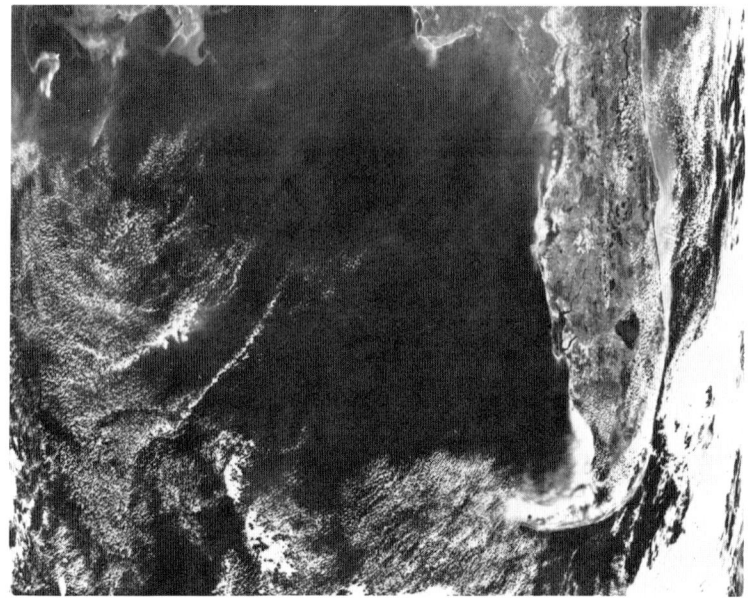

FIGURE 2c: NIMBUS-7, 1st November 1978, Coastal Zone Colour
Scanner images, channel 3, Gulf of Mexico.

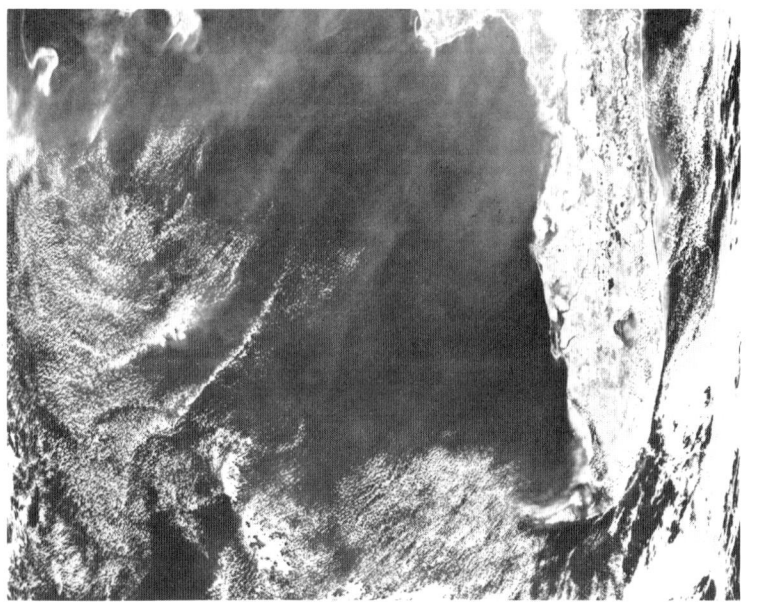

FIGURE 2d: NIMBUS-7, 1st November 1978, Coastal Zone Colour
Scanner images, channel 4, Gulf of Mexico.

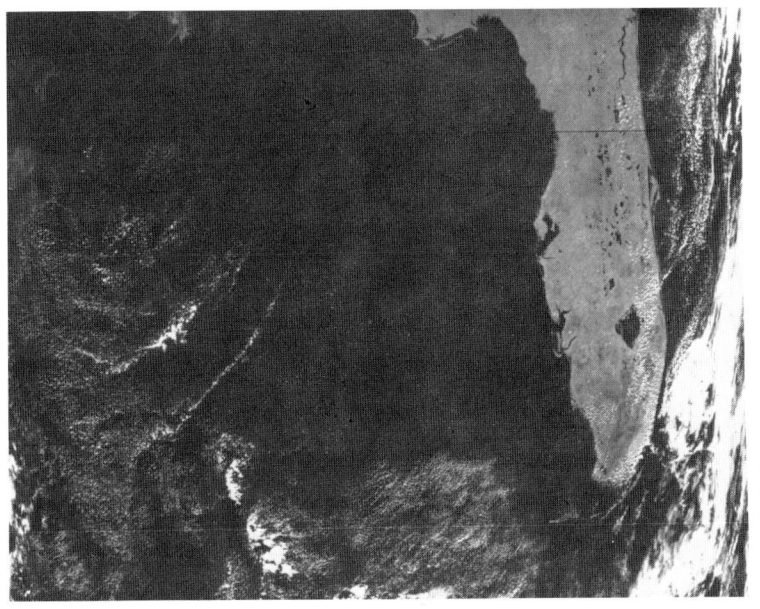

FIGURE 2e: NIMBUS-7, 1st November 1978, Coastal Zone Colour Scanner images, chanel 5, Gulf of Mexico.

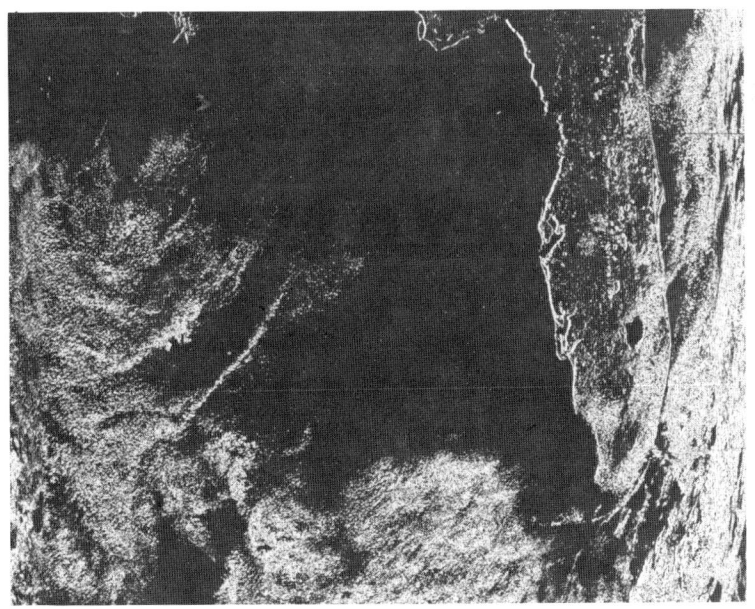

FIGURE 2f: NIMBUS-7, 1st November 1978, Coastal Zone Colour Scanner images, channel 6, Gulf of Mexico.

the satellite data to obtain temperature, chlorophyll and suspended sediment distributions for all European waters. At present my colleagues in Dundee are going out in their boat, about once every two weeks, coinciding with suitable tidal conditions and NIMBUS-7 overpasses, to collect sea-truth data on temperature, salinity, and suspended sediment and chlorophyll concentration. The data are being analysed and stored in anticipation of the day when CZCS tapes come through to us.

SOME EXAMPLES OF THE USE OF LANDSAT MSS DATA

I want to show you, in a qualitative way only, four examples of the use we are making of LANDSAT data for estuarine and coastal waters. More detailed quantitative results of this work will be published elsewhere later.

CONCENTRATED POLLUTION - A GROSS DUMPING EVENT

In Figure 3 there is shown a small extract from LANDSAT (Path 220, Row 21) MSS band 4 of 28th May 1977. The area covered is part of the coast of North-East England, the large river shown is the Tyne flowing through the city of Newcastle. An aircraft condensation trail can be seen over the water in the bottom right-hand corner of the picture. The gray scale used to make this picture has been chosen to maximize contrast in the water and, consequently, the detail over the land is almost all lost. The reason for which I have chosen this picture is the very sharp "hairpin" feature. This is clearly a trail of some kind of "pollutant" being dumped by a ship as it sailed away from the shore for several miles in a direction slightly East of North (the more "fuzzy" arm of the hairpin) dumping as it went. It then stopped twice and continued dumping (the two circular dots at the top of the hairpin) and continued dumping as it sailed back (the sharper arm of the hairpin).

The following points are worth making:

(a) Features like this have been observed elsewhere before, particularly off the German coastline in the vicinity of Hamburg.

FIGURE 3: An extract from LANDSAT (Path 220, Row 21), 28th May 1977, MSS band 4, coastline near Newcastle-upon-Tyne, England.

(b) The hairpin feature is also visible, though less clearly, in MSS band 5.

(c) It so happened that the previous day was also cloud-free over this area and, because of the overlap of successive days' swaths, there is also a LANDSAT MSS tape available for that day, namely 27th May 1977. We have examined the MSS band 4 scene for this area for 27th May 1977 and there is no sign of this feature on that day. We therefore conclude that the dumping actually occurred early in the morning of 28th May 1977 and was virtually completed

by about 10:00 a.m. GMT, the approximate time of the
LANDSAT overpass.

(d) We have made enquiries of the Ministry of Agriculture,
     Fisheries, and Food in London who are supposed to
     receive, from shipping companies etc., the details of all
     dumping events of this type. We asked for confirmation
     that this was an authorized, as distinct from an illegal,
     dumping event and for some indication of the nature of
     the material being dumped so that we could establish its
     spectral "signature". The receipt of our letter of
     inquiry was acknowledged, over six months ago, but so far
     no answer to our question has been obtained.

DISTRIBUTED POLLUTION

Within the EURASEP project there was a big pre-launch experiment carried out in the summer of 1977. This involved flights of
aircraft, mounted with scanners, over the Dutch, Belgian and north
French coastline, while simultaneous sea-truth voyages were made
from various ports in that locality. The Dundee University Tay
Estuary Research Centre was involved by providing a team to man one
of the nine boats involved in the sea-data collection. From the
aircraft scans it has been possible to construct maps of chlorophyll and suspended sediment distribution. This pre-launch training exercise has been described in some detail by Sφrenson (1978).
Recently we have been analyzing a LANDSAT scene which corresponds
to one of the days on which these aircraft-flown pre-launch experiments were performed. We show in Figure 4 a part of this scene and
in this picture one can see evidence of an enormous amount of
material being discharged into the sea by the Rhine and Schelde
rivers. Our task is then to interpret this general "murkiness" in
the water in terms of chlorophyll and suspended sediment concentrations, using the measured values along the tracks of the ships
involved in sea-data collection on that day.

FIGURE 4a: An extract from LANDSAT (Path 214, Row 24) 27th June 1977, MSS band 4, coastline of Low Countries.

FRONTS

We have heard quite a lot about fronts at this conference. In Figure 5 we show an extract from LANDSAT (Path 220, Row 21) MSS band 4 of 24th October 1976. The quality is not very good because, being October, the signal-noise ratio is much worse than in mid-summer. This is the only scene of this area in the British archives of LANDSAT, held at RAE Farnborough, corresponding to low

FIGURE 4b: An extract from LANDSAT (Path 214, Row 24) 27th June
1977, MSS band 5, coastline of Low Countries.

tide at Dundee. One can see sandbanks in the upper Tay Estuary and
at the mouth of the river. One can see whitecaps along a great
deal of the coastline and in the sandbanks at the mouth of the
river. There is also some cloud. However, there is a clear curve
passing in a gentle sweep from about Arbroath in the north (nearly
at the top of the picture) to St. Andrews in the south (three
quarters of the way down the picture, where there is a sharp right-
angle in the coastline). This picture has, indeed, been used by my

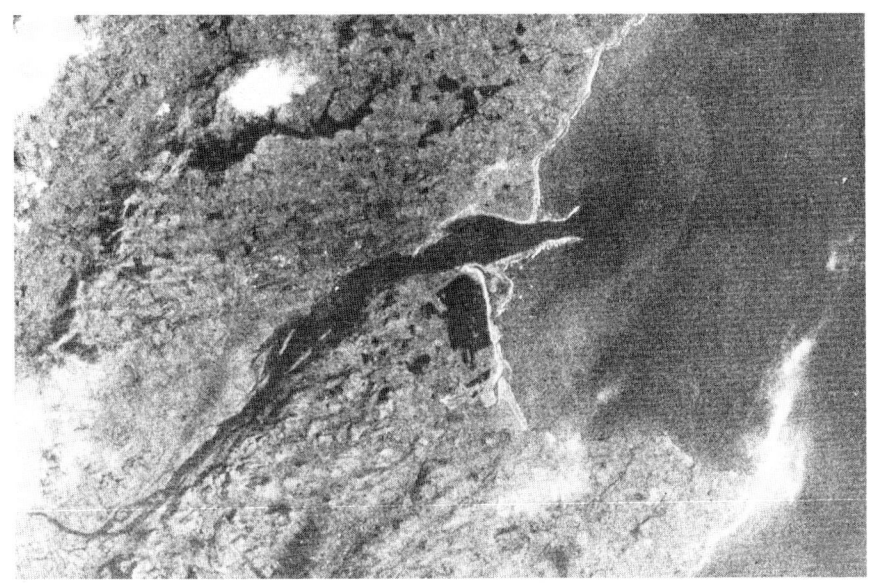

FIGURE 5: An extract from LANDSAT (Path 220, Row 21), 24th October 1976, MSS band 4, Tay Estuary, Scotland.

colleagues to help plan the course of their sea-truth collection cruises. Preliminary results indicate (J. A. Charlton, private communication) strong evidence of a front in almost exactly the position indicated by this LANDSAT picture. Many quantities vary across the front but, of course, the one which causes it to appear in the MSS band 4 imagery is the difference in suspended sediment (or possibly chlorophyll) concentration (not the difference in temperature or in salinity). In other scenes of this area which we have examined, and which correspond to high water at Dundee, there is no sign of this front in the open sea. This is, indeed, what we would expect because at high tide the front will have moved right back within the Estuary.

MAPPING OF SANDBANKS

Finally in Figure 6 we show the LANDSAT MSS band 7 picture for the same area and date as the picture in Figure 5. This scene corresponds, within a few minutes, to low-tide at Dundee. Thus in this picture we have an accurate way of mapping the sandbanks that are uncovered at low tide. What is particularly interesting is

FIGURE 6: An extract from LANDSAT (Path 220, Row 21), 24th October
1976, MSS band 7, Tay Estuary, Scotland.

that the shape of the Abertay Sands, on the south side of the mouth of the River Tay are changing all the time (on a time-scale of years). There are quite significant differences of detail between the outline of these sands, as seen from LANDSAT in October 1976 and as given on the latest Ordnance Survey map, which is based on aerial photographs of (probably) about ten years ago. If one goes further back in time, to Ordnance Survey maps of about 50 years ago, or to Admiralty charts, the differences from the October 1976 outline are very much more marked. We are currently investigating the outline of these sands quantitatively using this low-tide LANDSAT scene. We also hope, by ratioing on bands 4 and 5, to use high-water scenes to plot the present contours of these sandbanks.

ACKNOWLEDGEMENTS

The work described in this paper has been supported financially by the Carnegie Trust for the Universities of Scotland and by the Marine Technology Directorate of the Science Research Council. The LANDSAT image-processing work has been performed at the DFVLR, Oberpfaffenhofen, with the assistance of Dr. K. A. Ulbricht, and the NIMBUS-7 CZCS images were produced for us at RAE Farnborough.

REFERENCES

Hammond, M. J. 1977. A survey of Earth-surface observation satellites and the interface between remote sensor and attitude control system. ESA Journal 1: 327-43.

Johnson, R. W., I. W. Duedall, P. M. Glasgow, J. R. Proni, and T. A. Nelsen 1977. Quantitative mapping of suspended solids in wastewater sludge plumes in the New York Bight Apex. J. Wat. Poll. Contr. 49: 2063-73.

Morley, L. W. 1977. Satellites and Sovereignty. Report of the Interdepartmental Task Force on Surveillance Satellites. Ottawa: Canada Centre for Remote Sensing, Department of Energy, Mines and Resources.

Plevin, J., V. Hood, and T. D. Guyenne 1978. Earth observation from space and management of planetary resources. Proceedings of an international conference held in Toulouse 6-11 March 1978. Paris: European Space Agency.

Ulbricht, K. A. and D. Schmidt 1978. Mass appearance of blue-green algae in the Baltic: Evaluation of multi-spectral LANDSAT

scenes by image processing. In: Earth observation from space
and management of planetary resources. Proceedings of an
international conference held in Toulouse 6-11 March 1978. J.
Plevin, V. Hood and T. D. Guyenne. Paris: European Space
Agency.

Sϕrensen, B. M. 1978. Potential applications of the ocean colour
scanner North Sea experiment 1977. In: Utilization pour
l'Océanologie des satellites d'observation de la Terre, Brest,
6-8 Fevrier, 1978. Paris: Centre National Pour
l'Exploitation des Océans: 343-63.

Statistical Methods of Probable Use for Understanding Remote Sensing Data

Donald P. Gaver

ABSTRACT

This paper discusses several modern topics in statistical data analysis and probability modeling that may be useful in interpreting remote sensing data. These are robust regression, non-linear smoothing, and the jackknife in conjunction with stochastic point process modeling. Illustrative applications are to exploring a wind-speed, whitecap coverage relationship, to surface temperature measurement in the presence of cloud cover, and to estimating the probability distribution of ice-ridge sail heights.

INTRODUCTION

Statistical methodology has long been a familiar tool for use in understanding our natural environment. Classical examples of

263

applications of statistics are seen in weather forecasting, in evaluation of attempts at weather modification by cloud seeding, and in descriptions of the fluctuations in the sea surface. Now the accessibility of new and extensive data from a variety of remote sensing sources, such as earth orbiting and geostationary satellites, again calls for the development and application of appropriate statistical methodology. Classical methods of statistics and of probability modeling frequently must be adapted to the new needs. The process of adaptation will proceed most efficiently if statisticians work cooperatively with the scientists actually obtaining data and studying the associated natural phenomena. Conferences, such as PRIMARS I, are of great value in promoting the necessary interchange of information and the stimulus to approach novel and difficult problems in a realistic manner.

This paper describes new approaches to the analysis of data, in particular to quite "noisy" data of the sort that is likely to be encountered when observing the natural environment. The descriptions given will necessarily be brief, but an attempt will be made to show how the methods and viewpoints presented may be applied to problems arising in remote sensing.

ROBUST METHODOLOGY: REQUIREMENTS AND POSSIBILITIES

Many scientists who have closely examined real data have encountered occasional, or even frequent, anomalous behavior. Apparent anomalies in data may be with respect to either

(a) preconceptions as to "proper" data behavior, these perhaps being buttressed by (physical) theory, or

(b) the nature of the general pattern of the data, especially those data points in the immediate neighboorhood, e.g., in time or space.

Plots

In simple circumstances graphical plots will quickly reveal

those points that are blatant anomalies. For instance, suppose that one wishes to investigate data concerning the relationship between wind velocity and whitecap cover in the ocean. Theory may suggest a specific relationship, e.g. that whitecap cover, C, be nearly a cubic function of wind velocity v, so that it will be tempting to plot C _vs_ v and note an appearance as shown on Figure 1; there solid black dots represent (simulated) raw data. Since the eye finds it difficult to distinguish curves of the form $C = \alpha v^2$, $C = \alpha v^3$, $C = \alpha v^{7/2}$, etc., from one another, and yet is sensitive to departures from linearity, a graph of C _vs_ v^3 suggests itself, but is not included here. A plot on log-log paper may be still better. As presented, the date conforms in general to the theorized relationship or scaling, with the obvious exception of the circled point to the right. Such an anomalous point, or points, represents a challenge both to statistical technology and to the ultimate user of the data. Statistical technology assumes the responsibility for revealing the presence of such points, and, if possible, for providing a meaningful and useful summary of the remaining points. It falls to the consumer or ultimate user of the data, preferably with the help of a subject-matter specialist (physicist or oceanographer), to interpret the apparently anomalous maverick - or exotic, or outlying -- data point. The following are two possible options:

(i)   Is it an evidence of the failure of the relation $C = \alpha v^3$, say for large velocities?

(ii) Is an outright error in data recording, and to be disregarded?

Note that simple graphs are invaluable for pointing out extreme outliers in simple, one-explanatory-variable, situations. If more variables are required, informative plots are more difficult without the use of more statistical technology. We next show that classical, least-squares, technology may be quite misleading, but that replacements are available. See Mosteller and Tukey (1977), abbreviated MT hereafter, especially pages 353-365.

Fits and Residual Plots

Suppose that one wishes to summarize data, such as that in Figure 1, by fitting the relationship $C = \alpha v^3$, i.e., determining the parameter α from the data. The classical and automatic way of doing so is to apply least squares; computer programs are universally available, even for handheld calculators (the TI 59, or HP 67). What are we likely to find? A least-squares line (treating $w = v^3$ as the independent variable presents $C = \alpha w$; one can also plot and fit $C^{1/3} = (\alpha)^{1/3} v$, and there may be reasons for this choice) is quite apt to totally misrepresent the situation, responding much too sensitively to the single (here encircled) outlying value, and straying systematically away from the main body of the data; see the points represented by o in Figure 1.

An alternative method for fitting, described in MT, is less susceptible to outlier influence -- is far more robust to departures from basic assumptions--than is the ordinary least squares (OLS) method. This new method, termed biweight fitting, is carried out by a procedure that uses the OLS computation iteratively. In the course of the computations, weights are automatically developed that reduce the influence of the encircled value of Figure 1, permitting the fit to more closely approximate the main body of the data. We now describe and illustrate the biweight fitting procedure as it is adapted to the problem of determining the parameter α in the relation y_i vs αx_i.

Biweight Fitting Calculation

(1) Compute the kth ($k = 1,2,3,\dots$) iterative estimate of α, denoted by $\alpha^{(k)}$ by solving

$$\sum_{i=1}^{n} (y_i - \alpha^{(k)} x_i) x_i w_i^{(k-1)} = 0,$$

to obtain

$$\alpha^{(k)} = \frac{\displaystyle\sum_{i=1}^{n} y_i x_i w_i^{(k-1)}}{\displaystyle\sum_{i=1}^{n} x_i^2 w_i^{(k-1)}}$$

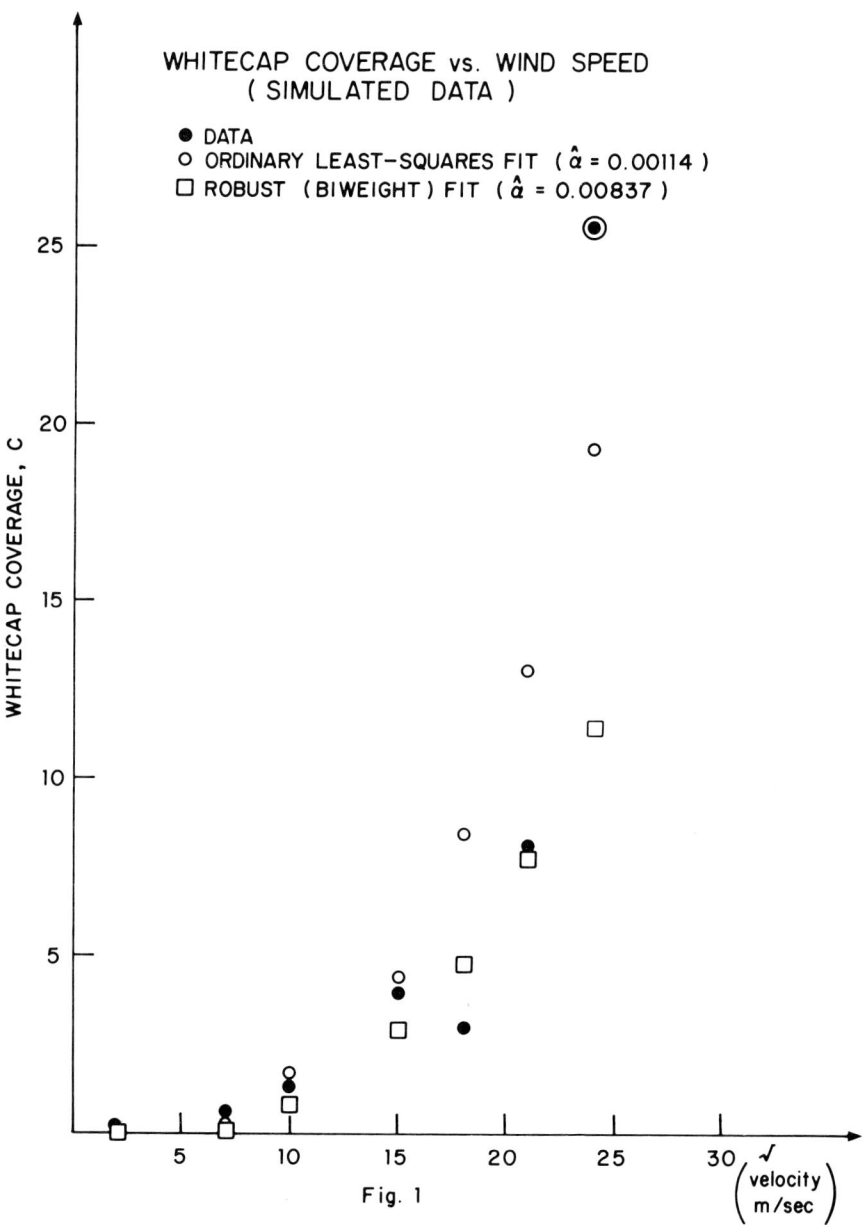

Fig. 1

(2) the weights, $w_i^{(k-1)}$, are of this form:

$$w_i^{(k-1)} = \left[1 - \left(\frac{y_i - \alpha^{(k-1)} x_i}{cS^{(k-1)}} \right)^2 \right]^2 \quad \text{if } (\cdot) \leq 1$$

$$= 0 \qquad\qquad\qquad\qquad \text{if } (\cdot) > 1$$

where $(\cdot)$ refers to the term $\{(y_i - \alpha^{(k-1)} x_i)/cS^{(k-1)}\}$; $S^{(k-1)}$ is a scale factor (robust replacement for the standard deviation) that may be computed in the following manner.

(3) The k-1st iterated value of the scale factor is
$$S^{(k-1)} = \text{median}\{|y_i - \alpha^{(k-1)} x_i|\},$$

c being a constant of value 6, or 9; the smaller value of c is the less tolerant of outliers, and large (infinite) c leads to OLS.

(4) the first value $\alpha^{(1)}$, of the iterative sequence can be obtained by equalizing all weights $(w_i^{(0)} = 1)$, which is equivalent to OLS; alternatively, one can utilize a "robust start," suggestions for which can be found in MT.

The iteration is carried on until the difference between successive values is small; usually 4 to 8 iterations are sufficient. The resulting α-estimate can be denoted by $\hat{\alpha}$.

Following the fitting it is informative to plot the residual values:
$$r_i = y_i - \hat{\alpha} x_i = y_i - \hat{y}_i, \quad i = 1, 2, \ldots, n.$$

$\hat{y}_i$ being shorthand for the predicted y value. In case there is a single outlier, as in Figure 1, the fitted line will tend to hug the major point cloud, and a histogram of the residuals will dramatically reveal the presence of the outlier, suggesting further investigation. A plot of r_i vs y_i is also useful. See MT for further suggestions.

Numerical Illustration

The following are a set of (simulated) whitecap percentages and corresponding wind velocities. Alongside are values for whitecap coverage estimated by OLS and by the biweight procedure.

Velocity	Cover ("Actual")	Cover (OLS Estimate)	Cover (Robust Estimate)
2	0.011	0.011	0.0067
7	0.63	0.49	0.29
10	1.30	1.43	0.84
15	3.89	4.82	2.83
18	2.89	8.33	4.88
21	8.16	13.2	7.76
24	25.7	19.7	11.6

It is clear from the above Figures, and perhaps clearest in Figure 1, that the OLS solution, in its attempt to fit the point $C(24) = 25$, systematically and considerably over-estimates the points at $v= 15$ and above. The biweight estimator performs much better, allowing a closer fit to all data other than $C(24)$. A residual plot brings attention to bear on that point.

Since the values of "Actual Cover" were actually constructed by forming $0.008v^3$ and adding Gaussian random noise with value proportional to $C(v)$, and since the sequence of values of $0.008v^3$ were 0.0064, 0.27, 0.80, 2.7, 4.67, 7.41, 11.06, we cannot fault the manner in which the biweighted procedure functioned in this example, and are encouraged to use it more widely.

Possible Application to Remote Sensing Data

In a paper in this conference proceedings by de Priest (p.51), and in Fleming (in press), a problem arising from partial cloud cover contamination of remote sensing data is described and addressed. This problem has the following origin. A series of measurements are made on a physical quantity (sea surface temperatures), but they are contaminated. That is, in the case of sea surface temperatures, if no clouds are present the measurements are approximately normally distributed around μ (the true temperature). However, if clouds are present, a fraction of the measurements is made artificially smaller, cloud temperatures being lower than those at the earth's surface. The problem is to estimate μ. Techniques for doing so are described by de Priest (p. 51) and by Fleming (in press). We describe a possible alternative approach that uses robust regression. Operational characteristics of the two procedures have not yet been compared.

(1) Arrange the measurements in order: $y_1 < y_2 < y_3 < \ldots < y_{n-1} < y_n$. The largest observations may well appear similar to the largest order statistics of a normal distribution with (unknown) mean μ and standard deviation σ (sometimes assumed known, although caution is in order), while the smaller ones are likely to depart systematically.

(2) Carry out a preliminary plot of

$$y_k \text{ vs } \Phi^{-1} \left(\frac{k}{n+1}\right), \qquad k = n, n-1, n-2, \ldots$$

where $\Phi^{-1}(p)$ is the inverse function of the unit normal: recall that if

$$\Phi(y) = \int_{-\infty}^{y} \exp(-\tfrac{1}{2}z)^2 \frac{dz}{\sqrt{2\pi}},$$

is the unit normal distribution, then the solution of the equation $\Phi(y) = p$ gives

$$y(p) = \Phi^{-1}(p);$$

Φ, and hence Φ^{-1}, are widely tabulated. Alternatively, use Arithmetic Probability Paper. If y_k is an ordered observation from a normal population, then the plot should appear straight, while a systematic departure from linearity indicates a departure from normality. Suppose departures begin to occur at $k = D$; sometimes D may be greater than $n/2$. One may first eye-fit a straight line to the points $k = n, n-1, \ldots, D$. Then $\hat{y}_{n/2} = \hat{\mu}$ (estimated temperature), i.e. the value of the fitted line at $n/2$ should give reasonable value for μ.

(3) Going further in a formal direction, one may wish to fit a line to the data points. Here biweight fit should behave well, tending to be oblivious to spurious (cloud contamination) points. One can proceed to fit the relation

$$y_k \text{ vs } \mu + \sigma x_k$$

with

$$x_k = \Phi^{-1} \left(\frac{k}{n+1}\right), \qquad k = n, n-1, n-2, \ldots;$$

a start using the eye-fit points $k = n, n-1, \ldots, D$ may be worthwhile. Finally, quote the estimate

$$\hat{\mu} = \text{med } \hat{y}_k = \frac{1}{2}(\hat{y}_{n/2} + \hat{y}_{(n/2)+1}), \qquad n \text{ even};$$

$$= \hat{y}_{(n+1)/2}, \qquad n \text{ odd}.$$

where

$$\hat{y}_k = \hat{\mu} + \hat{\sigma} x_k.$$

The above procedure seems worth further investigation and refine-
ment. One inportant step may be to adjust for the effect of correlation between order statistics when carrying out the regression.

SMOOTHING DATA

If one plots certain environmental data, e.g. monthly total
rainfall, or perhaps daily maximum temperature, at a particular
location, systematic regularities seem to appear, but may be masked
by noise. Often there is a seasonal pattern, i.e. one that is
roughly cyclic in nature. Attempts to fit such a pattern with
polynomials is doomed to failure, and selection of a set of sines
and cosines that does well (Fourier series) may lead to many terms.
Some method of smoothing the original series that lays bare the
regularities is to be desired. After this is made available, one
can study the residuals around it. Spectral analysis or some
formal procedure may then be of use.

Classical smoothing procedures involve some form of moving
average, and are susceptable to the "python-swallowing-the-pig"
difficulty. Imagine using the linear smoothing operation

$$Sy_t = \frac{y_{t-1} + y_t + y_{t+1}}{3}$$

on the y_t series

t	1	2	3	4	5	6	7	8	9	10
y_t	11	9	7	8	8	29	10	8	6	9
Sy_t	(11)	9	8	7.7	15	15.7	15.7	8	7.7	(10)
Ry_t	(11)	9	8	8	8	10	10	8	8	(10)

clearly the entry of 29 at t = 5 into the smoothed value series Sy_t
gives a serious distortion, travelling as it does in partially
digested form through the next two terms of the smoothed series.
If further smoothing is attempted, the bulge is reduced slightly,
but spreads out in time.

On the other hand, the non-linear operation of taking running
medians, as suggested by Tukey (1977), performs effectively (in

both cases circled end values are copied from the original series, more sophisticated procedures can be investigated as well). The last row in the table, labelled Ry_t, gives the result of this robust smoothing; note that it behaves in an intuitively appealing manner, essentially ignored the outlying value 29. Further steps can be taken to improve the "smooths," but we refer to Tukey (1977) Chapters 7 and 16 for details.

Two further points may be made. The first is that the analysis of a sequence of data points, and their projection or forecasting in space or time, should not end with providing a smoothed or averaged version. Examination of the remaining variation, e.g. the sequence $y_t - Ry_t$, called the "rough" by Tukey, may well be rewarding; presence and suggestiveness of various outliers is much more evident in the rough (residuals) sequence than in the original sequence of data points. Secondly, the procedure described for smoothing simple sequences of data points must be adapted to planar (two-dimensional) data; some work has been done, but much remains.

STOCHASTIC MODELING OF ICE PRESSURE RIDGES

The dynamics of ice formation in the Earth's cold regions results in the development of irregular ice pile-ups, or pressure ridges. These ridges occur in an apparently random fashion in space; in fact the following regularities are observed by remote sensing methods (Dr. W. Weeks, personal communication; and Weeks et al., in press):

(1) Along a sampling line (e.g. airplane flight path, or straight submarine track) ice ridges seem to appear in accordance with a stationary Poisson process, so if $R(x)$ is the number of such ridges encountered over a distance x, then approximately

$$P\{R(x) = n\} = e^{-\lambda x} \frac{(\lambda x)^n}{n!} \qquad n = 0,1,2,\ldots,$$

where $\lambda > 0$ is the density of ice ridges.

(2) The probability distribution of ridge "sail heights" (or "keel depths") may be approximated by the forms $F(y) = 1-e^{-\mu y}$, or $1 - e^{-\nu y^2}$; the best-fitting distribution may well depend upon the

method of observation (averaging properties). For further details
see work referenced in Weeks et al. (in press).

Now it may be of interest to compute the distribution of the
maximum sail height, or keel depth, that one is to encounter over a
course of length x. This is very simple, given the particular
distributions of sail number and size and furthermore assuming
independent between ridge heights. Let $\bar{H}(x)$ be the maximum sail
height; then

$$P\{\bar{H}(x) \leq y\} = \sum_{n=0}^{\infty} e^{-\lambda x} \frac{(\lambda x)^n}{n!} [F(y)]^n$$

since all of the Poisson-distributed heights must be less than y in
order for the maximum to be below y.

Sum out to obtain

$$P\{\bar{H}(x) \leq y\} = \exp\{-\lambda x[1-F(y)]\}$$

Depending upon which distribution is picked for ridge heights, we
get

a) $P\{\bar{H}(x) \leq y\} = \exp(-\lambda x e^{-\lambda y})$

b) $P\{\bar{H}(x) \leq y\} = \exp(-\lambda x e^{-\nu y^2})$

These closely resemble classical extreme value distributions. Note
that if logs are taken simplicity occurs:

a) $\ln P\{\bar{H}(x) \leq y\} = \exp(-\lambda x e^{-\lambda y});$

$\ln(-\ln P\{\bar{H}(x) \leq y\}) = \ln(\lambda x) - \mu y$

b) $\ell \nu \ell\{\bar{H}(x) \leq y\} = \exp(-\lambda x e^{-\mu y});$

$\ln(-\ln P\{\bar{H}(x) \leq y\}) = \ln(\lambda x) - \nu y^2$

If either of these formulas are to be used for practical purposes,
values of the parameters must be obtained. In order to estimate
parameters λ, μ, υ in the above models from data, one naturally
thinks of the method of maximum likelihood. Suppose that we have
observed $R(x) = n$ ridges of heights y_1, y_2 ..., y_n. Then the
maximum likelihood estimates are

$$\hat{\lambda} = \frac{x}{n} \, , \quad \hat{\mu} = \frac{1}{\bar{y}} \, , \quad \hat{\upsilon} = \frac{1}{\bar{y}^2}$$

where as usual we have put

$$\overline{y^k} = \frac{1}{n} \sum_{i=1}^{n} y_i^k$$

Hence our estimates are of the form

a) $\text{est}\ln(-\ln P\{\overline{H}(x) \le y\}) = \ln \hat{\Lambda} + \ln x - \hat{\mu}y$

b) $\text{est}\ln(-\ln P\{\overline{H}(x) \le y\}) = \ln \hat{\Lambda} + \ln x - \hat{\nu}y^2$

If rather large samples are available and if distributional assumptions are well satisfied, one may feel comfortable with conventional standard errors based on Fisher information and normality (see Cramér, 1946). On the other hand, it is of interest to apply the jackknife technique (see R. G. Miller, 1974) to obtain estimates of the variance of estimate due particularly to the ridge heights. To carry out the calculation, (i) compute $\hat{\nu}_{all} = n/\overline{y^{-2}}$; then (ii) compute

$$\hat{\nu}_{(-j)} = \frac{n-1}{y_1^2 + y_2^2 + \ldots + y_{j-1}^2 + 0 + y_{j+1}^2 + \ldots + y_n^2}$$

for $j = 1, 2, \ldots, n$; then

(iii) compute the pseudovalues $\hat{\nu}_j = n\hat{\nu}_{all} - (n-1) \hat{\nu}_{(-j)}$, and (iv) average to obtain a jackknifed point estimate $\hat{\nu}_{JK} = (1/n) \sum_{j=1}^{n} \hat{\nu}_j$, and its variance

$$\frac{1}{r} \left[\frac{1}{n-1} \sum_{j=1}^{n} (\hat{\nu}_j = \hat{\nu}_{JK})^2 \right] = s_{\hat{\nu}_{JK}}^2.$$

Then we can estimate the standard error of the probability prediction, (b") by comparing

$$\text{S.E.} = (\text{Var}[\text{est } \ln(-\ln P\{\overline{H}(x) \le y\})])^{\frac{1}{2}}$$

$$\underset{\sim}{} \frac{1}{\hat{\Lambda}_{JK}} s_{\hat{\nu}_{JK}}^2 + y^2 s_{\hat{\nu}_{JK}}^2$$

A similar calculation is easily performed for model a); details are omitted. From the above results, approximate confidence intervals may be constructed for the probability of encountering a (maximum) ridge sail height less than y in magnitude.

Recent theoretical results of Efron and Hinkley (1978) suggest that if a traditional maximum likelihood approach is taken, one is better off using observed Fisher information rather than expected Fisher information in order to establish an approximate standard error in either case a) and b). However, work of Reeds (1978) suggests that use of the jackknife in conjunction with maximum likelihood yields results that tend to be rather robust to choice of the basic model. Both of these suggestions must be validated by further work, a good deal of which will necessarily involve Monte Carlo simulation. Such work should be of great importance and interest to those who must assess the probabilities of extreme or rare events, and who furthermore wish to provide some reasonably valid estimates of the error of their estimates.

ACKNOWLEDGEMENT

The writer is much indebted to LCDR C. F. Taylor, Jr., for his assistance in the example of robust regression computations. He is also indebted to the Office of Naval Research for support of this research.

REFERENCES

Cramér, H. 1946. Mathematical Methods of Statistics, Princeton University Press, Princeton, N.J. (p. 503)

de Priest, D. 1979. "Considerations in using a truncated normal distribution for remote sensing data." This volume.

Efron, B., and D. Hinkley. 1978. "Assessing the accuracy of the maximum likelihood estimator: observed versus expected Fisher information," Biometrika. 65: 457-488

Fleming, H. E. 1980. "Application of the truncated normal distribution technique to the derivation of sea surface temperatures," In: Remote Sensing of Atmospheres and Oceans, ed. by A. Deepak; Academic Press.

Miller, R. G. 1974. "The Jackknife -- a Review." Biometrika. 61: 1-16.

Mosteller, F., and J.W. Tukey. 1977. Data Analysis and Regression. Addison-Wesley Publishing Co., Reading, Mass.

Reeds, J. A. 1978. "Jackknifing maximum likelihood estimates." Annals of Statistics 6: 727-739.

Tukey, J. W. 1977. Exploratory Data Analysis, Addison-Wesley Publishing Co., Reading, Mass.

Weeks, N. F., W. B. Tucker, M. Frank, and S. Fungcharoen. 1979. "Characterization of the surface roughness and floe geometry of the sea ice over the continental shelves of the Beaufort and Chukchi seas." In: "Sea Ice Processes and Models, Proc. AIDJEX/ICSI Sympos." (R. S. Pritchard, ed.), University of Washington Press.

Platform Kinematics and Attitude Control for Remote Sensors

Archie E. Roy

ABSTRACT

The forces shaping and governing the orbit of an artificial satellite are reviewed. Parameters available for designing suitable orbits for remote sensing programs are listed. Methods for station-keeping, attitude stabilization and control are described.

INTRODUCTION

This paper is divided into two major sections. In the first we list the forces acting on an earth satellite and the way in which these forces change relevant orbital parameters or elements describing the evolution of the orbit with time. We also consider in this section how suitable orbits can be chosen for particular remote sensing programs. In the second section a review is given of methods that have been adopted for station-keeping of the satellite, also for stabilization and control of its attitude.

ORBITAL MOTION OF AN ARTIFICAL EARTH SATELLITE

Forces Acting on an Artifical Earth Satellite

The earth's gravitational sphere of influence (the radial distance from the earth's center of a sphere within which any mass is effectively controlled by the earth's gravitational influence) is of order 10^6 km in radius. Although the moon is a rival mass in an orbit about the earth of semi-major axis 384,000 km, its mass of only 1/81.3 that of the earth ensures that outside its own sphere of influence (of order 20,000 km in radius) its gravitational field at most disturbs slightly the geocentric orbit of an earth satellite. The earth's gravitational field therefore dominates any earth satellite orbit, the effects of other planets being completely negligible. The sun's influence becomes important as a perturbation only if information about the values of the higher harmonics in the earth's gravitational potential from observations of artificial satellites is being sought, or other very high order effects. If desired, solar and lunar perturbations may be calculated by one of a number of theories giving the changes in the Keplerian elements of the satellite orbit.

The earth's atmosphere produces drag and lift forces on the satellite at any point of its orbit low enough for the satellite to be subject to the continual collision of air molecules, atoms and ions. The magnitude and direction of these forces depends upon a number of factors that vary with time such as velocity, altitude, latitude and attitude of the satellite, as well as diurnal, seasonal, and solar event changes in air density. But unless the satellite is below an altitude of 150 km, the drag and lift forces can be treated as perturbing forces.

Eddy currents can be induced in a satellite as it passes through the earth's magnetic field if the satellite has metal in its construction. The slight retardation acting on the satellite is very small.

Solar radiation can produce marked effects on a satellite orbit if the satellite mean density is small. For example the balloon satellite ECHO 1 suffered changes in perigee height of order 500 km in a period of ten months, the synodic period of the perigee point (the time it took perigee to make one revolution of the earth relative to the sun). Even for balloon satellites, however, such changes may be treated by perturbation techniques.

Uncharged particles such as neutral atoms, or meteoritic dust, encountered by a satellite must have a braking effect, as must charged particles, either of direct solar origin or contained within the atmosphere but any drag produced must be very high order and may be safely neglected.

Summing up, then, it is seen that for almost all earth satellites the major perturbations of the two-body Keplerian orbit are caused by the earth's gravitational field and atmospheric drag.

The Satellite Orbit and its Description

In Figure 1 the orbital parameters are displayed.

Five elements describe the size, shape, and orientation of the orbit; a sixth provides information enabling the satellite position and velocity to be found at any time by utilizing the well-known two-body problem equations. The first two elements, the right ascension of the ascending node, Ω, and the inclination, i, give the orientation of the orbital plane. The third element, the argument of perigee, ω, fixes the orientation of the elliptic orbit in the orbital plane while the fourth and fifth, namely the semi-major axis, a, and the eccentricity, e, give the size and shape of the orbit. The sixth, τ, the so-called time of perigee passage, is an epoch when the satellite was at perigee.

If the earth was a point-mass, or a sphere with a radially-symmetrical density distribution of matter, and no other forces such as atmospheric drag acted on the satellite, the six elements would be constant for the orbit would be a Keplerian orbit, fixed in size, shape and orientation. The other forces operating perturb the orbit so that the six elements describing the orbit and the "timetable" of the satellite in the orbit, vary with time. The differential equations of the elements may be set up and solved by the method of successive approximations to give analytical expressions providing descriptions of the elements' variations with time because of the perturbing forces.

These equations, known as the Lagrange Planetary Equations, are of the form:

$$\frac{d\sigma_i}{dt} = \phi_i, \quad i = 1, 2, \ldots 6 \tag{1}$$

where σ_i is any one of the elements and ϕ_i is a function of the elements and the disturbance. We sketch out the procedure in the two main disturbances, namely (a) the departure of the earth's gravitational potential from that of a point mass, and (b) atmospheric drag.

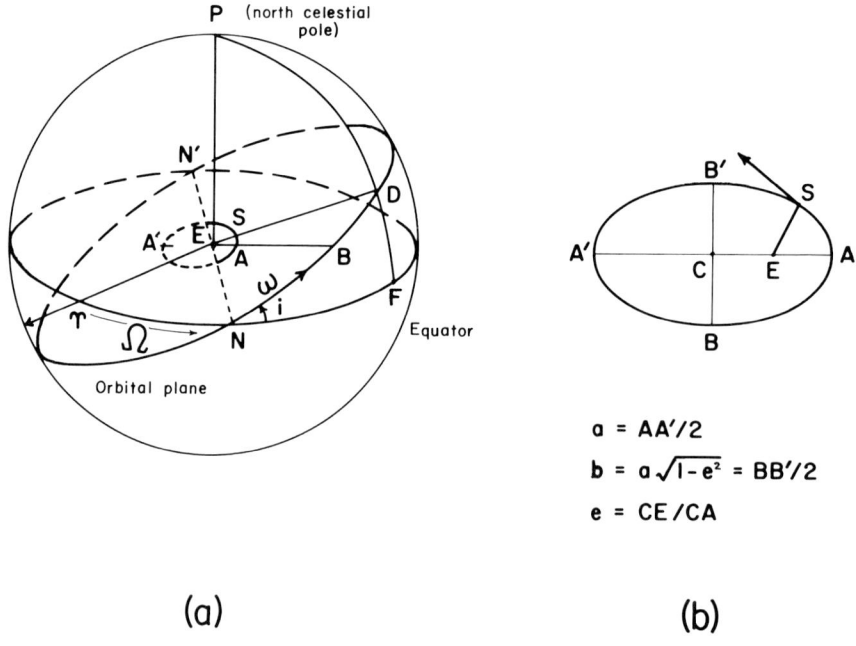

(a)   (b)

FIGURE 1: Elements of an earth satellite orbit

Earth's Gravitational Field

If the earth is assumed symmetric about its axis of rotation, its gravitational potential U is of the form

$$U = \frac{GM}{r} \{1 - \sum_{j=2}^{\infty} J_j \left(\frac{R}{r}\right)^j P_j(\sin \delta)\} \tag{2}$$

where   G is the constant of gravitation,

m is the earth's mass,

r is the distance of the satellite from the earth's

centre of mass

J_j are the harmonic constants of the earth's gravita-

tional potential,

R is the earth's equatorial radius,

δ is the satellite's latitude or declination and

$P_j(\sin \delta)$ is the Legendre polynomial of degree j in the

quantity sin δ.

Then the quantity $F = U - U_o$, where $U_o = Gm/r$, is the gravitational disturbing function due to the earth's field departing from
that of a point mass. Substitution of it into the Lagrange equations provides, on solution, the short period variations, the long
period variations and the secular behavior of the satellite orbital
elements due to this cause. The short period effects are of small
amplitude, with period of order the orbital period of the satellite
(usually a few hours or less); the long period effects can be of
observable and useful amplitude with periods of order a few months
and have been used to obtain values of the odd harmonics such as
J_3. The secular effects of interest occur in the elements Ω and ω,
being given in the first-order theory by the expressions

$$\Omega = \Omega_o - (K \cos i)t = \Omega_o - \Omega_1 t,$$

$$\omega = \omega_o + [K(2-\frac{5}{2} \sin^2 i)] t = \omega_o + \omega_1 t,$$

where

$$K = \frac{3}{2}J_2 R^2 n/[a^2(1-e^2)^2],$$

and

$$n^2 a^3 = GM, \tag{3}$$

n being the mean angular velocity (or mean motion) of the satellite.

There is a similar secular effect in τ but without interest. The other three elements, a, e and i suffer no secular changes but small periodic displacements about mean values. In higher order theories, of course, additional expressions involving J_j, $j \geq 3$, appear. It is found that $J_2 \sim 10^{-3}$ while $J_j \sim 10^{-6}$, $j \geq 3$, so that J_2 dominates the behavior of the elements of interest.

The effect of the secular behavior in the ascending node's right ascension Ω is to cause the orbital plane to precess backwards (i < 90°) at an almost constant rate. The exact rate is obviously a function of the values of a, e, i, n, J_2 and R. The parameters J_2 and R are given in value, n is a function of a through the relation $n^2 a^3 = Gm$ so that the value of Ω_1, the rate of precession of the node N, is governed by the choice of the values of a,e, and i.

The secular behavior of the argument of perigee, ω, is likewise governed by the choice of a,e, and i. In particular a critical value of i exists, given by

$$2 - \frac{5}{2} \sin^2 i = 0. \qquad (4)$$

The solution $i_{cr} \sim 63°26'$ of this equation is called the critical inclination. For $i < i_{cr}$ the orbit advances, ie $\omega_1 > 0$; for $i < i_{cr}$ the orbit precesses, ie $\omega_1 < 0$.

It may be remarked that for a close earth satellite orbit of low inclination, Ω_1 and ω_1 are of order 4 degrees per day.

Atmospheric Drag

In general a satellite of arbitrary shape moving with some velocity V in an atmosphere of density ρ will experience both lift and drag forces, both of which will vary with time if the satellite is changing its attitude with respect to the velocity vector as well as passing with varying velocity through regions of varying density. In the absence of precise knowledge of the satellite's attitude and of the atmospheric density at any instant it is not possible to predict exactly the changes in the satellite orbit.

For practical purposes, however, it is sufficient to assume that the lift forces average out to zero since the satellite's attitude is changing and to assume an average cross-sectional area for the satellite when computing the drag. If the satellite is spherical or nearly so, the cross-sectional area is constant or nearly so. The law of density change with altitude is sometimes taken to be a simple exponential fall-off of density with height such as

$$\rho = \rho_0 \exp(-(\eta-\eta_0)/H), \qquad (5)$$

where η is the altitude, η_0 is some standard altitude usually taken to be the altitude of perigee, ρ_0 is the density at the standard altitude and H is the scale height assumed constant. The scale height is that vertical distance in which the density changes by a factor e = 2.81 and depends upon the altitude. H is about 6 km at sea level reaching 40 km at a height of about 200 km.

If such a law is not assumed then a law of density based on some model atmosphere with parameters determined empirically from satellite observations may be used.

Let the satellite mass be m_s. The drag force per unit mass, F, caused by the satellite moving with velocity V through an atmosphere of density ρ is given by

$$F = \frac{1}{2m_s} C_D A \rho V^2, \qquad (6)$$

where C_D is the aerodynamic drag coefficient and A is the average cross-sectional area of the satellite. The coefficient C_D has a value between 1 and 2. It takes a value near the former figure when the mean free path of the atmospheric molecules is small compared with the satellite size and a value close to 2 when the mean free path is large compared to the size of the satellite.

Then for the simple case of spherical earth with a non-rotating atmosphere we can write down the Lagrange planetary equations of the problem (using their Gaussian form).

It is found that Ω = constant and i = constant, showing that the orbital plane's orientation is unaltered by drag, understandable since there is no component of drag perpendicular to the orbital plane. The symmetric form of the density function (it has the same value for θ degrees after perigee as for θ degrees before) means that in the first approximation to the solution for the other four elements only a and e exhibit secular changes. It is usual to compute the changes in perigee and apogee height over a complete orbit. Respectively, they are $\Delta[a(1-e)]$ and $\Delta[a(1+e)]$, given by

$$\Delta[a(1-e)] = - (\frac{A}{m_s}) \, C_D a^2(1-e) \int_0^{2\pi} \rho(1-\cos E) \, (\frac{1+e \, \cos E}{1-e \, \cos E})^{\frac{1}{2}} \, dE$$

$$\Delta[a(1+e)] = - (\frac{A}{m_s}) \, C_D a^2(1+e) \int_0^{2\pi} \rho(1+\cos E) \, (\frac{1+e \, \cos E}{1-e \, \cos E})^{\frac{1}{2}} \, dE$$

$$(7)$$

By inspection it is seen that unless e is very small, the apogee change is much larger than the perigee change. Thus the decay of a satellite orbit affected by air drag may be depicted as shown in Figure 2.

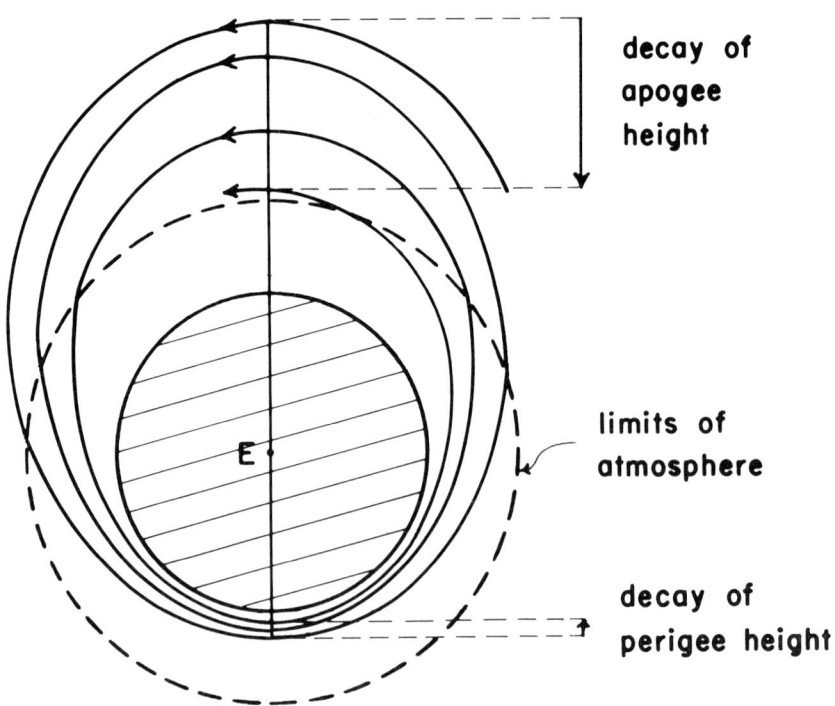

FIGURE 2: Mode of decay of earth satellite due to atmospheric drag.

Several additional comments may be made at this point.

A more accurate description of the earth's gravitational potential involves sectorial and tesseral harmonics since the earth is not quite symmetrical about its axis of rotation. The harmonic constants are more general and the formula for U is more complicated but their effects are of small amplitude and in the first order are purely periodic. When the period of one of these effects is commensurable with that of the satellite orbit, enhanced effects are obtained enabling measurable effects to be observed. The slow decay of a satellite orbit under atmospheric drag can 'drift' the orbit through one of these resonances.

Finally, the more general theory of atmospheric drag takes into account the oblateness of the atmosphere over a non-spherical earth, also its rotation, as well as the diurnal, seasonal, and solar cycle variations that take place. Solar cycle activities and their effects on air density can only be predicted with rough accuracy as evidenced by the decay of Skylab's orbit some two years before the scheduled time for that event.

Use of Orbital Theory to Design Orbits for Specific Purposes

The sidereal orbital period of the satellite is T, given by T = $2\pi/n$. Other periods may be defined, for example the nodical and the anomalistic, respectively the time intervals of successive passages of the satellite through the ascending node or perigee. The fourth period, the synodic, is the time interval between successive similar configurations of earth E, satellite S and sun Θ, for example the time between two conjunctions (i.e. E-S-Θ). All these periods are nearly equal. We now consider some special cases.

Geostationary Orbits

If a satellite is placed in a circular, equatorial orbit at a height of 33500 km (22322 mi), it will have a sidereal period of $23^h56^m04^s$, which is the length of the sidereal period of rotation of the earth (the sidereal day). The satellite will therefore remain above a particular point of the earth's surface. Such an orbit is called a geostationary orbit and has been used extensively for communication and surveillance purposes. In theory three such

satellites at the corners of an equilateral triangle will provide
world-wide coverage. Among the satellites orbited in geostationary
orbits are INTELSAT, DSCS (Defense Satellite Communication System),
SKYNET and METEOSAT.

Soviet Highly-Elliptic Satellite System (Molniya)

Russia put about 60 of these communication satellites into
orbit. Their orbits are designed to utilize Kepler's second law
(the radius vector in an orbit sweeps out equal areas in equal
times). With perigee and apogee heights of 400 and 4000 km respectively, the orbital eccentricity e is high and the satellite makes
2 revolutions of the earth per day. The inclination i is of order
64° and apogee is over Soviet territory. These criteria ensure
that the satellites may be used for communication purposes for 8
hours of each 12-hour period (Figure 3).

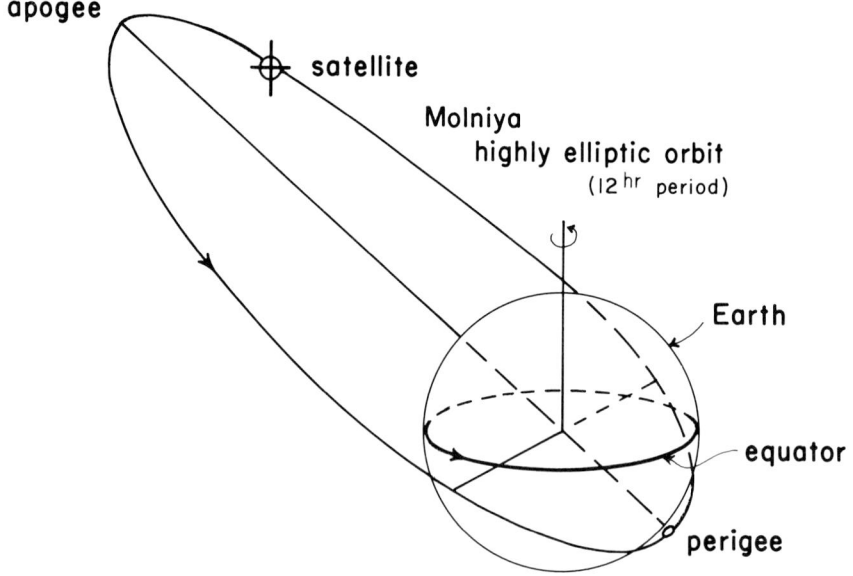

FIGURE 3: Molniya high-elliptic orbit for communications satellite.

Near Polar Orbits

The higher the inclination the closer to complete earth cover
the satellite sensors approach. Nevertheless, the inclination

value i is one of the three parameters governing the behavior of Ω and ω (equations 3 and 4) but this is no disadvantage since it is often desirable to make use of the secular behavior of these elements (see below).

Sun-Synchronous Orbits

Suppose an altitude (or semimajor axis a) is chosen for a satellite such that the orbital period T is an integer divisor of the earth's sidereal rotation period T_E. Thus $qT = T_E$ where q is an integer known as the <u>recursion number</u>. The satellite will therefore reappear over the same point of the earth's surface once each day when q revolutions have been completed. To achieve this, however, the value of the inclination i must be suitably chosen. Examination of equation 3 shows that if i = 90°, then $\Omega_1 = 0$ and a value of q and T may be chosen so that $qT = T_E = 23^h56^m04^s$. The value of T is, by $T = 2\pi/n = 2\pi(a^3/Gm)^{\frac{1}{2}}$, a function of a but not e. If these criteria are satisfied, the satellite will reappear over the same point on the earth's surface in a true inertial recursion manner.

The <u>mean solar day</u> is longer than T_E by about four minutes and is the time between successive passages of the mean sun (a fictitious body moving in the equator in the direction of the sun's apparent yearly orbital motion around the earth) across the observer's meridian. The mean sun's constant rate of increase in right ascension being 0.985 degrees/day, it takes the earth about 4 minutes to rotate through the extra 0°.985 before the mean sun is back on the meridian (Fig. 4).

If the satellite is put into an orbit of slightly higher altitude (and semimajor axis a) with therefore a slightly longer period of revolution T_D' the same recursion number may be used but now $qT' = T_M = 24^h$ = one mean solar day. If the value of i is suitably chosen, so that i is a few degrees greater than 90°, Ω_1 is positive and can be made equal to 0.985 deg./day. The satellite is therefore put into a sun-synchronous orbit in an orbital plane that will keep pace with the sun. Uniform lighting conditions are also ensured every time the sensors scan a particular point on the earth's surface.

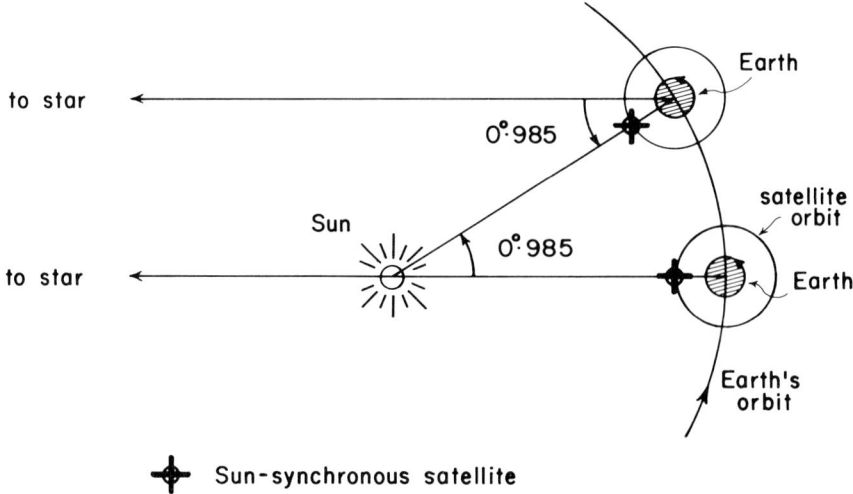

FIGURE 4:   Sun-synchronous satellite orbit.

TABLE 1:   Sun-synchronous recursion orbits.

Recursion number	Period (min)	Altitude (km)	Comments
13	110.8	1262	high for sensors
14	102.9	897	acceptable
15	96.0	569	acceptable
16	90.0	278	Low for life

Table 1 illustrates four sun-synchronous orbits with recursion numbers 13 to 16.   Those for numbers 14 and 15 would be sensible for platforms with remote sensors aboard.   In Figure 5a we see the relation between values of altitude and inclination if the orbits are to be sun-synchronous.   It is useful, however, in some types of survey to provide a slight drift of the earth track in that after 14 or 15 revolutions the satellite does not pass exactly over the

previous track. If therefore a 185 km wide swath is being surveyed
by the sensors aboard the satellite and an 18.5 km overlap seems
desirous between corresponding satellite passes on successive days,
this may be arranged by shifting the recursion track some 167 km
per day. To do this, the chosen altitude and therefore period is
decreased to just below true sun-synchronous recursion as indicated
in Figure 5a. In Figure 5b is shown the result for the sun-
synchronous orbit with recursion number 14. It may be noted that a
slight change in the inclination i is also required.

The LANDSAT series of artificial satellites were flown in
circular sun-synchronous near-polar orbits at a height of 917 km.
This choice of orbit enabled them to cover the entire surface of
the earth every 18 days. Their primary sensor was a Multi-Spectral
Scanner which detected emitted or reflected electromagnetic energy
using an oscillating mirror that produced a 2300-line image of an
185 km square.

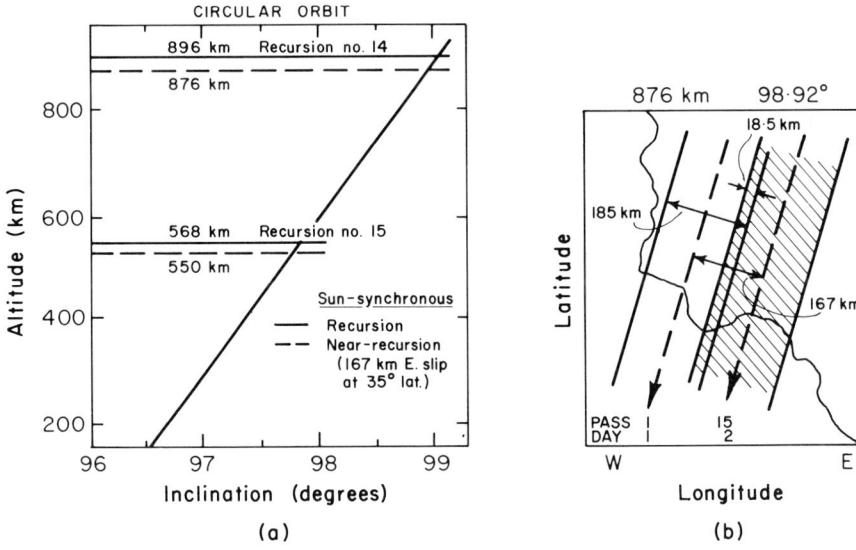

FIGURE 5: (a) Relation between altitude and inclination for a
sun-synchronous satellite orbit. (b) Satellite track with recursion number 14 on successive days. (After Alfred I. Sibila).

STATION KEEPING ATTITUDE CONTROL AND STABILIZATION

Methods of Attitude Control and Stabilization

Without attitude control and stabilization, the sensors aboard an artificial satellite would wander in their orientation with respect to the earth's surface as their platform tumbled through space. No systematic surveying or sensing program could be carried out and any data collected by chance could not be correlated with any certainty with respect to location.

Various methods of attitude control and stabilization have been tried out, assessed in their efficiency, modified and improved. The major ones are (i) gravitational, (ii) magnetic, (iii) spin-gyro effect, (iv) angular momentum flywheel, (v) inertial platform with accelerometer, gyros and thrusters, (vi) sun and star trackers, (vii) infra-red sensors. We consider these briefly in turn, giving examples of satellites utilizing these methods.

Gravity Stabilization

If a mass is raised in the earth's gravitational field its weight decreases. If therefore a uniform massive rod is free to rotate in a vertical plane about a frictionless axle through its center of mass, the stable position of equilibrium for the rod is vertical, the lower half of the rod being weightier than the upper half. The unstable position of equilibrium is horizontal; any angular displacement produces an unequal distribution of weight leading to an oscillation about the vertical position.

The moon, a triaxial ellipsoid, illustrates this action of the gravity gradient by the fact that it always turns the same face towards the earth in its revolution about the planet, a consequence of the fact that the longest axis of the moon points towards the earth's center. Similarly a number of the satellites of Jupiter and Saturn also have rotational periods equal to their periods of revolution.

Any elongated artificial satellite will therefore tend to take up this position and if sensors are placed at one end of the long axis, they will always point downwards towards the earth's surface. Among those artificial satellites using gravity stabilization were the NASA Geodetic Explorer spacecraft GEOS-A, -B and -C and the Transit spacecraft. These spacecraft had long gravity gradient booms. EOLE, built by the French space agency and launched by a Soviet rocket in 1971, had a 10-meter long gravity stabilization boom while NIMBUS -D's attitude control system also contained an extensible rod with a mass at its tip to improve the inertial distribution to take advantage of the gravity gradient torque. Indeed the elongated design of NIMBUS and LANDSAT satellites itself provides some measure of attitude control by gravity stabilization.

Magnetic Stabilization

A crude example of this is the preferred direction of a compass needle in the earth's magnetic field. If applied to an artificial satellite, it has the disadvantage that as the satellite pursues its elliptical orbit inclined at some angle to the earth's geographic equator, irregular changes in the direction of the magnetic lines of force are experienced by the satellite. However, the method has been used. For example the Anna geodetic satellite had a permanent magnet fixed to the satellite with its axis parallel to the satellite axis. This maintained its axis parallel to the geomagnetic field to within plus or minus 3 degrees of the earth's field.

The Beacon-Explorer satellites were also magnetically stabilized, a pyramid reflecting cap being placed on the north-seeking end of the spacecraft.

A more sophisticated and versatile form of attitude control by magnetic stabilization was flown in the ITOS satellites (originally TIROS M (Improved Tiros Operational Satellite)). Six ITOS were flown between 1970 and 1976 to monitor weather systems. They also measured the amount of heat radiated into space by the earth and studied near space proton and electron flux. The ITOS were three-axis stabilized. To keep the ITOS scanners permanently pointed towards the earth, a momentum flywheel rotated at 150 r.p.m. causing the platform to roll through one revolution per orbit. The other two axis' stability was governed by a magnetic coil wound round one end of the package. A current in the coil interacted with the earth's magnetic field providing a force capable of changing the

satellite attitude. This type of attitude control allowed continuous data-gathering throughout each orbit, the current's magnitude and direction being controlled by commands from the ground.

ESRO IV Scientific Satellite also used the same magnetorquer
device while UK-6, although spin-stabilized at 60 r.p.m. also used
a magnetic coil for attitude control.

Spin Stabilization

In this form of attitude control use is made of the gyro
effect. A spinning object in free fall will maintain its axis of
rotation's direction. Many satellites have been stabilized by
spinning them. It has already been mentioned that UK-6 is spin-
stabilized at 60 r.p.m., utilizing small gas thrusters at the end
of its solar cell wings to maintain the satellite's spin rates.

The IDSCS (U.S. Initial Defense Satellite Communication
System) of 26 satellites were also spin-stabilized. They were
placed in orbits just below geostationary altitude so that each
satellite drifted some 30 degrees per day with respect to the
ground. SKYNET and NATO geostationary communication satellites
were also spin-stabilized as was ANIK, operated by Telesat Canada.

The first eight TIROS weather satellites, designed to operate
in near-circular orbits between altitudes of 680 and 940 km, in
inclinations from 48° to 59°, were spin-stabilized. They contained
two TV cameras and infra-red scanners. The cameras were mounted
parallel to the spin axis therefore useful pictures were obtained
only when the satellite axis pointed at the earth. On either side
of this point in each orbit increasingly oblique angle pictures
were taken.

The last two TIROS satellites used a 'cartwheel' control
method to widen the TV coverage, having two TV cameras as before,
but each looking outwards from opposite sides of the satellite
rather than directly down from one end. TIROS 9 and 10 made a 10
r.p.m. roll with the axis of rotation at right angles to the orbital plane. The cameras therefore viewed the earth's surface twice
in each spacecraft rotation about its axis. These two satellites
were also in polar orbits.

NASA's SMS/GOES (Synchronous Meteorological Satellite/ Geostationary Operational Environmental Satellite) system are, as the name implies, in geostationary equatorial orbits. They are also spin-stabilized at 100 r.p.m.

Momentum Flywheel

This method of continuously altering the attitude of a satellite has already been mentioned in connection with the ITOS series where the ITOS scanners were kept permanently pointed earthwards by the action of a momentum flywheel rotating at 150 r.p.m.

Inertial Platform With Accelerometers, Gyros and Thrusters

A schematic of an inertial platform is given in Figure 6a together with the basic principle of an accelerometer (Fig. 6b). Essentially, an inertial platform consists of a platform to which is attached three gyroscopes, each gyro having a single degree of freedom so that all three rectangular axes are taken care of. The platform is suspended in gimbals. When the spacecraft swings about an axis, a torque acts upon the corresponding gyro so that it precesses. Its angular velocity is sensed and governs a servomotor which opposes the disturbing torque, maintaining the platform in its reference attitude. Attached to the platform are three accelerometers, again one for each axis of reference. When the spacecraft thrusters act, accelerating the vehicle, inertia causes the accelerometer mass to move against one of the springs. A slider attached to the mass moves across a potentiometer giving an output at C (Fig. 6b). Double integration gives velocity and distance changes during a 'burn' by the motor, data that may be compared with the computer midcourse correction program to design a new midcourse correction if required. Obviously thrusters are required for attitude changes if no magnetic coil is used. These motors may be small hydrazine thrusters or ion motors.

Sun and Star Trackers

These are extensively used for attitude control not only in artificial satellites of Earth and Mars but also in interplanetary space probes. In general the procedure is to acquire suitable celestial objects and then lock on to them thus maintaining the correct attitude of the spacecraft. Objects used are the sun and

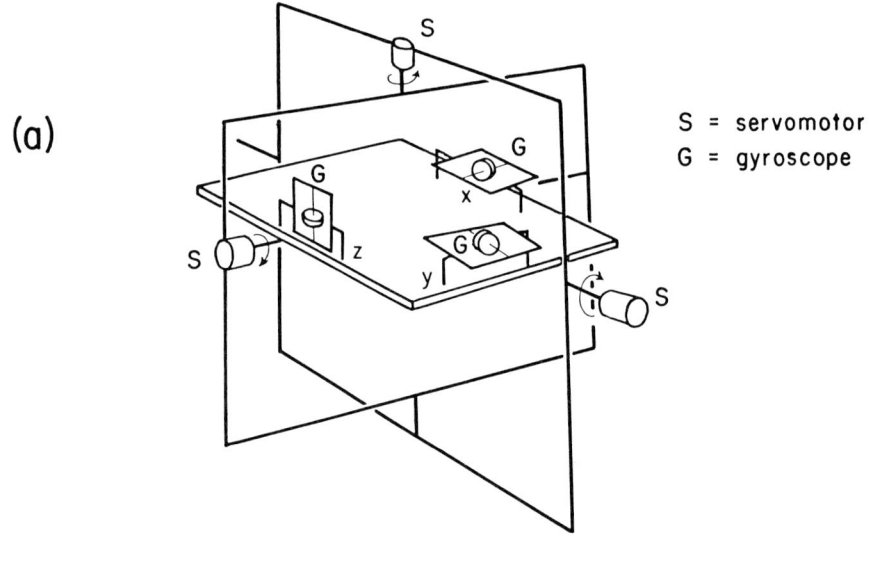

(a)

S = servomotor
G = gyroscope

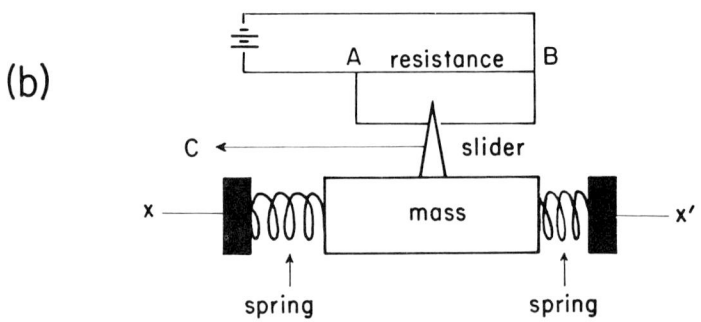

(b)

FIGURE 6: (a) Principles of stabilized platform (b) Principles of accelerometer.

suitable bright stars such as Polaris and Canopus. Usually the sun is acquired first followed by a search for the star.

For example in the attitude control system for the Hawker Siddeley Dynamics X-4 sun-synchronous artificial satellite, the acquisition sequence utilizes for sun acquisition (1) a coarse secondary sun sensor array providing full cover (2) a coarse primary sun sensor with a field of view of ±80° and an accuracy of ±2°

(3) a fine sun sensor with a field of view of ±8° and an accuracy of ±0°.05. Star acquisition (on Canopus) then follows. Once acquired, a limit cycle of 1° amplitude and rate of 8°/hour about the locked star begins. In this system the attitude is maintained by gas jets.

Infra-Red Sensors

These may be used for attitude control and have been flown in various satellites launched. Accuracies of order 0.°1 are easily obtained by this method which involves the detection of the infrared output from the earth's surface. Once the satellite has achieved an attitude that acquires the earth, a drift in that attitude will cause the infra-red (or albedo) sensor's field of view to slide off the earth (that is, off the satellite horizon) into space. The sudden change in the intensity of infra-red radiation received by the sensor operates a gas jet to oppose and correct the change in attitude. Infra-red sensors have the advantage of providing reference data in earth coordinates whereas celestial sensors, though capable of much more accurate performances, have the disadvantage of greater complexity and the need for transformation from celestial to earth coordinates.

Station - Keeping for Geostationary Satellites

It might be thought that if a satellite is put into a perfectly circular, equatorial geostationary orbit at the correct altitude it would remain over a particular point on the equator, that is, it would maintain the geographical longitude of that point. In general this is not so. For example the sun and moon try to take the satellite out of the earth's equatorial plane. Again, because of irregularities in the earth distribution of matter, the Earth is not quite spin-axially symmetric with the result that at the geostationary orbit altitude there are slight geopotential 'hills' and 'valleys' as one goes from geographic longitude 0° east and west to 180°. For example a geostationary satellite at 0° longitude is on a geopotential slope and will therefore drift slowly in an easterly direction. The Meteosat weather satellite stationed at 0° longitude is therefore corrected by ground control when it has drifted to longitude one degree east.

On command the radial thrusters provide an impulse for the space-craft to move westwards. The spacecraft then coasts up the geo-potential 'hill', slowing until it comes to a stop about one degree west. It subsequently drifts down the 'hill' eastwards again until a new station-keeping thruster impulse is ordered by ground control to repeat the cycle, the duration of which is about five months.

The Meteosat satellite is spin-stabilized, rotating at 100 r.p.m. about an axis parallel to the earth's rotation axis. Scan-ning of the whole hemisphere of the earth's surface below the satellite is achieved by (1) east-west scans by virtue of the satellite's spin and (2) south-north sequences of these scans by stepping of the sensors one increment northwards after each space-craft rotation.

Attitude measurement is achieved by four infra-red earth sensors. Two scan the earth's disc as the satellite rotates, each providing a pulse at the transitions from earth to space and vice versa. Two sun slit sensors give one pulse each per revolution and two accelerometers monitor satellite nutation. All these data are telemetered to ground control, giving a continual check on atti-tude.

CONCLUSION

It is evident from this review of platform kinematics and attitude control methods that by the end of the seventies, we have moved into a phase of earth satellite state of the art where the technology is certainly capable of solving the problems connected with the provision of a satellite platform for any proposed remote sensing program of the eighties.

REFERENCES

Attitude and Orbit Control Systems, 1977. Paris. European Space Agency.
Barrett, E. C. and L. F. Curtis, 1976. Introduction to Environmental Remote Sensing. London: Chapman and Hall.
Roy, A. E. 1978. Orbital Motion, 489 pp. Bristol: Adam Hilger.
Sibilia, A. I. 1970. Small Payload Potential for Earth Resources. Journ., Brit. Interplan. Soc., 23, No. 4., p. 307.

Satellite-Borne Microwave Sensors

D.P. Thomas and C.R. Francis

ABSTRACT

The role of satellite-borne active and passive microwave
sensors in the monitoring of the marine environment is considered.
In particular, passive microwave radiometry, scatterometry and
radar altimetry are described. The synthesis of algorithms for the
extraction of oceanographic parameters from measurements made by
such instruments is outlined.

SATELLITE-BORNE MICROWAVE SENSORS

INTRODUCTION

Within the family of microwave sensors, there are two main
groups: passive and active sensors. Active microwave sensors

provide their own illumination, while passive systems measure radiation upwelling from the earth's surface.

Active microwave sensors are basically radars - they operate by transmitting electromagnetic waves and then measuring the signal reflected from the target, such as the sea surface.

In the case of a passive microwave radiometer, the very low power thermal microwave radiance produced naturally by the sea surface and atmosphere is detected.

Probably the greatest asset of microwave instruments - both active and passive sensors - lies in their ability to sense the sea surface through overlying cloud. These sensors are also independent of daylight and operate equally well in complete darkness.

In the following sections, passive microwave radiometry and active scatterometry and altimetry will be considered.

PASSIVE MICROWAVE RADIOMETRY

In passive microwave radiometry, the antenna picks up thermal microwave radiation emitted by the ocean surface and overlying atmosphere. In general, the antenna smooths and averages the brightness distribution of the source observed. An accurate knowledge of the antenna reception pattern allows a partial correction for antenna effects and restoration of the actual brightness distribution. The radiometer is calibrated by switching the input of the receiver from the antenna to an accurately known standard such as a noise source. With careful design, it is possible to measure the antenna temperature to an absolute accuracy of less than $1°K$. Passive microwave radiometers have been used both on the ground and on aircraft before being installed on NASA spacecraft.

The Electrically Scanning Microwave Radiometer (ESMR) was carried on Nimbus 5 which was launched in 1972. This makes use of a phased array antenna with linear polarization and operates at 19 GHz. A second ESMR was carried on Nimbus 6, launched in 1975. In this case both vertically and horizontally polarized components were measured at a frequency of 37 GHz. The Electrically Scanning Microwave Radiometers have been particularly useful for polar ice mapping. It is possible to differentiate between new ice and snow, and older ice.

Nimbus 6 also carried the Scanning Microwave Spectrometer (SCAMS). This uses three horn antennae allowing measurements at 5 different frequencies. SCAMS has produced global images of water vapour and liquid water.

The latest development in satellite-borne passive microwave radiometers is the Scanning Multichannel Microwave Radiometer (SMMR). This is carried on SEASAT-1 and Nimbus-7, both of which were launched in 1978. The SMMR is a ten-channel instrument delivering orthogonally polarized antenna temperature data at the five microwave frequencies: 6.6, 10.7, 18, 21 and 37 GHz (Fig. 1). Six conventional Dicke-type radiometers are utilized, those operating at the four lowest frequencies measure alternate polarizations during successive scans of the antenna, the others, at 37 GHz, operate continuously for each polarization. A two-point reference signal system is used, consisting of an ambient RF termination and a horn antenna viewing deep space. A switching network of latching ferrite circulators selects the appropriate polarization of cali-

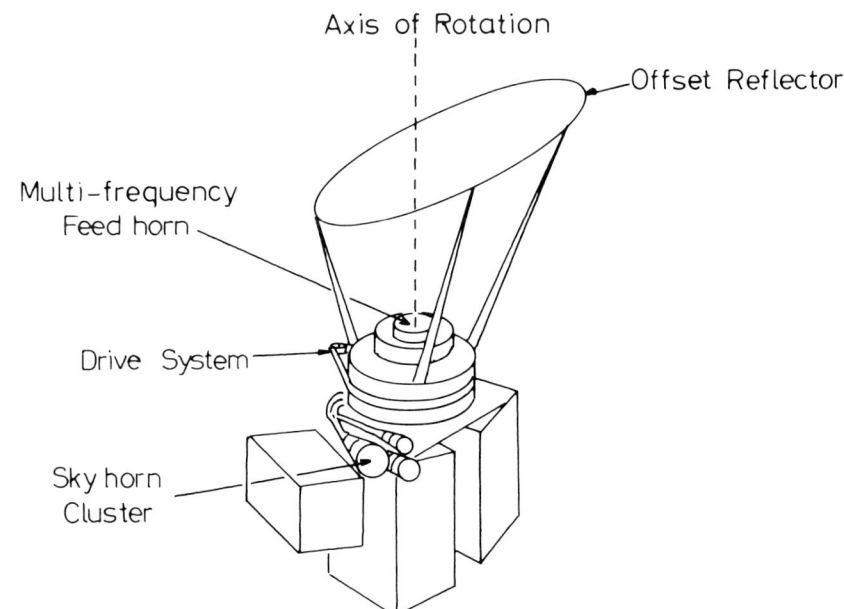

FIGURE 1: Scanning Multichannel Microwave Radiometer (SMMR).

bration input for each radiometer. The antenna is a 42° offset
parabolic reflector illuminated by a single feed horn covering the
entire range of operating frequencies and providing coaxial antenna
beams for all channels. Scanning is achieved by oscillating the
reflector about an axis coincident with the axis of the feed horn.
The instrument is installed on the spacecraft (Fig. 2) in such a
manner that this axis is parallel to the local vertical, resulting
in a conical scan pattern with the angle of incidence constant on
the surface of the earth, near 50°. In the case of Nimbus 7, the
scan-pattern is forward - viewing and scans equally to either side
of the orbital track so that the swath is centered on that track.
Further details of the SMMR design are given by Gloersen and Barath
(1977).

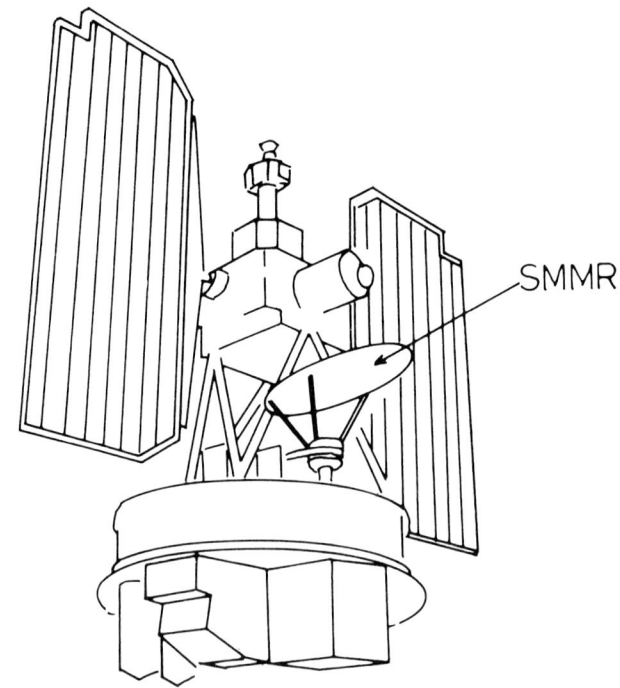

FIGURE 2: Nimbus-7 Satellite.

The SMMR allows a relatively wide swath of measurements to be
made at the earth's surface: 800 km for Nimbus 7, 600 km in the
case of SEASAT. The radiometric resolution varies from 1°K at 37

GHz down to 0.4°K at 6.6 GHz. Since all the SMMR channels utilize
a common reflector, the resolution at the earth's surface varies
with frequency. In the case of the SMMR carried on Nimbus 7, the
largest is ∿100 x 150 km at 6.6 GHz while the smallest is ∿30 x 20
km at 37 GHz.

Passive microwave radiometry is the detection of thermally
generated emission at microwave frequencies from the earth and
atmosphere. The characteristics of the received radiation (intensity, polarization, frequency variation, etc.) depend on the substance and form of the emitting surface. In the case of passive
microwave radiometry applied to oceanography, the surface - the
sea - is relatively simple compared with the land surface. The
brightness temperature of the sea surface is the product of the
emissivity and the physical temperature, the emissivity being
itself a function of physical temperature as well as windspeed and
salinity. Microwave radiometers are sensitive to salinity at
frequencies of ∿1 GHz. Due allowance has also to be made for the
overlying atmosphere particularly in the case of the higher frequency channels of the SMMR. The atmosphere acts both as an
attenuator to radiation upwelling from the sea surface and an
emitter in its own right. The total upwelling thermal radiation is
a function of the sea surface temperature, surface windspeed,
atmospheric water vapour content, cloud liquid water content, and
rainfall rate. To extract all of these parameters it is necessary
to utilize a multi-frequency instrument such as the SMMR.

An iterative algorithm has been produced for the extraction of
ocean and atmospheric parameters from SMMR measurements of brightness temperatures. An initial estimate of the required parameters
is made; this is used in a model of the ocean and atmosphere to
calculate brightness temperatures. These are then compared with
actual observations and the differences found. Differential correction is then applied to the initial guesses of the parameter
values, and the corrected values used in the model. This process
is repeated until the calculated brightness temperatures are within
the required error bounds of the measurements, yielding the
required parameter values.

SCATTEROMETRY

A scatterometer is essentially a radar that does not give range information - it measures the radar scattering coefficient of the surface being sensed.

The SEASAT Scatterometer consists of 2 pairs of orthogonal 'stick' antennae (Fig. 3). The $\frac{1}{2}$-power beam width is $0.5° \times 25°$, yielding a ground resolution of approximately 50 km. A fixed fan-beam design is employed, combined with electronic Doppler filtering. The operating frequency is 14.6 GHz. The primary viewing zone is for angles of incidence between 25° and 55°, giving measurements of windspeed and direction. Measurements are also carried out of high wind speeds in the zone 55° to 65° angle of

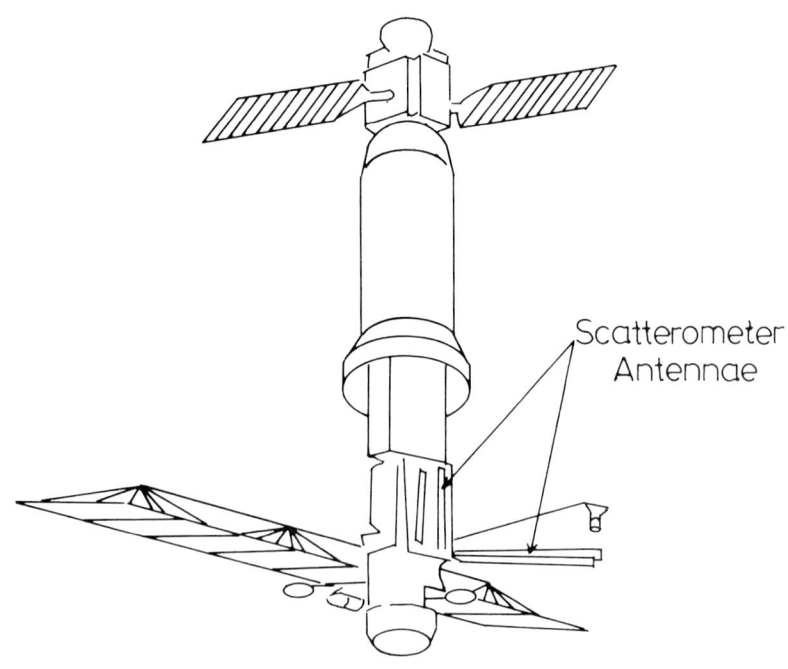

FIGURE 3:   SEASAT-1 Satellite.

incidence. Further details of the scatterometer design are given by Grantham et al. (1976 and 1977).

The Scatterometer antenna configuration allows measurements to be made within two swaths at the surface, one to each side of the subsatellite track. Each swath has a total width of 750 km: within the outer 250 km only high winds are registered. The ground resolution is 50 x 50 km. Two orthogonally placed antennae observe each ground cell, yielding values of the radar cross-section for two orthogonal views of the area. This information is particularly useful for deriving the wind direction. The objective of the SEASAT scatterometer is to measure the winds over the ocean within the swath of the instrument to an accuracy in speed of $\pm 2m.s^{-1}$ or 10% whichever is worse, and in direction to within about $\pm 20°$.

The calculation of the vector winds from the scatterometer measurements of radar backscattering cross-section depends on the relationship between the windspeed and direction and the radar cross-section. Such a function has been given by Pierson et al. (1974) and includes both theoretical knowledge and experimental measurements carried out using scatterometers carried on aircraft. Having obtained equations for radar cross section in terms of wind magnitude and direction for each incidence angle, it is possible to assemble a lookup table from which the windfield values can be extracted from the measured orthogonal radar cross section.

The algorithms produced for the SMMR and Scatterometer will be applied to data collected by Nimbus 7 and SEASAT in the near future. The oceanographic and atmospheric parameters extracted will then be compared with weather and sea condition reports collected by the Meterological Office and Oil Company platforms in the North Sea at approximately the same time and location.

Both SMMR and Scatterometer algorithms can be applied in combination in the case of SEASAT, particularly since the SMMR and Scatterometer ground tracks show a degree of overlap. This combined algorithm should have synergestic properties allowing better results for the windfields to be obtained then in the case where the measurements made using the two sensors are analyzed separately.

ALTIMETRY

A satellite-borne radar altimeter is able to make two principal measurements over the ocean. These are the absolute altitude
above the average sea surface, and the height of the ocean waves,
i.e. the sea state. These measurements are primarily of use in
geodesy and oceanography, and for offshore design purposes.

It is assumed that the mean ocean surface lies on a surface of
equal geopotential, so that determination of this surface enables
the geoid to be defined. This can be done by combining the measurements of absolute altitude with accurate orbit determination.
Of course many factors other than the shape of the geoid affect the
position of the sea-surface. Among these are tides, currents and
winds. Repeated measurements of the same area enable these components to be resolved.

Waveheight information is useful in its own right, without
further analysis. However, proper analysis of the altimeter return
pulse enables the waveheight distribution function over the footprint to be determined.

THEORY OF OPERATION

The radar altimeter is a pulsed radar operating in a pulse-
width limited mode. This means that the pulse is sufficiently
short so that not all the target is illuminated simultaneously.
This is illustrated in Figure 4. Here we are assuming that the
sea-surface is flat, and we introduce the fiction that the pulse
penetrates the sea-surface, for illustrative purposes. We also
assume that the surface is rough on the scale of the wavelength
employed; this is not unreasonable.

The illuminated area grows rapidly as the pulse intersects the
surface, and so therefore does the returned power (which is proportional to the illuminated area, for a rough surface and within the
antenna beamwidth). When the illuminated area becomes annular
however, the area (and hence returned power) remains constant as
the annulus expands. This illumination of successive annuli at

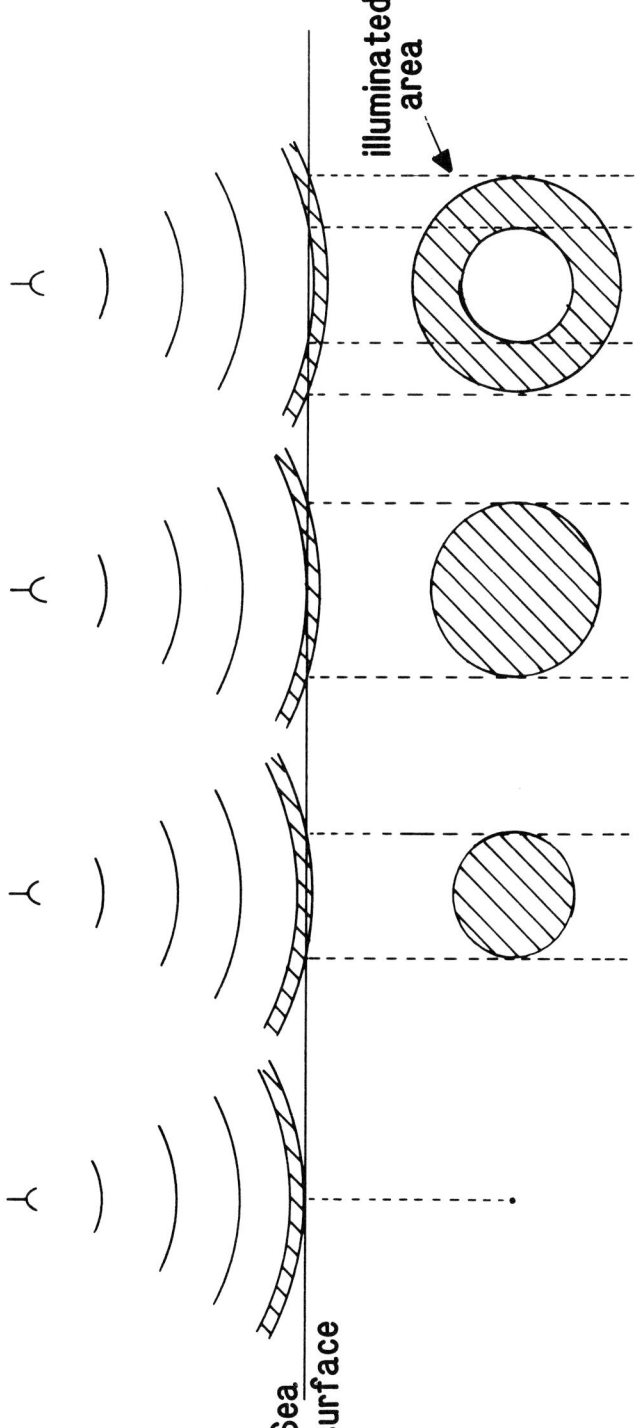

FIGURE 4: Intersection between the radar pulse and the sea surface.

different times is characteristic of the pulse-width limited mode as we defined it above.

Eventually the annulus expands beyond the footprint defined by the antenna beamwidth, and the returned power starts to decrease, as the gain of the antenna falls off.

The average return therefore shows a sharp rise, a plateau, and finally a long fall-off. This is illustrated in Figure 5, together with a typical individual return. In this diagram the time scale is non-linear; the initial rise is very short compared to the plateau or the fall-off. The power in an individual return fluctuates due to interference between parts of the wavefront reflected from different points on the ocean surface (for a surface which is not now flat). Note that a number (of the order of 100) of individual pulses must be combined to form the average pulse. This does not degrade the spatial resolution because the pulse repetition frequency is of the order of 1000/sec.

In the case of reflection from a surface which has waves of height greater than the width of the incident pulse, the return pulse becomes stretched. This is intuitively obvious and is shown in Figure 6. The effect of this on the average return is to introduce an additional slope on the leading edge, which is a function of the ocean-waveheight. This is shown in Figure 7 as a function of waveheight.

By differentiating this leading edge the waveheight distribution function may be obtained; however, an estimate of waveheight may be found by simply evaluating the slope of the linear part. The position (in time) of the mid-point of the leading edge is invariant with waveheight, and defines the return time from the average sea-surface.

The leading edge of the return pulse contains the altitude and sea-state information, as we have described, and it may therefore be examined in detail at the expense of the rest of the return pulse, which may be discarded. The measurement footprint is therefore much smaller than the footprint defined by the antenna beamwidth.

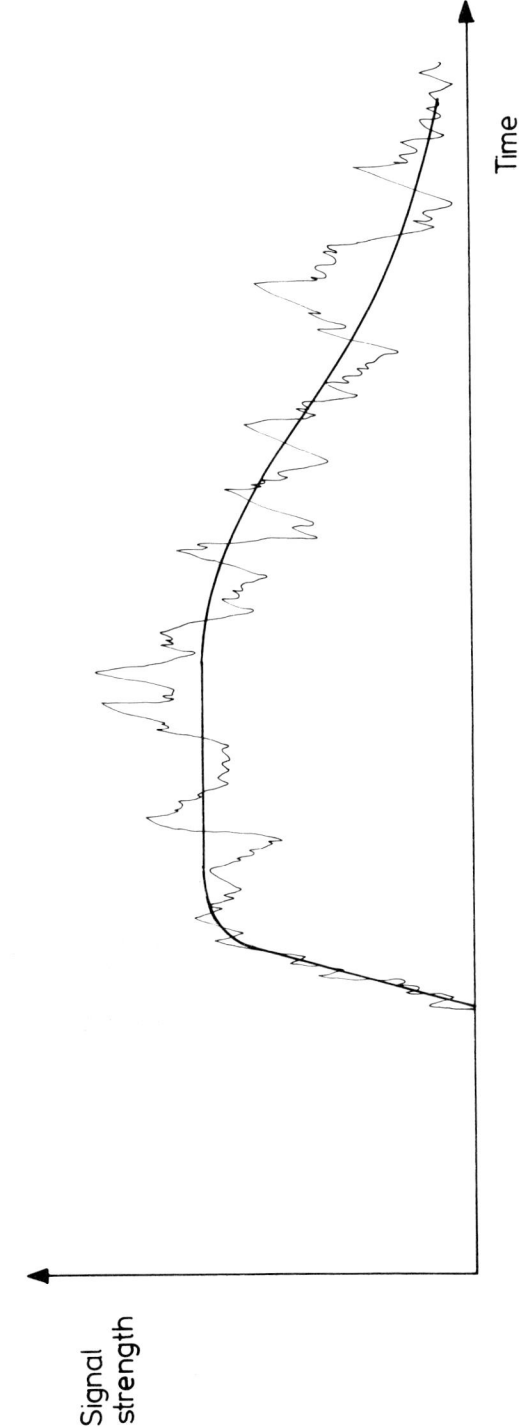

FIGURE 5: Average return pulse (heavy line) and an individual return pulse (light line).

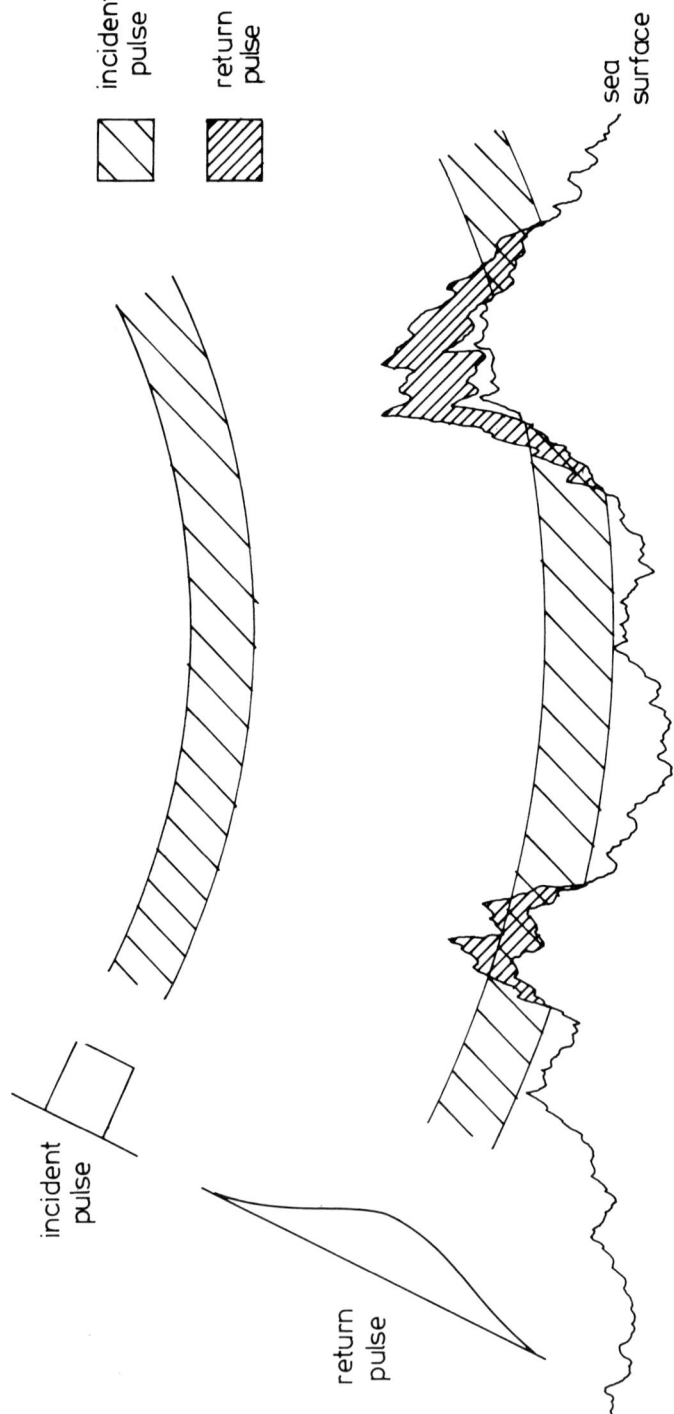

FIGURE 6: A short incident pulse becomes stretched on reflection from a rough sea-surface.

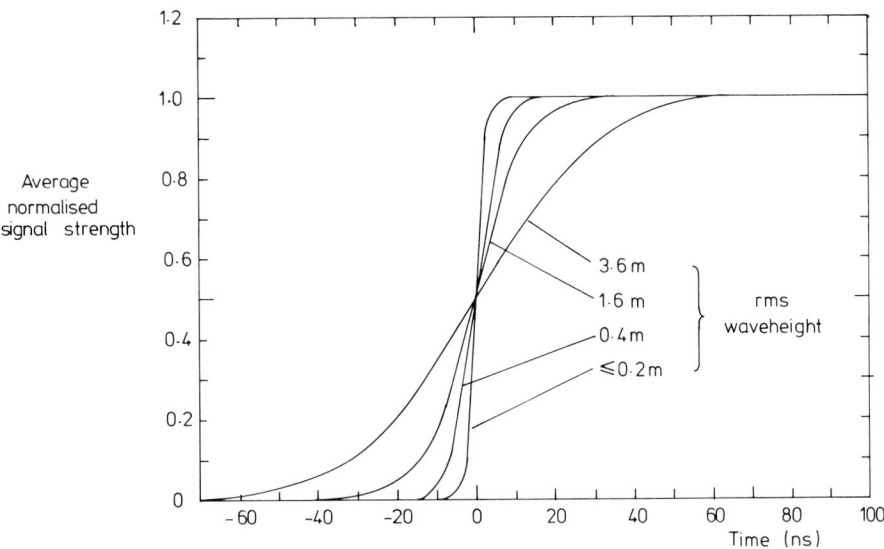

FIGURE 7:   Average normalized return signal as a function of rms waveheight.

CURRENT INSTRUMENTS

Three radar altimeters have been flown to date: the S-193 experiment on Skylab, and instruments on GEOS-3, (Geodynamics Experimental Ocean Satellite) and SEASAT-1. It is proposed that the National Oceanic Satellite System (NOSS) should also carry an altimeter. Some of the characteristics of these instruments are shown in Table 1.

There are four options for the NOSS instrument. Option A is similar to the SEASAT-1 instrument. Option B uses a single antenna with multiple feeds to produce a nadir beam plus two extra beams at 25km separation either along or across the track. Option C uses two dishes in an interferometric mode to produce a nadir beam and two side beams at 50km each off nadir, while option D uses three dishes to achieve the same separation along the track as well as across, by the interferometric method, in addition to the nadir beam.

A rough determination of the geoid was established by GEOS-3, but this spacecraft has now failed. The precision available with

TABLE 1

	Skylab	GEOS-3	Seasat-1	NOSS-A	Option B	Option C	Option D
Number of beams	1	1	1	1	1 + 2	1 + 2	1 + 4
Antenna diameter	1.1 m		1 m	1 m	1 m	2x1.1→1.8m	3x1.1→1.8m
Footprint diameter	8km	3.6km	> 1.7km	> 1.7km	> 1.7km +	> 1.7km +	> 1.7km +
Beamwidth	1.5°	2.6°	1.6°				
Pointing Accuracy		1°	0.2°				
Pulsewidth (uncompressed)	100 ns	1μs	3.2μs				
Pulsewidth (compressed)	-	12.5ns	3.125ns				
Repetition frequency	250 Hz	100 Hz	1020 Hz				
Frequency	13.9GHz	13.9GHz	13.5GHz				
Peak RF Power	2 kW	2 kW	2 kW				
Average RF Power	0.05W	0.24W	6.5W				
DC Input Power			164 W	168 W	168 W	168 W	168 W
Electronics, Volume			0.75 m^3	0.4m^3	0.4m^3	0.4m^3	0.4m^3
Electronics, Mass			93.8 Kg	164 Kg	170 Kg	200 Kg	220 Kg
Data Rate	10kb(max)		8.5 kb	8.5 kb	17 kb	20 kb	20 kb
Altitude Precision	< 1m	< 50cm	< 10cm	< 10cm	7cm + 35cm	7cm + 18cm	7cm + 18cm

the SEASAT-1 instrument would allow a good determination of the geoid, but limited coverage was achieved before the spacecraft failure. The precision available with this instrument would also have allowed useful oceanographic work to be done.

There are no satellite radar altimeters working in orbit at present.

REFERENCES

Gloersen, P. and F. T. Barath. 1977. IEEE Journ. Ocean Eng., OE-2,2,172.

Grantham, W. L., E. M. Bracalente and W. L. Jones. 1976. Oceans '76, MTS-IEEE, 10D 1.

Grantham, W. L., E. M. Bracalente, W. L. Jones and J. W. Johnson. 1977. IEEE Journ. Ocean. Eng. OE-2,2,200.

Pierson, W. J., V. J. Cardone and J. A. Greenwood. 1974. The Applications of SEASAT-A to Meteorology. The University Institute of Oceanography of the City University of New York.

An Objective Commentary of the Current Use of Satellite Data for Ocean Applications

James J. Gallagher

ABSTRACT

A commentary on some of the critical problems impeding use of quantified satellite imagery data is offered. This discussion is directed principally to uses of satellite data for open ocean water mass delineation and characterization; i.e., ocean fronts and eddies. Further, since the bulk of the experience in oceanography has been with visible and infrared imagery the discussion focuses on these aspects; however, similar problem areas associated with microwave sensor systems are addressed.

The discussion distinguishes between global and the regional users, their scope, and their requirements. The need for coordination of meterological and oceanographic satellite data analyses, and of on-going oceanographic and marine meterological in situ data acquisition and analyses is stressed. Lastly, the problems associated with image data processing requirements and capabilities at

small or remote oceanographically oriented laboratories are briefly
discussed.

INTRODUCTION

During the last five years, the oceanographic community has
been exposed to the unique satellite infrared pictures revealing
dynamic ocean surface features. Certainly, this synoptic imagery
information has been instructive, stimulating, and has provided new
insights into ocean dynamics. As a consequence, many oceanographers developed a general knowledge of satellite sensor technology and data products. Historically, the bulk of these photographic products were provided by the National Oceanic and
Atmospheric Administration's National Environmental Satellite
Service (NOAA-NESS). These products were enhanced non-linearly,
and while they contained valuable qualitative reconnaissance
information they appeared to have little appeal as oceanographic
data. Further, oceanographers soon learned to be cautious, perhaps
even skeptical, about the value of these pictures; for the data to
be of scientific and operational value it must be quantifiable and
reliable. Recently, digital magnetic tapes of the imagery data
have been made available by NOAA's Satellite Data Service (SDS).
However, the availability of, and experience with, digital computer
interactive image processing techniques at most marine user facilities has been largely non-existent.

Two factors served to maintain a low level of perceived reliability of satellite data for ocean application. First, until
SEASAT-A was launched more attention and fiscal support were
usually paid to atmospheric and terrestrial sensor developments
than to evaluation of existing sensor systems to meet ocean
requirements. Secondly, there has been a general lack of concur-
rent, integrated meterological analyses to remove interference
aspects of the atmospheric column for ocean surface analysis.
Until very recently, assessments of how well atmospheric corrected,
quantitative satellite derived ocean surface data could characterize in situ surface and related subsurface ocean conditions, and

under what limiting conditions, had received inadequate attention; of course, atmospheric features, not oceanic features, were of primary interest in most satellite related programs.

To date, most of the on-going ocean user activities have been generated piecemeal, and resulted from personal interests and initiatives rather than as a result of top-down direction within long-term structured programs. The exception to this situation is the exponential growth in the awareness of the applications potential of satellite technology that resulted from the organized focus on the SEASAT-A program. But, we are still rather low on the learning curve, and even in the SEASAT-A program could not readily apply an advanced knowledge from quantitative infrared data experiences to microwave systems analyses. There is much research and development work to be done before quantitative satellite data can be used with high confidence in an operational mode.

DISCUSSION

The following discussion is directed primarily to open ocean water mass delineations and characterizations using satellite technology. In particular, how well can satellite data be used to locate, track, and characterize ocean fronts and eddies, and circulation patterns? The discussion largely addresses infrared and visible imagery data since these data represent the bulk of our experience; however, similar considerations for emerging microwave and laser sensors data use are strongly implied. With respect to infrared, we are asking some of the same questions today that were asked ten years ago; i.e., how well can we discern ocean features relative to atmospheric features, particularly boundary layer atmospherics, and, how well does the satellite observed radiated expressions of the upper ten micrometers (microns) reveal subsurface structures, at least within the seasonal thermocline?

It is commonly assumed that the major categories of current ocean users of satellite data are Research and Development, and Operational. I contend that the real division is between institutional Global and Regional interests; Research and Development and

Operational users are found in both Global and Regional categories. Global here is defined as attempting to provide climatological data to characterize the world's oceans synoptically; unfortunately, in most large institutions that are responsible for satellite data handling and use operational means global. In order to satisfy the attendant requirements of global data using a reasonable amount of data, the incoming satellite and in situ data tend to be averaged over fairly large grid areas; i.e., on the order of 60 to 100 km^2, to provide climatological data. The regional user is generally interested in viewing an area of the ocean less than 1000 km^2, equivalent to about two days steaming time, and is interested in using high spatial resolution data; i.e., full pixel resolution. To federal program managers these two interests have been incompatible in terms of data volume, storage, handling, and processing requirements.

Can central or regional satellite receiving and processing facilities adequately support multiple marine users having diverse applications, interests, and requirements? While that is a major multifaceted problem, and beyond the scope of this paper, it should be noted that until very recently the central facilities were operationally oriented and essentially not receptive to marine regional users' needs, particularly R & D needs. Regional users are becoming more active, and are attempting to pool resources to satisfy their data needs; i.e., Scripps Institution of Oceanography, U.S.A.; University of Dundee, Scotland; Centre d'Etude Meterologie Spatiale, France, and the U.S. Northeast Satellite Users Consortium, USA. Regional users are becoming more active, and are attempting to pool resources to satisfy their data needs, i.e., Scripps Institution of Oceanography (USA), University of Dundee (Scotland), Centre d'Etude de Meteorologie Spatiale (France), and the U. S. Northeast Satellite Users Consortium (USA). Processing of satellite magnetic tape data is less well developed in the ocean community because many small (field) ocean laboratories are generally incapable of, or perhaps cannot afford to purchase and maintain the necessary equipment and software, digital processing of imagery data, particularly fisheries related users.

Among the questions frequently posed by potential ocean-related users are: What is the minimum size computer system required to view an area of 1000 km^2 or less, with individual pixel resolution? What associated peripheral equipment is necessary? What are typical maintenance costs? To what extent is on-site software development required? If a direct-readout is not possible, can such user sites expect to receive calibrated tape data within 48 hours of the satellite pass? On this last note it should be recognized that many scientists are wary of receiving fully pro-cessed "final" data, particularly when there are elements of inter-pretation involved in producing the final data set to the user. I maintain that the largest potential user market, in terms of numbers of users, is the Regional user category. It is fair to say that the type of data that the regional Research and Development user requires has not been readily available to him.

Consider the concerns of fisheries, offshore petroleum, or tactical Navy operators, or, technical decision makers who must assess the relative technical merits and cost effectiveness of a major commitment to the use of satellite technology. For the Regional users two questions immediately come to mind:

1. How well can synoptic quantitative satellite data be used to improve data input to dynamic oceanography/acoustic prediction models; i.e., time and space varying condi-tions.

2. Will the use of synoptic satellite data, (augmenting sparse data coverage, including remotely relayed in situ data) significantly influence technical management deci-sion making, relative to the combined use of climat-ological data and present standard data acquisition methods?

There is little accrued experimental evidence to provide definitive answers. With respect to the above two questions there are several implied considerations to be addressed by the oceano-graphic community:

1. Determine the repeatability of accurate relative (and absolute) satellite-derived values of ocean surface

properties for specific geographic areas. Determine the limiting environmental conditions for achieving required accuracies; in the vicinity of ocean fronts and eddies even relative gradients may be suspect because of atmospheric variability in the marine boundary layer.

2.   Determine the degree of reliability of interpreting critical in-water vertical profiles from satellite imagery. Correlate surface values with near-surface values, or those within the seasonal thermocline. Surface layer thicknesses associated with radiative sea surface temperature (SST) values must be defined. When strong, near-surface vertical gradients are present, the in situ SST values obtained are largely a function of the measurement technique used. Measurement techniques used for comparison with satellite data should be evaluated and standardized.

3.   With respect to real time ($\leq$ 24 hours) satellite data user requirements: When does the Research and Development community really need calibrated and corrected (atmospheric and geographic) quantified satellite data, in what forms are these data to be disseminated to remote institutional users or research vessels, and how well can these data be used directly by most resource limited remote users? Are requirements for frequent quick-look or prompt update information more or less stringent for Regional Operational and Research and Development users? How acceptable are satellite derived graphic data that are telemetered via existing facsimile methods such as enhanced but uncalibrated qualitative gray shade images, annotated line drawings, automated alpha-numeric or contour charts, etc?

   The needs and current deficiencies of satellite related ocean programs, from a regional users' standpoint, are outlined in Figure 1.

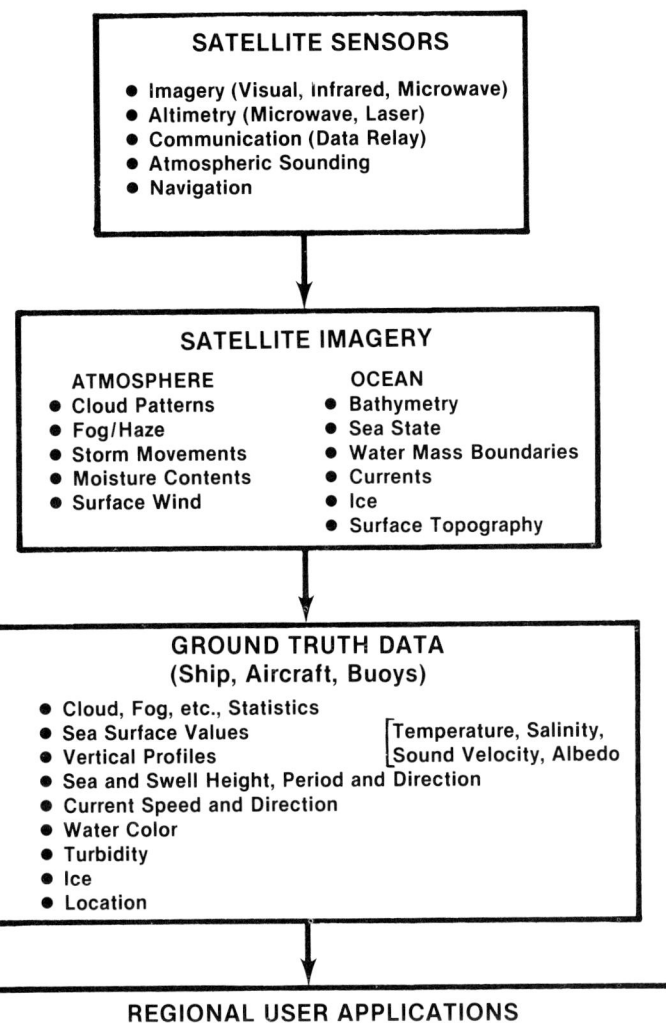

FIGURE 1

Sensor Box

Cloud cover continues to plague ocean experiments using satellite infrared imagery. The "footprints" of atmospheric sounder sensors are too large for valid corrections of high resolution infrared data in dynamic ocean areas. Use of all-weather microwave sensors for accurately monitoring water mass boundaries produces some different problems. For example, the Scanning Multifrequency Microwave Radiometers (SMMR) spatial resolution of approximately 100 km precludes accurate detection of frontal boundaries, particularly in coastal areas. The Synthetic Aperature Radar (SAR) data look intriguing but analyses and interpretation evaluations are complex and incomplete. Radar altimetery provides the most immediate promise, however, it is not an imaging system, and perhaps, should be used in combination with infrared and visual imagery to establish a data base for water mass boundary detection. It should be noted that there has been very little interdisciplinary intercourse among various satellite sensor data analysts; i.e., between infrared and microwave analysts for the same ocean area. The notable exceptions to this statement are the SEASAT-A Ground-Truth experiments and the coincident JASIN experiment. Increased emphasis must be directed to combining analyses from various satellite sensors, such that the best features of each data set being used to complement each other in analyses of the data.

Imagery and Ground-Truth Boxes

A tri-component data set represented by atmospheric imagery, ocean imagery, and ground-truth data is the critical focus of Figure 1, and of this paper. There is an immediate need to coordinate these three ingredients in the ocean application of satellite data if the necessary reliable satellite data bases, under varying ocean environmental conditions, are to be established. Satellite related meterological analyses over ocean areas cannot continue to be conducted totally independent of corresponding oceanographic analyses in the same geographic areas. Similarly, satellite derived ocean data should not be analyzed without regard to the transmission through the atmospheric column. Commonly, satellite derived ocean data analyses are conducted independent of on-going at-sea measurements in coincident geographic areas. If, indeed, creation of coincident time and space satellite data and ground-truth data sets are essential to evaluate the stand-alone utility of satellite data, and they are, then the above deficiencies reflect a woeful lack of planning and control at management levels. If valid standardized satellite derived ocean data bases are to be established, the three components of this data set must be working in unison and toward mutual goals. It is important to recognize that new programs, and new money are not necessarily required, but coordination of on-going programs is absolutely necessary.

In situ data obtained at remote, and even unmanned, platforms (ships, buoys, aircraft, etc.) can be relayed between these platforms and shore based analyses facilities via meterological satellite data communications channels. These channels are currently underutilized.

Aircraft used as multi-sensor, quasi-synoptic remote sensing and ground-truth data collection platforms, in conjunction with ships and buoys, provide rapid spatial coverage necessary for satellite data verification in dynamic ocean areas. Rejuvenation of aircraft oceanography deserves consideration, particularly within 400 km of shore.

Applications Box

The importance of the applications box in Figure 1 is the commonality of requirements of diverse Regional users; i.e., spatial and parametric resolutions, surface to subsurface correlations, and atmospheric corrections.

An idealized experimental design configuration is shown in Figure 2. It emphasizes integration and coordination of available existing data collection resources, provides for the data correction and verification necessary for satellite data applications evaluations, and, builds a data base for continued sensor developments. Implementation of such an approach in conjunction with increased capabilities in computer interactive imagery processing by local and Regional ocean oriented users, and, in high density data telemetry, should provide the critical path to standardization

and increased utilization of quantitative satellite data in the
marine community.

**INTEGRATED OCEAN DATA
COLLECTION AND COMMUNICATION**

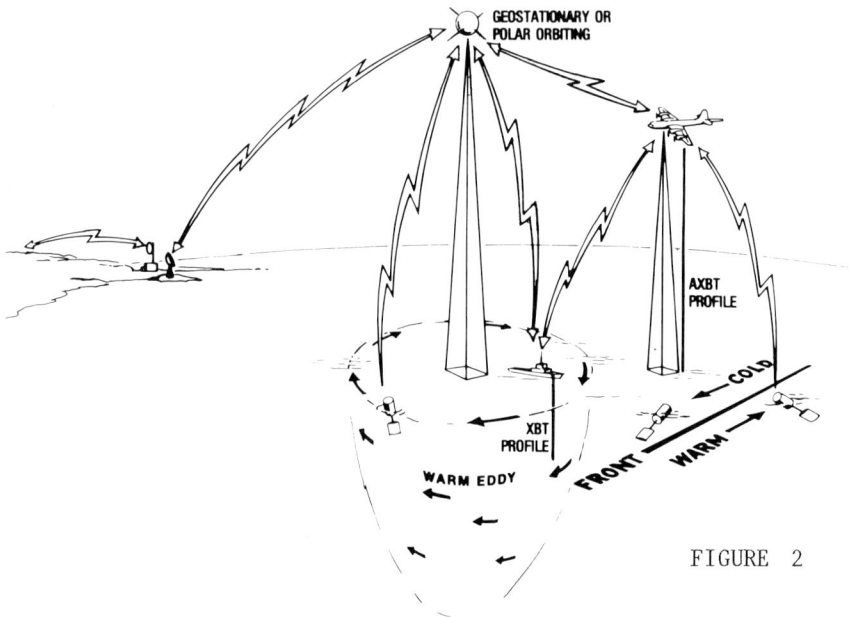

FIGURE 2

Acoustic Characterization of Sediments

Dr. N.G. Pace

INTRODUCTION

The final objective of the work described in this paper is the classification of sedimentary sea bottoms using remote, water-bourne, acoustic techniques. Such classifications may be of direct use in the present context as ground truth data in certain specialized remote sensing methods: for example three techniques used from aircraft which in some instances can provide sea-bed information are laser and acoustic sounding and synthetic aperture radar.

A background to the properties of sediments relevant in the acoustic context is first presented. This is followed by a brief account of work in progress on a normal incidence broad-band approach to small area classification. Finally, the main part of this paper reports in detail a study into the possible use of the ubiquitous side scan sonar method in a new role as an objective sedimentary sea bottom classifier.

ACOUSTIC PROPERTIES OF SEDIMENTS

Apart from geometrical spreading losses, three attenuation mechanisms are acknowledged (McLeroy and DeLoach, 1968; McCann and McCann, 1969; Hamilton, 1972; Thomas and Pace, 1980) for sound propagation in unconsolidated, non-cohesive sediments, namely Rayleigh scattering, viscous (or particle-fluid) losses and solid friction (or particle-particle) losses. Rayleigh scattering varies as the fourth power of frequency and is significant when the wavelength and particle sizes are comparable. Recent experimental measures of the attenuation as a continuous function of frequency (Thomas and Pace, 1980) in non-cohesive sediments together with a large body of data at spot frequencies (McLeroy and DeLoach, 1968; Smith, 1973) demonstrate a linear variation of sound attenuation with frequency, thus indicating the predominance of the solid friction losses over those due to viscous losses the latter requiring a square root dependence on frequency (McCann and McCann, 1969). An implication of this is the existence of a matrix formed by sediment particles in contact. Such a matrix provides what is termed the frame bulk modulus as a component of the total bulk modulus of the saturated sediment and can allow the sediment to support shear stresses.

The reflection of sound waves at a plane water/sediment interface depends on the contrast between the sound velocity-density product of the water and sediment. If, in addition to the interface being plane, the sound wavelength is many times greater than the sediment mean particle size, the reflection coefficient is independent of the sound wave frequency. As there are strong inter-relationships (Hamilton, 1975; Smith, 1975) between sound velocity, density, sediment particle size and porosity, such reflectivity measurements can provide information valuable in a sediment classification scheme (Breslau, 1965). However in practice the water/sediment interface is not plane and bottom topography plays a role, the significance of which increases as the scale of the topography and the acoustic wavelength approach each other. The problem then becomes one of relating statistics of the return signal to sediment properties (Breslau, 1965, 1967; MacIsaac and Dunsinger, 1978; Simpkin, 1978).

NORMAL INCIDENCE MEASUREMENTS

An underway experimental program at Bath University which is supported by the Procurement Executive of the Ministry of Defence employs wide bandwidth acoustic pulses confined to narrow beam-widths for gathering data which is evaluated in the context of sedimentary sea bottom classification. The acoustic pulses are generated by the self-demodulation (Berktay, 1965; Ceen and Pace, 1979) in water of Gaussian modulated carrier ($\sim$ 1MHz) waves and typically have Q-factors around unity centered on 50kHz and 3db energy beamwidths of $\pm 3°$. As the area of insonification is small, the bottom topography can, to a large extent, be determined by sediment particle size, in such cases a measure of bottom roughness is the ratio of sediment mean particle size to sound wavelength (d/λ). If (d/λ) $\ll$ 1, then the interface appears smooth whilst if (d/λ) $>$ 1 the interface appears rough. The amplitude of the signal returned from plane sea bottoms at a particular frequency is normally distributed for (d/λ) small and tends towards a Rayleigh distribution as (d/λ) increases (Beckmann and Spizzichina, 1963; Medwin, 1970). The ratio (m/σ) of the mean signal level to its (variance)$^{\frac{1}{2}}$ is usually large for a normal distribution but has a value of 1.92 for a Rayleigh distribution. The rate of change of (m/σ) with frequency, obtained from the returns of low-Q pulses incident on water/sediment interfaces, shows promise as a correlate of sediment particle size in the region where (d/λ) $\leq$ 0.25 (Pace, 1978; Williams, 1978).

In the regime (d/λ) $>$ 0.25, measures of the time spreading of the return of the short time duration acoustic pulse incident on water/sediment interfaces can be correlated with the sediment particle size (Pace, 1978).

However, although this normal incidence approach to sediment classification using low-Q, narrow beamwidth pulses shows promise, it has a number of limitations one of which is the small rate of area coverage possible.

SIDE SCAN SONAR

With an eye on the rapid area coverage possible (13 sq km per hour) with side scan sonar, the remaining part of this paper examines the usefulness of such a technique for quantitative sediment classification. In the side scan method (Chesterman and Heaton, 1977) use is made of the backscattering, at low grazing angles of incidence to the sea bed, of acoustic waves radiated in a fan shaped beam from a transducer contained in a towed body. The fan beam is oriented so that the wider angle is in the vertical plane and the narrower angle in the horizontal plane with the axis of the beam normal to the ship's track. The display of the returns is usually realized as a series of closely spaced intensity modulated lines on a paper recorder (see for example Fig. 1). In this way a two dimensional picture of the sea bed is built up. The interpretation of side scan sonar records by experienced observers can supply information on the nature of the sea bottom in terms of geological features, bottom topography, and sediment types. However, it is in the latter category that the most subjective opinions are offered and it is here that attention is focused. For the purpose of this study sea bed types are split initially into two categories: grossly flat sedimentary sea bottoms on the one hand and on the other hand geological and structured sea bottoms. Such features as sand waves will for example be described as a structured sea bottom.

The acoustic return from a grossly flat sedimentary sea bottom due to an acoustic pulse incident at a low grazing angle depends on the bottom roughness, and thus in part on sediment particle size (Wong and Chesterman, 1968). Hence the distribution of signal levels over an area of side scan sonar record will contain information on the statistics of the bottom roughness which in turn gives to the usual side scan sonar display, the appearance of texture.

By drawing on methods of texture quantification (Harlick et al., 1973), this paper explores the potential of using a number of textural features to characterize small areas of side scan sonar record with the view of producing, via a pattern recognition pro-

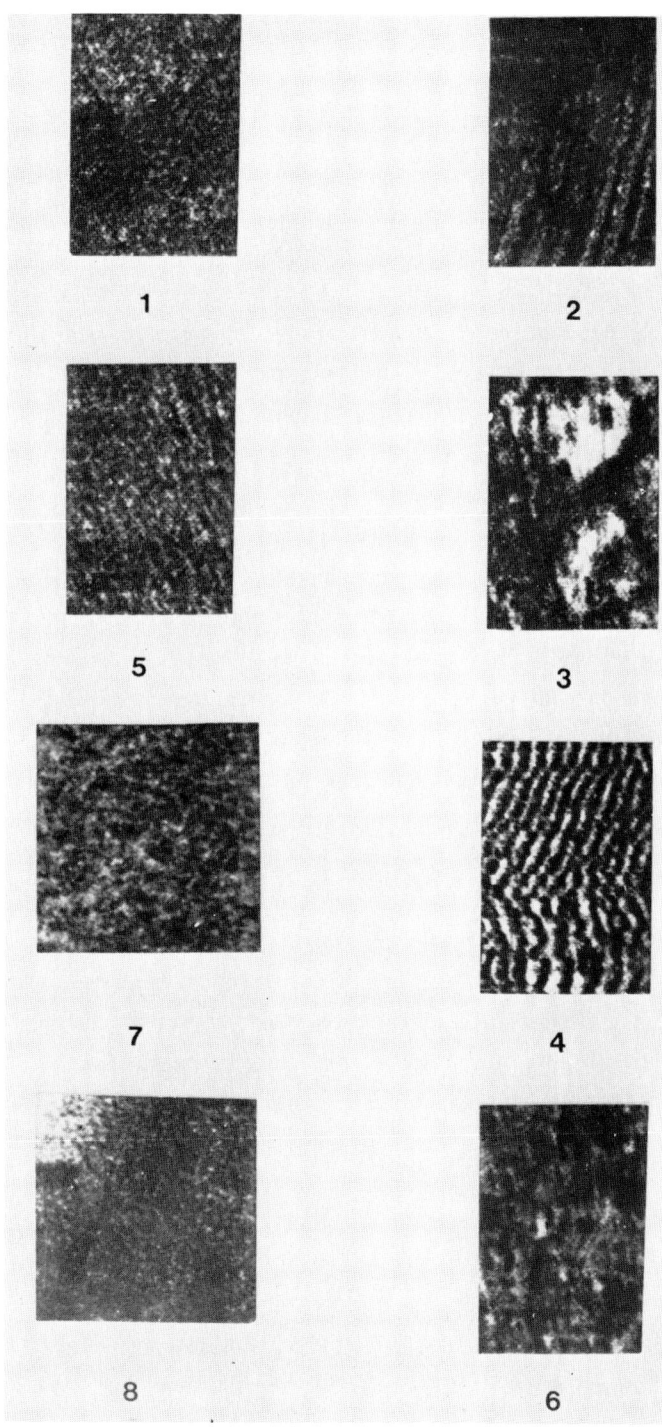

FIGURE 1: Side scan sonar records of eight different areas of sea bottom. The area of sea bottom involved is about 100 x 240 meters in all cases. Areas 1, 5, 7, 8 collectively form class A (see text) while areas 2, 3, 4, 6 collectively form class B.

cedure, classification decisions in the context of sedimentary sea
bottom identification.

Overview of Method

Side scan sonar techniques have been in use at Bath University
for many years and a library of side scan records stored on magnetic tape has been built up. The particular side scan system of
interest here has been described in detail elsewhere (Chesterman,
1974).

The principal parameters of the system are a transmitted pulse
length of approximately 1 m/sec at a 48 kHz carrier frequency. The
return signals are stored in analogue form on magnetic tape by
frequency modulation having a nominal 2 kHz bandwidth and a 48db
dynamic range.

Two sets of four areas of sea bed as detailed in Table 1 were
chosen from the library as being representative of a range of, on
the one hand, grossly flat sedimentary sea bottoms and, on the
other hand, of geological and structured sea bottoms. Side scan
records from each of the eight areas are shown in Figure 1, each
record represents an area of sea bottom approximately 100 m x 240
m. After digitization and some pre-processing, each of the eight
records were divided into twenty sub-areas corresponding to a sea
bed area of approximately 60 m x 20 m. Each of the 160 sub-areas
consisted of 50 consecutive samples taken from the same time window
in each of the returns of 20 successive transmissions. The pre-
processing was done because in practice when side scan sonar
surveys are being undertaken, subjective adjustments of both the
received signal level and the details of the time-varied gain are
made from time to time. If it is assumed however, that any given
settings of the equipment give rise to side scan records which are
monotonic transformations of records which could have been obtained
for different settings, then record normalization by equal probability quantization guarantees that the same quantized record will
be produced irrespective of the settings (Harlick et al., 1973).
All 160 sub-areas were equal probability quantized to 16 levels
thus ensuring that relative differences between sonar records not
originating in sea bottom differences were minimized. It should be

TABLE 1: Location and interpretation of the side scan records in Figure 1.

| AREA | POSITION | | INTERPRETATION |
	LATITUDE	LONGITUDE	
1	50° 16.8'N	0° 33.8'E	40% coarse sand 60% fine gravel
2	43° 28.8'N	3° 4.4'W	Rock ridges
3	61° 54.3'N	6° 42.2'W	Rock explosives
4	50° 18.7'N	3° 35.3'W	Sand waves
5	50° 21.2'N	0° 8.8'W	40% fine gravel 60% coarse sand
6	50° 46.5'N	4° 49.7'W	Rock
7	54° 59.3'N	7° 46.8'E	Coarse sand, with 20% small stones
8	54° 55.2'N	7° 6.6'E	Coarse sand, with 10% gravel and 10% broken shell

noted that absolute signal level information is absent from the equal probability quantized signal.

The quantification of the texture in each sub-area is equivalent to a data reduction procedure in that the texture generated by the 1000 numbers is represented by only N numbers termed collectively the feature vector. Using simple pattern recognition techniques on the 160, N-dimensional feature vectors the usefulness of some textural features for sediment classification were examined.

The abstraction of textural features from two dimensional digital arrays has seen an increased interest in recent years. A number of such features are becoming well known (Harlick et al., 1973) but as yet there is no obvious method for selecting optimum textural features for particular tasks. Eleven textural features are employed in this paper, and full details of their definition can be found in the reference Pace and Dyer, 1979.

Classification Experiments

For this initial study of sedimentary sea bottom classification, the following simple decision rule was employed. If d_i (X)
represents the Euclidean distance between a feature vector
$X(x_1, x_2, x_3 \cdots, x_n)$ and the mean feature vector M_i $(m_1, m_2,$
$m_3, \ldots, m_n)$ for class i, then the feature vector X is classified
into the ith class when d_i (X) $\leq d_j$ (X) for all j. A contingency
table of true class versus assigned class according to the above
decision rule for the eight class situation is shown in Table 2a.

TABLE 2a: Contingency table of true class versus assigned class
for an eight class situation.

ASSIGNED CLASS

		1	2	3	4	5	6	7	8
	1	11	0	0	0	7	0	2	0
T	2	1	14	2	0	0	0	0	3
R	3	1	6	10	0	0	1	0	2
U E	4	0	0	0	19	0	1	0	0
C	5	3	1	0	0	12	0	3	1
L A	6	0	2	0	6	0	12	0	0
S S	7	1	0	0	0	2	0	16	1
	8	2	1	0	0	2	0	3	12

A prerequisite for the present approach to sedimentary sea
bottom classification is to be able to decide which side scan sonar
records represent sedimentary bottoms and which do not. The first
test grouping of the 160 sub-areas was thus into two classes representing on the one hand grossly flat sedimentary sea beds (Class A)
and on the other hand structured or geological sea beds (Class B).
The probability of correctly classifying feature vectors representing sub-areas in this, two class situation, is given in the
first column of Table 2b.

TABLE 2b: Correct classification probabilities. Results are given
for three different groupings of the eight sea-bottom areas.

	Correct Classification Probabilities Achieved	
A,B	1,5,7,8	1,5,7,8,B
94	65	78

Of those sub-areas classified correctly as belonging to class
A, how many can be correctly placed into one of their four classes
of origin (1, 5, 7 or 8)? This question is answered in the second
column of Table 2b where it is seen that a 65% probability of
correct assignment is achieved.

Classification results for a grouping of the original eight
classes into five classes, namely 1, 5, 7, 8, B provides an overall
picture of the success with which both the initial discrimination
between class A and class B is achieved, together with the all
important detailed discrimination within class A; a correct classification probability in this context of 78% is obtained.

SUMMARY

A brief introduction to the acoustic properties of non-
cohesive marine sediments was followed by a mention of some on-
going normal incidence studies designed to aid the remote acoustic
classification of sediments.

A report on a study of the potential use of side scan sonar
for the rapid classification of areas of sedimentary sea bottoms
showed that considerable success can be achieved with a simple
approach. That fairly high levels of correct classification probabilities were achieved using small numbers of features derived
from relative signal level variations rather than absolute ones
together with an uncomplicated decision rule is an indication of
the performance that could be expected in a more sophisticated
implementation of the method.

REFERENCES

Beckman, P. and Spizzichino. 1963. The scattering of electro-
    magnetic waves from rough surfaces. New York, Pergamon.
Berktay, H. O. 1965. Possible exploitation of non-linear acous-
    tics in underwater transmitting applications. J. Sound Vib.
    2, 435-461.
Breslau, L. R. 1965. Classification of sea floor sediments with a
    shipborne acoustic system. La Petrole et la Mer No. 132,
    Woods Hole Oceanographic Inst. Contribution 1678, pp. 1-9.
Breslau, L. R. 1967. The normally incident reflectivity of the
    sea floor at 12 kHz and its occurrence with physical and
    geological properties of naturally occurring sediments. Woods
    Hold Oceanographic Inst. Ref 67-16.
Ceen, R. V. and N. G. Pace. 1979. Acoustic signals in marine
    sediments due to water-borne parametric arrays. Proc. Inst.
    Acoustics, Underwater Applications of Non-Linear Acoustics,
    University of Bath, England, 10/11th September.
Chesterman, W. D. 1974. The Ocean floor. Contemp. Phys. Vol. 15,
    501-516.
Chesterman, W. D. and M. Heaton. 1977. Sonar studies of the
    continental shelf. Proc. Ultrasonics Internat., Hotel
    Metrople, Brighton, pp. 1-18.
Hamilton, E. L. 1972. Compressional wave attenuation in marine
    sediments. Geophysics 37, 620-646.
Hamilton, E. L. 1975. Acoustic properties of the sea floor: a
    review. Oceanic Acoustic Modelling, Pt. 4, Conference
    Proceedings No. 17. Saclant ASW Res. Centre, La Spezia,
    Italy.
Harlick, R. M. and K. Shanmugam. 1973. Computer classification of
    reservoir sandstones. IEEE Trans. Geoscience Electronics,
    Vol. GE11, 171-177.
Harlick, R. M., K. Shanmugam, and I. Dinstein. 1973. Textural
    features for image classification. IEEE Trans. Man and
    Cybernetics Vol. SMC - 3, 610-621.
MacIsaac, R. R. and A. D. Dunsinger. 1978. Ocean sediment proper-
    ties using acoustic sensing. 4th Internat. Conf. on Port &
    Ocean Engineering under Arctic conditions POAC Memorial
    University of Newfoundland.
McCann, C. and D. M. McCann. 1969. The attenuation of compres-
    sional waves in marine sediments. Geophysics 34, 882-892.
McLeroy, E. G. and A. DeLoach. 1968. Sound speed and attenuation
    from 15 to 1500 kHz measured in natural sea-floor sediments.
    J. Acoust. Soc. Amer. 44, 1148-1150.
Medwin H. 1970. Scattering from the sea surface. In: Underwater
    Acoustics. Ed. R.W.B. Stephens, pp. 57-89. London Wiley
    Interscience.
Pace, N. G. 1978. Underwater acoustic backscattering of wideband
    signals from sedimentary materials. University of Bath,
    School of Physics Report.
Pace, N. G. and C. M. Dyer. 1979. Machine classification of
    sedimentary sea bottoms. IEEE Geoscience Electronics. GE17,
    52-56.

Simpkin, P. G. 1978. Evaluation of broad-band high resolution seismic data for sea floor sediment classification. Proc. Oceanology Internat. Conf., Brighton, England.

Smith, D. T. 1973. Acoustic and mechanical loading of marine sediments. In: Physics of Sound in Marine Sediments. Ed. L. Hampton. Plenum Press.

Smith, D. T. 1975. Geophysical assessment of sea floor sediment properties. Proc. Oceanology Internat. Conf., Brighton, England pp. 320-328.

Thomas, P. R. and N. G. Pace. 1980. Broad-band measurements of acoustic attenuation in water-saturated sands. Ultrasonics. 18, 13-17.

Williams, J. P. 1978. Quantitative sediment identification using remote acoustic techniques. Ph.D. thesis, University of Wales.

Wong, H. K. and W. D. Chesterman. 1968. Bottom backscattering near grazing incidence in shallow water. J. Acoust. Soc. Amer. Vol. 44, pp. 1713-1718.

On the Use of Infra-Red Imagery to Study the Surface Temperature of the Shelf Seas

J.H. Simpson and D.C. Bowers

ABSTRACT

Infra-red imagery has been used to study the distribution of sea-surface temperature (SST) in the shelf seas. A series of fronts, which form boundaries between mixed and stratified areas, are found to coincide with the predictions of the h/u^3 criterion. The variability of the fronts about their mean position is shown to have a significant component due to tidal advection, but the adjustment due to variations in tidal mixing between neaps and springs is found to be small. Other sources of variability, including frontal instabilities, are discussed.

The role of bathymetry in influencing SST is described with particular reference to winter conditions.

INTRODUCTION

The shallow seas of the continental shelf have a large heat
capacity which enables them to act as a "thermal flywheel" and damp
variations in atmospheric temperature. This storage capacity is
rendered more efficient by the agencies of wind and tide stirring
which can rapidly mix surface heat input down through the water
column without large changes in sea surface temperature (SST). At
the same time the processes of horizontal advection and diffusion
serve to redistribute the heat in the horizontal plane moving, for
example, large quantities of heat from shallow to deeper water
regions during the positive phase of the seasonal heating cycle.

Our understanding and knowledge of such processes is limited,
at least in part, by the sparse data available from ship surveys.
The advent of satellite infrared imagery offers a modest but significant improvement in our ability to study these problems, since it
provides synoptic pictures of SST, data which are not readily
available from any other source. There are a number of well known
weaknesses with such data; it is not absolute, it applies only to
the top millimetre of the ocean and is severely limited in temperate latitudes by cloud cover. In spite of these limitations,
considerable use is being made of it to elucidate the important
processes in the shelf seas. As an example we describe, in this
note, the use of analogue infrared imagery in the study of strong
frontal features in SST.

THE CRITERION FOR STRATIFICATION

These fronts arise from spatial variations in the level of
stirring which in turn are due to the dynamic response of the shelf
seas to forcing by the ocean tide. Where tidal streams are sufficiently large, complete vertical mixing is maintained throughout
the year whereas in low energy regions strong stratification
develops in the summer months and wind stirring takes over as the
main agent of vertical mixing. Fronts occur as the transition
between these two regimes. A balance between surface heating

(producing stratification) and tidal stirring (tending to destroy it) is defined by the parameter Qh/u^3 (Simpson & Hunter, 1974) where

Q   =   net rate of heat input to the sea surface

u   =   tidal stream amplitude

h   =   water depth.

For a limited spatial region and given time of the year, Q is approximately constant so the controlling parameter is simply h/u^3.

In view of the paucity of measurements of u, the value of h/u^3 is perhaps best determined from a vertically integrated numerical model of the shelf seas and this has recently been done for the whole European shelf (Figure 1) by Pingree and Griffiths (1978). Fronts should occur on a critical contour of h/u^3 which is found to be $\log_{10} h/c_D u^3 = 1.5$ in the notation of Figure 1 or $h/u_1^3 \simeq$ $70m^{-2}s^3$ if we use the surface tidal amplitude at springs u_1 from observations.

A verification of the relevance of this parameter has been given (Simpson, Hughes and Morris, 1977) based on available records of density structure from data banks, but the existence of the surface manifestations of the fronts may be readily confirmed from remotely sensed data. Figure 2a illustrates the positions of some of the fronts on a specimen TIROS N image of the U.K. shelf. The fronts are visible as more or less abrupt light to dark (colder to warmer) transitions which lie approximately parallel to the h/u^3 contours (Figure 1). In addition to extensive features like the Celtic Sea (B) and Western Irish Sea front (A), the criterion also successfully predicts a number of smaller scale features like those on the south-west coast of Scotland (H, J & K) and that in Cardigan Bay (F). This image, and almost all the satellite data we have used, is received and processed at University of Dundee satellite receiving station which is described elsewhere in this volume (Baylis and Brush, 1979).

The imagery is produced as 25cm square black and white prints, one visible and one infrared for each satellite pass. At least two

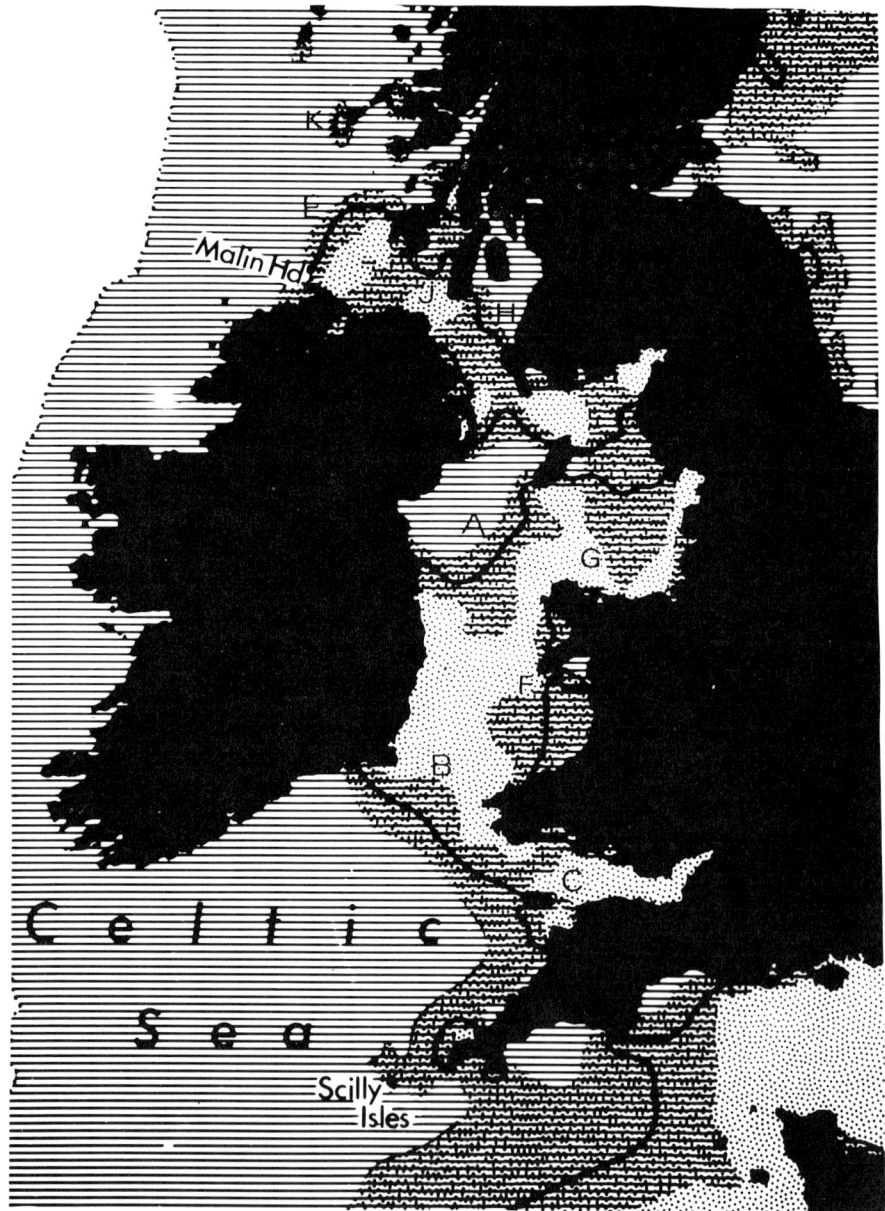

FIGURE 1: Distribution of the parameter h/u³ as computed from the
numerical model of Pingree and Griffiths (1978) showing the predicted areas of stratified water (striped) and vertically mixed
water (dotted) around the British Isles. The position of fronts is
shown here as dark black lines embedded in the darker shading,
which represents transitional water. (Reproduced with permission
from "The effects of vertical stability on phytoplankton distributions" by R. D. Pingree, P. M. Holligan, and G. T. Mardell, Deep-
Sea Research 25, 1011-1028".)

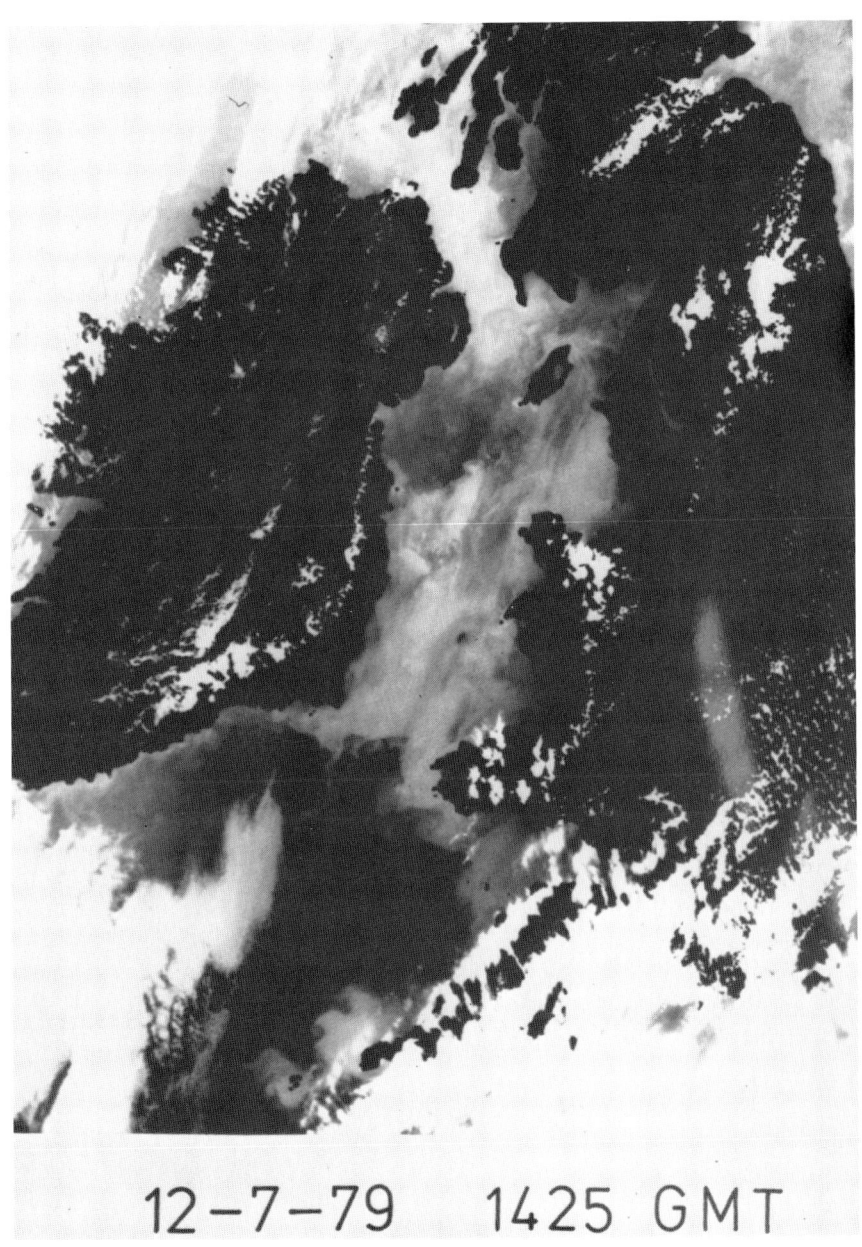

FIGURE 2a: Infrared satellite images in summer (TIROS-N satellite). Shading represents temperature with white being coldest and black warmest.

4-3-78   1028 GMT

FIGURE 2b: Infrared satellite images in winter (NOAA 5, satellite). Shading represents temperature with white being coldest and
black warmest.

passes a day are usually available. The shades of grey which the
photographic print is capable of showing are spread over the range
of sea surface temperatures, so that on a summer day the land,
which is warmer than the sea, appears a uniform black, and clouds

appear white. The prints are corrected at Dundee for distortion due to slant angle viewing of the earth's surface by the satellite.

Figure 3 represents a composite picture in which lines of maximum gradient have been traced from all available images for the period. The subjective element in determining the position of the front does not introduce large uncertainties because the transition is usually sharp and well defined. There is however a case for using digital imagery on CCT to make this procedure completely objective and this we hope will be possible with TIROS N data in the near future.

In addition to the previously reported fronts A-E, Figure 3 also shows evidence of a persistent feature in Liverpool Bay (G) which is a shallow water area where h/u^3 is marginal. It is thought that in this situation the input of buoyancy as freshwater from rivers, particularly the Mersey, is of importance in controlling the occurrence of stratification. The associated front has a strong salinity contrast and has been observed during the winter months when the other fronts are absent.

MOVEMENTS OF THE FRONTS

It is clear from Figure 3 that the fronts are close to, but not exactly coincident with, a single h/u^3 contour. Several factors may contribute to this variability. Part of it will be due to tidal advection which causes the fronts to oscillate with the tidal frequency and an amplitude of $\sim$ 5km. There are also possibilities of adjustments due to variations in the mixing and heating rates. The former of these could involve the restabilization of marginal areas during the fortnightly springs-neaps cycle in which the intensity of mixing (αu^3) varies by a factor $\sim$ 6. Such regular changes of stratification, if they occur, could strongly influence phytoplankton production.

Satellite imagery offers a means of answering questions about these adjustments, though there are difficulties. Apart from the problem of cloud cover already mentioned, the satellite samples at two almost fixed times of day roughly 12 hours apart. This can lead to an ambiguity in which the tidal advection may be confused with neaps-springs adjustment. It is therefore desirable to remove

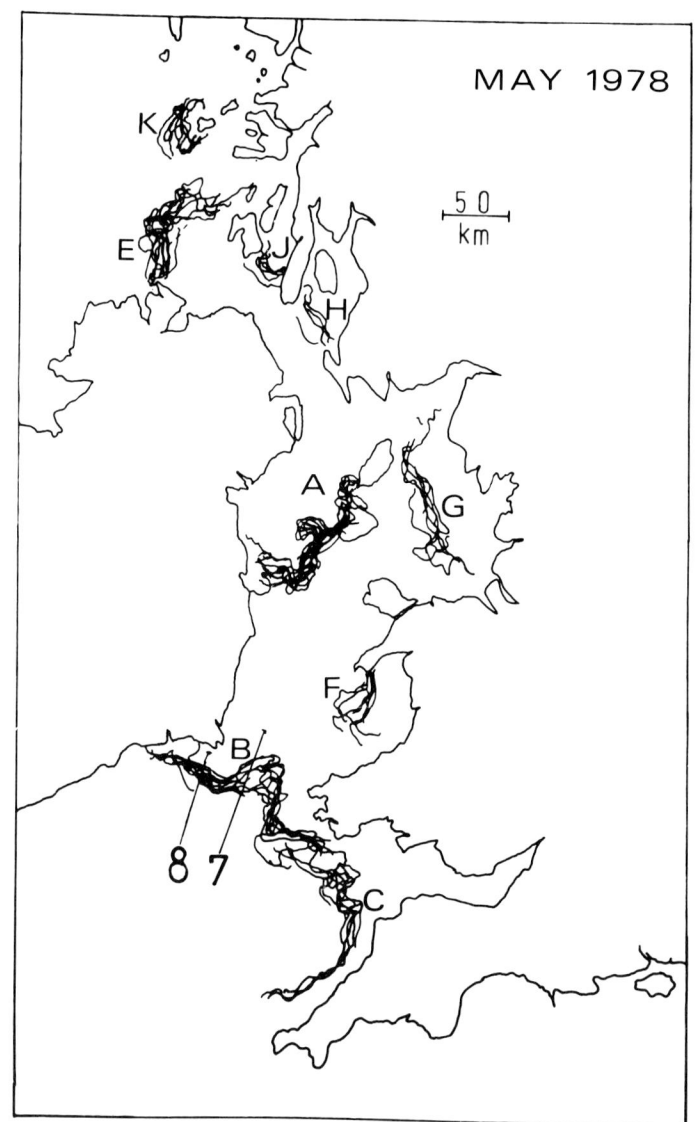

FIGURE 3: Composite showing front positions on 17 images during
May 1978. The sections 7 and 8, along which measurements were made
are also shown. The fronts are labelled as follows:
      A : Western Irish Sea Front
  B & C : Celtic Sea Front
      E : Islay Front
      F : Cardigan Bay Front
      G : Liverpool Bay Front
      H : Clyde Sea Front
      J : Kintyre Front
      K : Tiree Front

the advection signal by the use of observed or computed tidal velocities. This point was emphasized in an earlier paper, Simpson and Bowers (1979), which considered the movements of the Islay and western Irish Sea fronts. Here we present more recent results on the adjustment of the Celtic Sea front.

The displacement Z of the observed frontal boundary along lines perpendicular to the mean orientations of the front (Figure 3) was measured from all useful images obtained between August 1976 and August 1978. We consider Z to consist of a tidal contribution and a residual associated with longer period adjustments.

i.e.
$$Z = A_o + \zeta + \Delta z\ (t)$$

$$= A_o + F\ A_1 \sin (\omega t + \mu) + \Delta z\ (t)$$

A_o represents the mean position of the front and the factor F allows for increases in tidal range from mean neaps (F = 1). The amplitude A_1 and phase μ of the tidal motion may be obtained from the particle displacements X and Y near the front at neaps. It can be shown that if

$$X = a \sin (\omega t + \phi); \quad Y = b \sin (\omega t + \varepsilon)$$

Then
$$A_1^2 = \frac{a^2 \cos^2 \alpha + b^2 \sin^2 \alpha - a\ b \sin 2\alpha \cos(\phi - \varepsilon)}{\sin^2 \Upsilon} \tag{1}$$

$$\tan \mu = \frac{a\ \sin\phi\ \cos \alpha - b\sin\varepsilon\sin\alpha}{a\ \cos\phi\ \cos \alpha - b\cos\varepsilon\sin\alpha} \tag{2}$$

where

α = angle between the front and true north

Υ = angle between the front and the reference line

Plots of Z versus ζ are shown in Figure 4. It is clear that the tidal movement makes a significant contribution. For the data points shown, which represent ~30 observations over a total of 9 months in the summers of 76, 77 and 78, a linear regression accounts for 42% of the variance on line 7 and 54% on line 8.

The dotted lines in Figure 4 represent estimates of the tidal movement based on relations 1 and 2 using tidal excursion data from a numerical model (Pingree and Griffiths, 1978). The slope in both

cases is less than that of the regression line. This difference in
slope is significant at less than the 10% level, but some exaggera-
tion of the tidal signal may be expected from the effects of frontal curvature and changes with time in orientation of the front.

CHANGES BETWEEN NEAP AND SPRING TIDES

To test whether the front is moving in response to changes in
tidal stirring, we plotted ΔZ against F in Figure 5. ΔZ was calculated from Figure 4 by determining the displacement of each point
from the model prediction, and F is the ratio of tidal range at the
time of the satellite pass to the tidal range at mean neaps. No
appreciable increase in ΔZ is apparent with increasing F and the
slopes of both lines are found to be statistically insignificant.
The maximum possible adjustment at the 95% confidence level on line
8 is 5.0 km, which is consistent with estimates for the western
Irish Sea front (Simpson and Bowers, 1979). The corresponding
upper bounds for line 7 is 15.7 km which reflects the large scatter
of the data from this line.

Why does residual variance change so markedly between these
two lines? An obvious difference is in the gradient of the tidal
mixing parameter h/u^3 (Figure 1) which is greater on line 8 by a
factor of ~2. This enhanced gradient may serve to limit the
development of instabilities whose occurrence on fronts has previously been reported on the basis of satellite I.R. imagery (e.g.
Simpson, Allen & Morris, 1978).

These eddies, which have scales of ~20km, could represent the
major source of variability of frontal positions during the summer
months. No conclusive evidence of seasonal variation with Q has
yet been obtained although investigation of the autumnal phase when
the fronts should retreat has been restricted by the paucity of
cloud free images in September and October. Attempts to establish
a relationship to meteorological factors have also proved negative,
so instabilities remain as a likely candidate explanation of frontal variability. They are also probably the principal mechanism of
cross-frontal mixing and their further study by I-R and other
methods seems desirable.

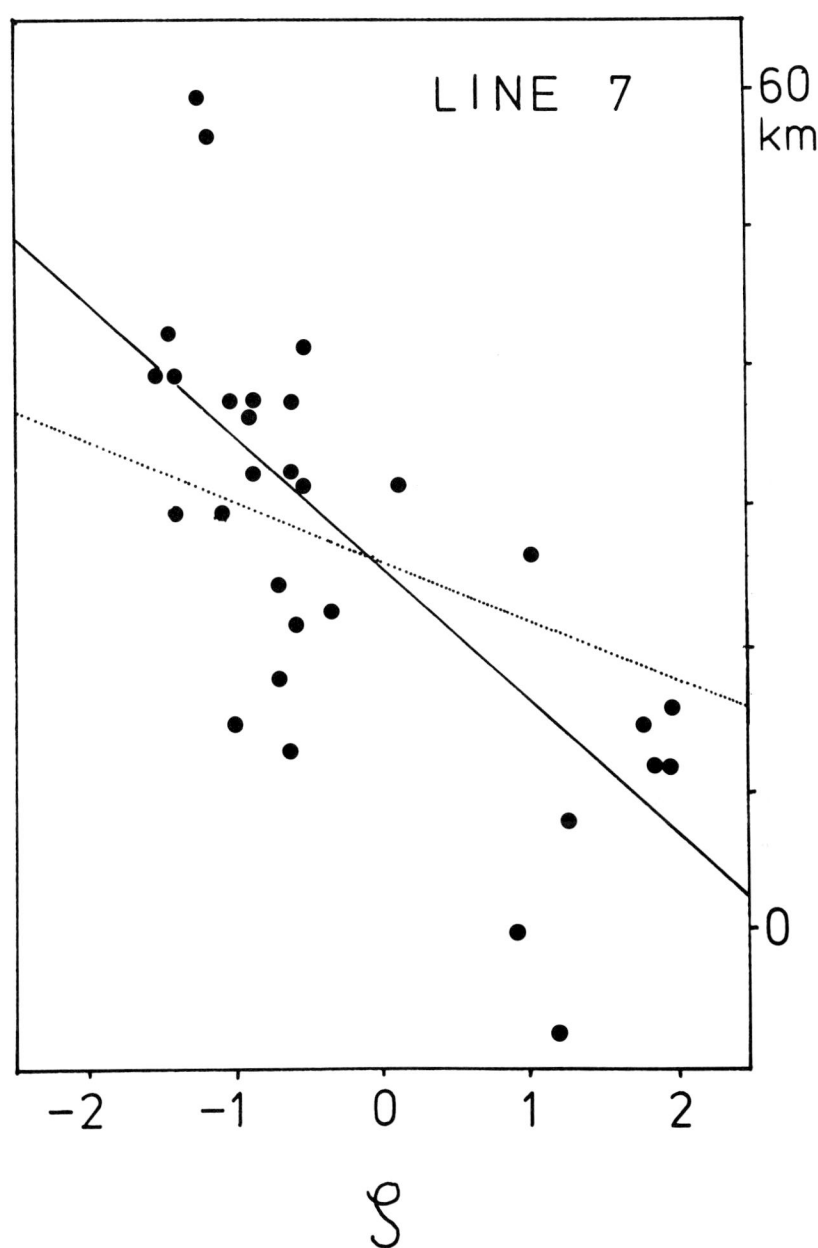

FIGURE 4: Front displacement, Z, plotted against ζ for (a) line 7
and (b) line 8. The solid lines show the regression slopes and the
dotted lines the slopes predicted by a tidal model.

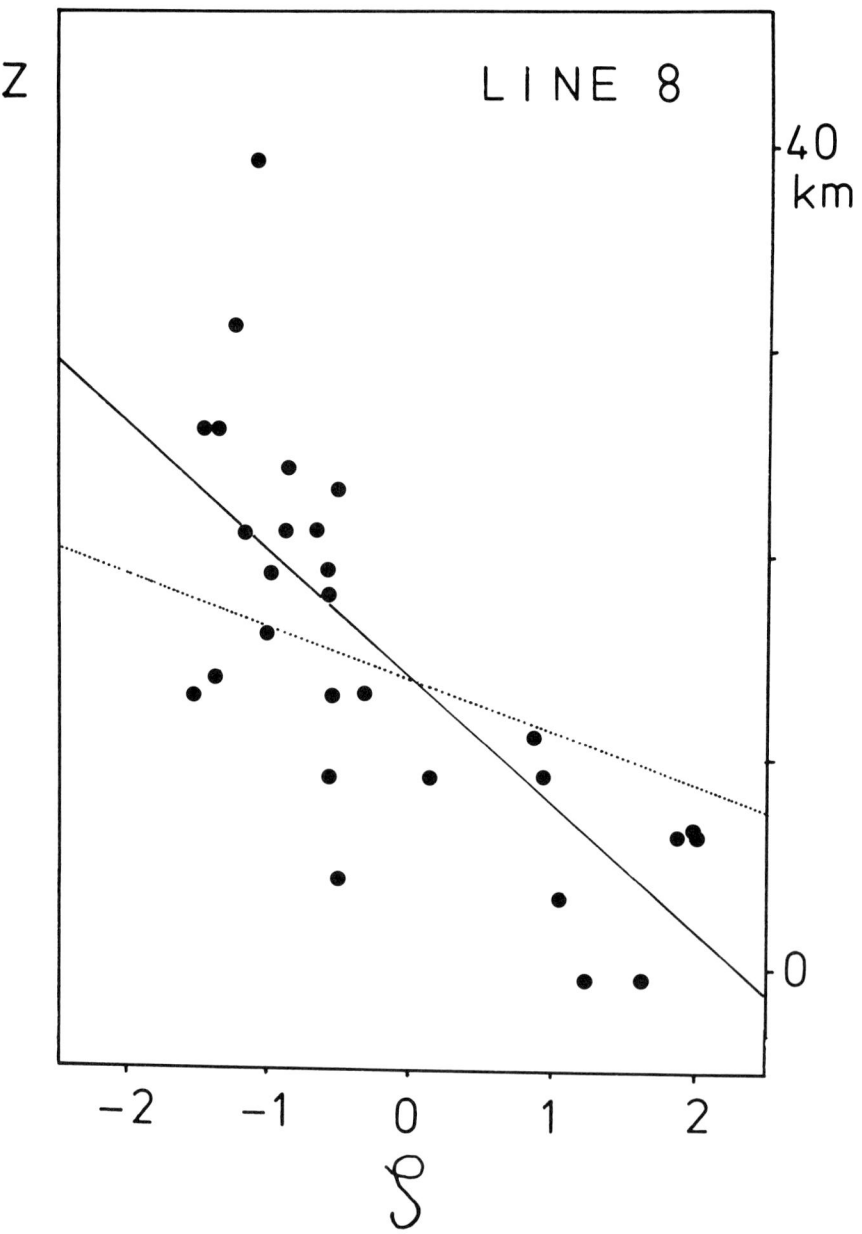

SHALLOW WATER

    The rise in surface temperature in the stratified areas during
the heat input phase is, of course, a consequence of the limitation

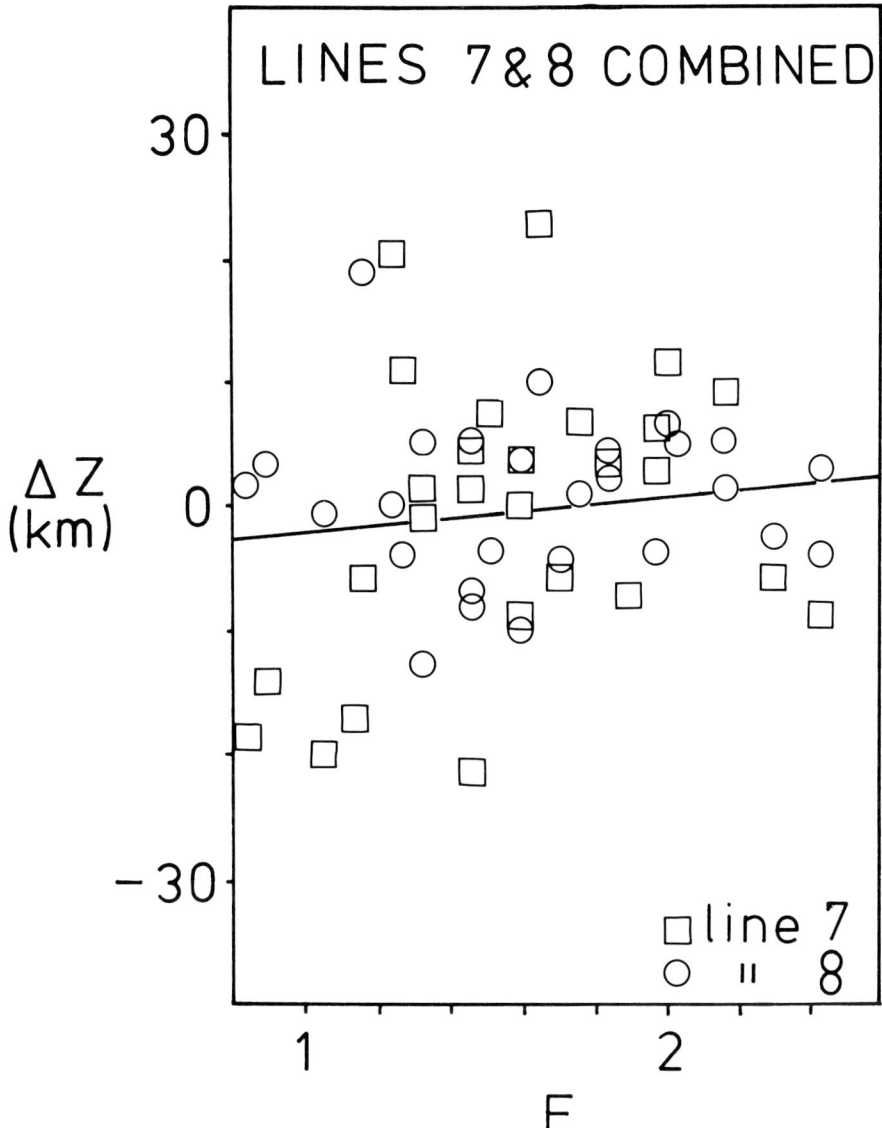

FIGURE 5: Non-tidal movement of the Celtic Sea Front plotted against non-dimensional tidal range factor, F. The solid line represents the result of a regression.

of the depth of mixing by the pycnocline. The same effect arises where the water depth decreases and heat input accumulates in a shorter vertical column. Consequently there is a tendency for

surface temperature to be inversely related to depth in mixed
areas. This can be seen in the areas of warm water visible near
the coast in Figure 2a though the situation is complicated by local
stratification which may also contribute to the temperature rise.
The high temperatures occuring in Liverpool Bay for example, are
only partly due to this 1/h effect, much of the surface heating
being associated with the stratification of part of the bay by heat
and freshwater inputs.

The inverse depth effect also operates in winter when the
shallow water areas cool rapidly while the deep water, with greater
heat capacity per unit area, remains relatively warm. The resulting fringing of the land by cold water is apparent in the very few
cloud free winter images which are so far available. Figure 2b is
one of the best examples from the NOAA-5 archive which shows cold
water margins around the coast with indications of sharp boundaries
occurring in places between cold and warmer deep water. These
putative "winter fronts" invite direct investigation.

CONCLUSIONS

Satellite infrared imagery has provided strong evidence for the
persistence of the shelf sea fronts in positions defined by the
h/u^3 criterion. It has also made possible studies of the variability of frontal positions which indicate that springs to neaps
adjustment is small, probably not exceeding a few kilometers.

Bathymetric influence on surface temperature is also apparent
in the images especially in winter months when thermal stratification is absent and shallow regions become substantially colder than
deeper water areas. Although the cloud cover problem is restricting, the careful study of long term archives of satellite data
should further improve our knowledge of the behaviour of fronts.
With the accumulation of more data points, it should be possible to
put tighter bounds on springs to neaps and seasonal adjustments.
We might also hope, in time, to obtain more insight into the distribution and evolution of the large scale eddy motions which, at

the moment, appear to be the major contributor to frontal variability.

Satellite imagery now also plays an important part in guiding
the effort of conventional ship surveys and this role too will
probably expand in the future. Indeed the successful study of eddy
type processes may only be possible through well co-ordinated
exercises in which aircraft remote sensing as well as satellite
play a part in supporting ship surveys.

ACKNOWLEDGEMENT

We are grateful for the continuing assistance of Mr. P. Baylis
and Mr. J. Brush of the University of Dundee in supplying satellite
imagery.

REFERENCES

Pingree, R. D. and D. K. Griffiths (1978). Tidal fronts on the
    shelf seas around the British Isles. Journal of Geophysical
    Research, 83(9): 4615-4622.
Simpson, J. H., C. Allen, and N. C. G. Morris (1978). Fronts on
    the Continental Shelf. Journal of Geophysical Research,
    83(9): 4607-4614.
Simpson, J. H., and D. Bowers (1979). Shelf sea fronts adjustment
    revealed by satellite IR imagery. Nature, 280: 648-651.
Simpson, J. H., D. Hughes and N. C. G. Morris (1977). The relation
    of seasonal stratification to tidal mixing on the continental
    shelf. In: A Voyage of Discovery pp 327-340. Ed. by M.
    Angel. Pergamon Press.
Simpson, J. H., and J. R. Hunter (1974). Fronts in the Irish Sea.
    Nature, 250: 404-406.

Potential of Satellite Remote Sensing in Coastal Geomorphology

Paul M. Mather

ABSTRACT

The coastal geomorphic system can be considered in terms of (i) active processes of wave, wind and currents; (ii) landforms and their morphological features, and (iii) flows of sediment and energy. This is shown by the Gibraltar Point Study. Of these system components, the third one (sediment flow) is most amenable to satellite monitoring. Requirements are systematic sea truth observations and an image processing capability, together with rapid access to satellite-acquired data. The development of suitable image processing and pattern recognition software is a necessary preliminary, and represents a considerable investment of time and effort. Rapid access to satellite-acquired data is essential if experiments are to proceed heuristically, that is, by incorporating lessons learned from previous experiments. At the moment,

351

both Landsat and Nimbus imagery is difficult to obtain quickly (indeed, at the time of writing, CZCS data are still unavailable).

The application of remotely-sensed data in coastal geomorphic studies is at a comparatively early stage of development. Results reported so far are encouraging. Nevertheless, there are two important points which must be borne in mind: firstly, remote sensing technology is advancing rapidly. The quality of data, in terms both of spectral and spatial resolution and also of weather-independence, is likely to improve considerably during the coming decade, so that many of the difficulties presently experienced may disappear (to be replaced, no doubt, by others). Secondly, remote sensing is not a substitute for field observation; at present it can only augment rather than replace traditional methods of data collection. Even so, it gives the coastal geomorphologist the potential to project his observations over a region, and it is this aspect which may well be considered to be its major attribute.

INTRODUCTION

Remote sensing can be defined as the acquisition of data about an object without the establishment of physical contact with that object. Aerial photography, visual observation, and radar sensing of sea-state all fall within the scope of this definition, but for the purposes of this paper the term "remote sensing" will be limited to data collection from satellite-mounted sensing systems. Data are not of any value, however, until they have been incorporated into an explanatory framework (a model or theory) which describes and accounts for particular features and relationships in a portion of the real world. In this paper, remotely-sensed data are considered in terms of the coastal landforms, the processes which operate to produce these landforms, and the flows of sediment and energy which maintain the equilibrium of the system.

According to Huh and Noble (1977, p.1) the growth of remote sensing is attributable to four factors: defence requirements for early warning reconnaissance and surveillance, rapid expansion of aerospace technological capabilities, rapid development of large-capacity computers and numerical methods, and widespread political awareness of high-priority environmental problems. The validity of the first point is clearly demonstrated by the sponsorship of this workshop by the United States Navy, and details of the Office of Naval Research interest are given in Mr. Badgley's contribution. Military applications are not considered here, however, and attention is given to the implications of the availability of data from satellite-borne sensors from the point of view of an academic geomorphologist. This point of view encompasses the environmental problems of the coastal area which are, as Huh and Noble (1977) point out, of high priority - particularly in a densely-populated country such as England. Technological advance in both the aerospace and computer industries is obviously relevant to the development of remote sensing, but this paper will be restricted to a discussion of the implications of such technological advances where they are relevant to the use of the end product (the raw data) in the context of coastal geomorphology.

THE COASTAL ZONE

The coastal zone can be considered to be the interface between the terrestrial and marine environments. It contains a variety of landforms which are formed and maintained by meteorological, biological, and marine processes, such as wind, plant growth and development, and wave and current action. Flows of material and energy take place within the coastal zone, being represented principally by the transport of sediment and the movement of waves and currents. The nature of the coastal area, its landforms, the magnitude and frequency of the predominant processes, and the rates of change brought about by the flux of material and energy are of concern to several groups. Environmental protection agencies are concerned with maintenance or improvement of the coastal zone from an aesthetic point of view, although economics enter here, for the tourist industry is adversely affected by deterioration in environmental conditions, such as might be brought about by oil spillage or sewage pollution. The monitoring of the position and movement

of contaminated water in the coastal zone is one of the areas in
which the availability of sequential imagery from orbiting satellites has made some impact. The second group with an interest in
the coastal geomorphic system is the military; the successful
prediction of nearshore and beach conditions is essential for the
accomplishment of operations, such as the Normandy landings in
World War II. Lastly, academic geomorphologists have an obvious
interest in the coastal zone for they are concerned with the
dynamic relationships between coastal processes and coastal landforms.

The nature of this interest is perhaps best explained with
reference to the investigations into the geomorphology of the
south-east coast of Lincolnshire by researchers from the University
of Nottingham. The area studied is shown in Figure 1, from which
it will be seen that the main characteristic of the study area is
its location on the north west corner of the large estuary of the
Wash. This estuary is typical of the bays and estuaries of the
east coast of Britain in that sandbanks and channels are present.
These sandbanks and channels are created by the movement of tidal
currents and the availability of large amounts of unconsolidated
and heterogeneous sediments which were deposited during the glacial
epochs of the Pleistocene period. This sediment is transported,
reworked, and deposited by the action of waves and currents, so
that there is a close relationship between the location, morphol-
ogy, and sedimentary composition of the sandbanks and the flow of
tidal currents entering and leaving the Wash. These currents flow
from north to south on the flood tide; this direction is reversed
on the ebb of the tide. As the ebb currents diverge on leaving the
estuaries, the morphological response has been the generation of a
complex set of intertidal sandbanks, the surface of which are
formed into characteristic patterns of sand ripples and waves.
Between the sandbanks are tidal channels, which can be classified
into flood-residual or ebb-residual depending upon the relative
magnitude of current velocities attained during the flood tide and
ebb tide periods. Sediment movement is in a southerly direction in
the flood-residual channels and in a northerly direction in the

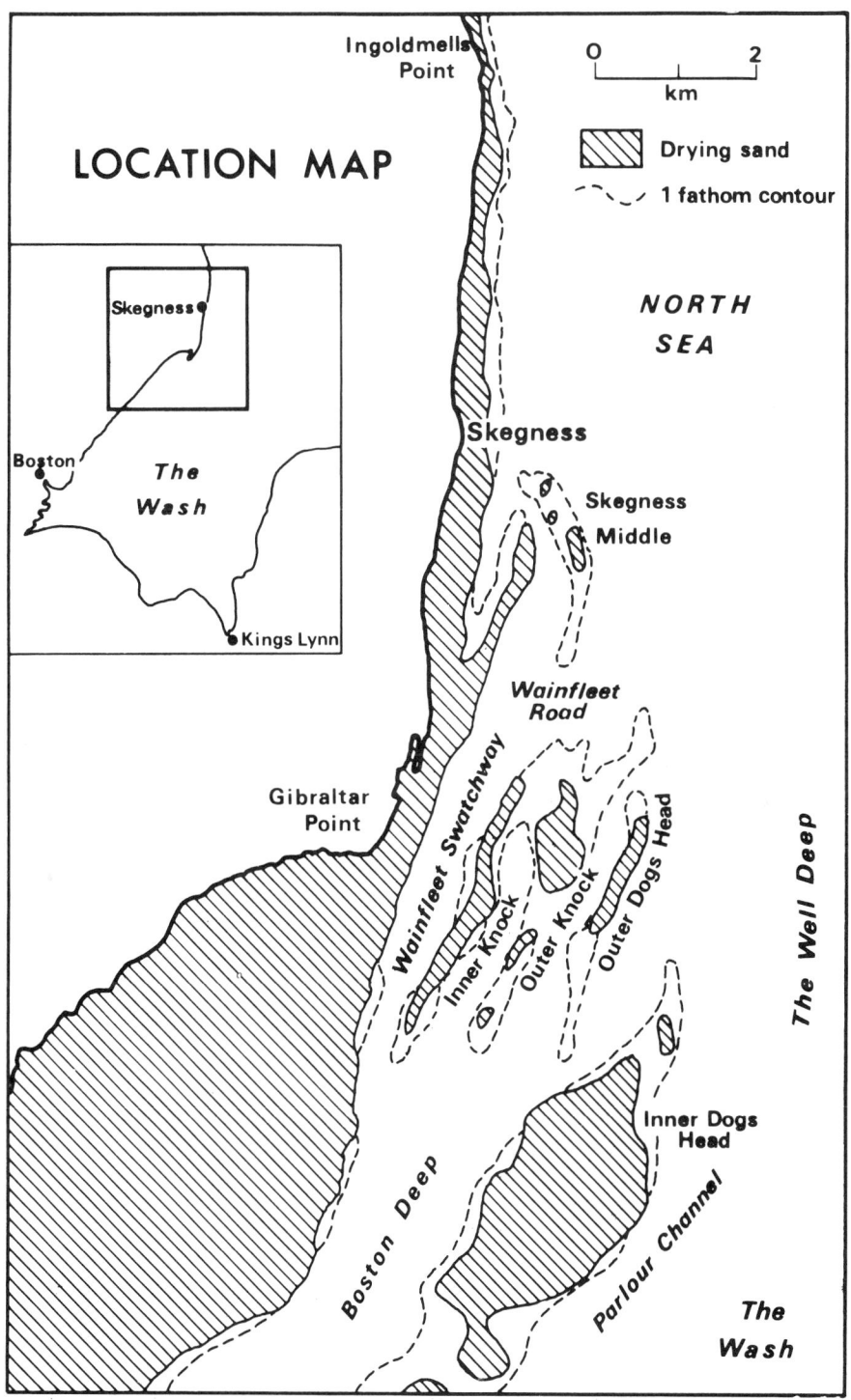

FIGURE 1:   The Gibraltar Point area and the South Lincolnshire Coast.

ebb-residual channels. Although these movements are mainly
parallel to the coastline, sediment transfers take place between
the nearshore and foreshore zones at a point where a sandbank lies
at an acute angle to the shore, forming a ness at the point of
contact between the sandbank and the foreshore, resulting from
onshore movement of sediment. Such a ness is located 2km north of
Gibraltar Point, where the Skegness Middle sandbank lies at an
angle of approximately 20° to the coastline. Sand is transferred
from the nearshore circulation system to the foreshore at this
point. Once on the foreshore, the sand is moved southwards by wave
and tidal current action.

A close association between the magnitude and direction of the
processes operating in this area, the resulting landforms, and the
movement of material has thus been demonstrated by the work of
Dugdale et al. (1978) and Dugdale (1977).

REMOTELY-SENSED DATA

To be of use in coastal geomorphic studies, remotely-sensed
data must contain information which can be applied to the understanding and explanation of the three components of the coastal
geomorphic system, namely: the processes operating, their magnitudes and frequencies; the landforms present and the patterns of
movement of sediment which are generated by the operation of geomorphic processes. Various sensors provide differing types of
data, but not all are of value in the context set out above. Thus,
records of sea-surface temperatures are not of direct interest to
geomorphologists working at the scale of the Gibraltar Point investigation, described previously, although they are of much
greater relevance in larger-scale studies. The properties of the
lower atmosphere again are not of direct interest. Microwave
remote measurement systems have the major advantage of being independent of weather conditions, which is an important factor in an
area such as the British Isles. Sea-state monitoring by radar
imagery is one application of microwave techniques. Thus, the
radar altimeter carried by Seasat-1 was capable of monitoring wave

heights to within 0.5 to 1 meter (see the contribution by Cartwright, this volume). At the present time, however, records of visible and near-infrared radiation reflected from the earth's surface are the most useful source of remotely-sensed data for the purposes of coastal geomorphology. In particular, the multi-spectral scanners of Landsat and Nimbus provide data which are essentially spatial averages of reflected radiation in particular wavebands. The Landsat Multispectral Scanner (MSS) has four wavebands, covering the ranges 0.5 to 0.6μm, 0.6 to 0.7μm, 0.7 to 0.8μm, and 0.8 to 1.1μm. The reflectance of the target area in each of these four wavebands is recorded and the resulting data are available for unit areas (pixels) of 79 x 57 metres.

Nimbus-7 carries on board the "Coastal Zone Colour Scanner" or CZCS, which is the first instrument to be specifically designed for the purpose of measuring ocean colour from a spacecraft. It is a conventional multichannel scanning radiometer, like the multi-spectral scanner of Landsat. Since the CZCS is designed to measure ocean colour (and hence the content of sea water) the spectral channels have been selected to cover specific absorption bands; thus channel 1 (0.433 to 0.453μm) is intended to observe chloro-phyll absorption, channel 2 (0.510 to 0.533μm) records chlorophyll correlation, channel 3 (0.540 to 0.560μm) "yellow stuff", channel 4 (0.660 to 0.680μm) chlorophyll absorption, channel 5 (0.700 to 0.800μm) surface vegetation, and channel 6 (10.5 - 12.5μm) surface temperatures. The CZCS channel 5 has the same spectral response as Landsat channel 6. The Nimbus CZCS has, obviously, a higher spectral resolution than the Landsat MSS. On the other hand, its pixel size is 875m x 875m compared to the 79m x 57m of Landsat's MSS. Another consideration is frequency of coverage; for Landsat this is 18 days, whereas for Nimbus 7 it is roughly three overflights every five days (full account of CZCS is given by Cracknell, this volume).

From the point of view of the coastal geomorphologist, the high spectral resolution of CZCS and the frequency of Nimbus over-flights are preferable to the broad-band spectral coverage and 18-day overflight period of Landsat. Depending on the scale of the

study, the pixel size will generally be the dominant factor. For
the Gibraltar Point investigation, the 875 metre square pixel size
of CZCS makes the data of marginal value because the spatial scale
of the study area is such that high-resolution data is required
(see also Figure 1).

Landsat MSS imagery produces data for four spectral channels
with a pixel size of 79m x 57m. The information required in
coastal geomorphic studies is that which relates to variations in
suspended sediment under the influence of waves and tidal currents.
The radiance values in the four channels must, therefore, be correlated with records of the actual suspended sediment concentrations at the time of overpass. Collection of sea-truth data is the
key to the successful use of MSS data in coastal geomorphic
studies; radiance values alone may reveal interesting patterns, but
deductive reasoning by itself cannot produce insights into the
significance of such patterns. In the Gibraltar Point area, it is
proposed to use a grid sampling scheme, with automatic data loggers
attached to buoys at the nodes of the grid. The instruments, which
are currently under development, are described elsewhere in this
volume by Dugdale. A further development, which is technically
possible but currently beyond the resources of the group at
Gibraltar Point is the direct interrogation of the data logging
instruments by satellite. The Argos system, which is a cooperative
project between the French Centre National d'Etudes Spatiales
(CNES), NASA, and NOAA, is carried onboard the Tiros-N spacecraft,
and similar data collection systems will be carried by the NOAA
series of satellites.

Argos allows for the transmission of analogue or digital
information from each user platform to the onboard data collection
system; these messages are stored and subsequently re-transmitted
to a ground telemetry station. The messages are received by the
CNES coordination centre at Toulouse, France, via NESS (National
Environmental Satellite Service) centre at Suitland, Maryland.
Users then receive their data in one of several forms, for example,
print-outs are forwarded by post on a weekly basis. Since the
number of satellite passes a day at 55°N is 18, on average, with

170 minutes visibility time over each 24-hour period, the prospects are attractive to the user of remotely sensed data products. This is especially true in the coastal environment where poor or hazardous conditions and/or the time-consuming nature of data collection may limit severely the amount of sea-truth data that can be collected in a given time by a small team of investigators.

Results of correlating sea-truth measurements with MSS data so far reported in the literature are encouraging. Johnson and Harris (1980) indicate that the research area of estuarine and continental shelf sediment transport dynamics can benefit from the use of remote sensing techniques. Examples include the work of Alfoldi and Munday (1978), Anderson et al. (1973), Bowker et al. (1973), Bowker and Witte (1977), Hunter (1973), Klemas et al. (1973, 1977), Weisblatt et al. (1973), and Williamson and Grabau (1973). Bowker and Witte (1977) report, for example, that they found a correlation of 0.96 between Landsat Band 5 radiance and sediment concentration up to 50 mg/1. Weisblatt et al. (1973) also reported a high correlation (0.92) between Landsat Band 5 radiance and turbidity.

REFERENCES

Alfoldi, T. T. and J. C. Munday, Jr. 1978. Water quality analysis by digital chromaticity mapping of Landsat data. Canad. Jnl. Remote Sensing, 4: 108-126.

Anderson, D. M., L. W. Gatto, H. L. McKim, and A. Petrone 1973. Sediment distribution and coastal processes in Cook Inlet, Alaska. Sympos. on Signif. Results Obtained From the E.R.T.S.-1. NASA Goddard SFC, March 1973: 1323-1339.

Bowker, D. E., P. Fleischer, T. A. Gosnick, W. J. Hanna, and J. Ludwick 1973. Correlation of ERTS multispectral imagery with suspended matter and chlorophyll in Lower Chesapeake Bay. Sympos. on Signif. Results Obtained From the E.R.T.S.-1. NASA Goddard SFC, March 1973: 1291-1297.

Bowker, D. E. and W. G. Witte 1977. The use of Landsat for monitoring water parameters in the coastal zone. Satellite Applications to Marine Technology - a collection of technical papers. Proc. AIAA Joint Conference, New Orleans, Louisiana, November 1977: 193-198.

Dugdale, R. E. 1977. Movement of sediment in the nearshore zone, Gibraltar Point, Lincolnshire. Unpubl. Ph.D. thesis, University of Nottingham.

Dugdale, R. E., H. R. Fox and J. R. Russell 1978. Nearshore sandbanks and foreshore accretion on the south Lincolnshire coast. East Midland Geographer, 7: 49-63.

Huh, O. K. and V. E. Noble 1977. Remote sensing of environment: achievements and prognosis for the future. Naval Research Reviews, 30: 1-18.

Hunter, R. 1973. Distribution and movement of suspended sediment in the Gulf of Mexico off the Texas coast. Sympos. on Signif. Results Obtained From the E.R.T.S.-1. NASA Goddard SFC, March 1973: 1341-1348.

Johnson R. W. and R. C. Harris 1980. Remote sensing for water quality and biological measurements in coastal waters. Photogramm. Engineering and Remote Sensing, 46: 77-85.

Klemas, V., R. Srna, M. C. Treasure 1973. Applicability of ERTS-1 imagery to the study of suspended sediment and aquatic fronts. Sympos. on Signif. Results Obtained From the E.R.T.S.-1. NASA Goddard SFC, March 1973: 1275-1290.

Klemas, V., J. L. Davis, W. Whelan, and G. Tornatore 1977. Satellite, aircraft, and drogue studies of coastal currents and pollutants. IEEE Trans. Geoscience Electronics, GE-15: 97-108.

Weisblatt, E. A., J. B. Zaitzeff, and C. A. Reeves 1973. Classification of turbidity levels in the Texas marine coastal zone. Proceedings, Sympos. on Machine Processing of Remotely Sensed Data, Purdue University, Lafayette, Ind., 3A-42 to 3A-59.

Williamson, A. N. and W. E. Grabau 1973. Sediment concentration mapping in tidal estuaries. Third E.R.T.S.-1 Symposium, NASA SP-351: 1347-1386.

Application of Image Processing to Remotely Sensed Phenomena

Klaus A. Ulbricht

ABSTRACT

Summaries of investigations of several remotely sensed phenomena, investigated with the help of DFVLR's digital image processing
facility DIBIAS, are given. Topics described show the application
of image processing on remotely sensed data. Investigation of
LANDSAT scenes of North Africa via image processing has been supported and verified by recent ore exploration mission. LANDSAT
scenes of the Lake of Constance have been studied with the help of
limnologists. Thematic mapping of the Rhinegraben resulted in a
land use inventory, not treated here. Finally, evaluation of
LANDSAT scenes of the Baltic, showing the mass appearance of blue-
green algae, may lead to a research project investigating the part
of the nitrogen budget of the Baltic Sea due to blue-green algae.

361

DIBIAS, DFVLR's Digital Interactive Bavarian Image
Analysis System

Central point of the image processing system is the 85 k-bite
interdata processing computer (Figure 1). Images from computer
compatible tape (CCT) can be read in by 800 and 1600 bpi magnetic
tapes, from slides, coloured, and black and white, and from paper
prints. Output is achieved in the same form. An alphanumeric
terminal with keyboard allows communication between DIBIAS and the
user. From the keyboard it is possible to

- start the different image processing programs,
- to read parameters into the program,
- to verify the actual status of the system, and
- to receive system messages.

A colour TV-screen enables the user to verify the results of each
processing step. This way, cost effective and time-consuming use
of colour films is avoided. Additionally, there is the direct
interaction possibility with the displayed image via track ball and

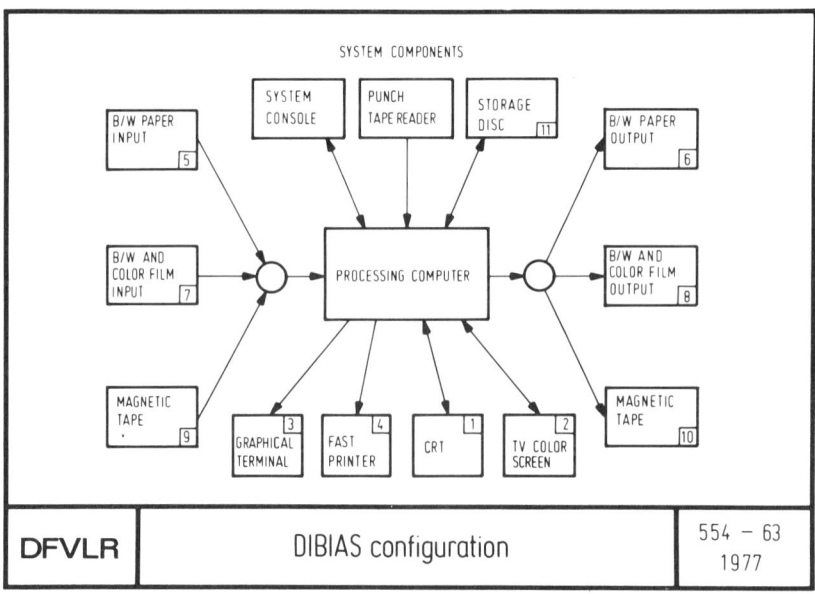

FIGURE 1: DIBIAS configuration.

light target. A graphic terminal is used for display of graphic information, for instance histograms, grey level distribution, clusters, etc. A mechanical drum scanner (Muirhead) is used to digitize black and white paper prints up to a size of 20 cm times 20 cm. A drum recorder produces black and white photographs.

A Dicomed flying spot scanner is used to digitize black and white, as well as colour slides of 35 mm and 70 mm film. Recording of black and white and colour film, results of the processing activities, is done by an electronic filmrecorder, also Dicomed, which can expose polaroid films, too.

The user concept shows the basic concept for the DIBIAS software (Figure 2). The user reads image data into the computer with the corresponding input module. The system has been built in modular form, so newly developed programs can be added any time. A multitude of processing programs of all kinds serves to bring out the contents of the image. Programs are divided into input, output, utility, and (mainly) processing programs. Very important is the possible display of the processed image after each processing step. Direct interaction with the processed image is possible via track ball and light target. The system has been constructed to be easily run by experimenters of different scientific disciplines after short explanation phase. The approximately 150 programs corresponding to about 10 man years, are divided mainly into

- image enhancement,
- multispectral classification,
- digital filtering, and
- texture analysis programs.

Image processing at Oberpfaffenhofen has been applied to a multitude of subjects, ranging from purely geological topics as forwarded by the German Federal Institute for Geosciences and Natural Resources (BGR), looking for long lineaments, ore deposits or oil, over university departments, e.g. the marine biology department of the University of Kiel in their investigation of the blue-green algae appearance in the Baltic Sea, the forestry department of the Freiburg University in their studies of the black

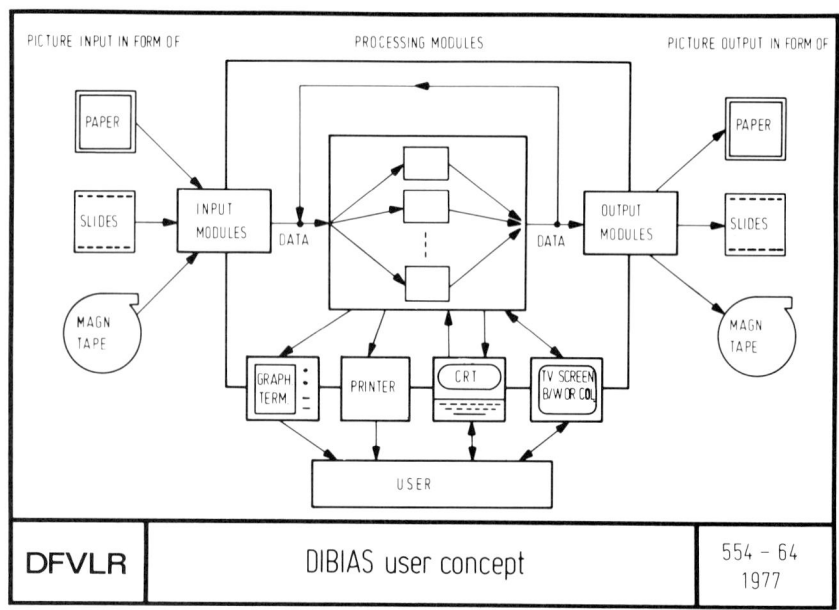

FIGURE 2: DIBIAS user concept.

forest region, government agencies in a study to use LANDSAT imagery for cartographic mapping, atmospheric research institutes in
their study of aerosols in the atmosphere, over environmental
research institutions to all kinds of users with specific
interests.

APPLICATION OF DIGITAL IMAGE PROCESSING MODULES ON LANDSAT
SCENES OF NORTH AFRICA

For the last 3 years DFVLR and the German Federal Institute
for Geoscience and Natural Resources (BGR), have cooperated in
testing the digital image processing facility DIBIAS and its applicability to geological problems and questions posed by the BGR
(Hoppe, Ulbricht, 1978). The basic material for the tests were
computer compatible tapes of the Earth Resources Technology
Satellite LANDSAT. With this satellite it has been possible for
the first time, to gain multispectral ortho-images of larger areas

of the earth. Images were taken repetitively in 7- to 14-day
intervals at the same time in the morning, so with the same sun
azimuth the same direction of the shadows was obtained. This
allows combination and comparison of images from different orbits
for geoscientific interpretation.

The evaluation of the distribution of linear features and
distinct superficial patterns, given by different grey levels in
the four multispectral channels of the LANDSAT spacecraft, provides
the base for geoscientific maps of different types. These are for
example:

a)   geomorphological maps, classifying the superficial forms,

b)   structural maps, analyzing the construction of the
     earth's crust,

c)   geological maps, showing the distribution of rock types
     and lithological units,

d)   hydrological maps, indicating probable reservoirs of
     water resources both on the surface and underground, and

e)   maps of areas, investigated for oil and ore exploration.

In contrast to aerial photography used previously, LANDSAT
images allow surveying of areas up to 35 000 km^2 within one scene.
This fact provides a new knowledge of the fracture patterns in
tectonic analysis, especially for long lineaments. Many of these
all over the world were discovered this way, unknown on older maps
and invisible in the field. Long lineaments reflect the border-
lines of blocks of fractured earth crust as well as trends in their
lateral movements, originating in ancient history of the earth and
moving again during later mountain building periods. Such linear
structures, mostly the deeper ones, are sometimes accompanied by
groundwater and ore deposits.

LANDSAT images can also be helpful in detailed mapping. In
this case, image quality and resolution of the smallest picture
element are a stronger limiting factor to the evaluation than would
be in small scale reconnaissance mapping. At this point, image
processing systems like DIBIAS improve LANDSAT images, as handed

out by the EROS Data Center in Sioux Falls. Using computer compatible tapes, such processed images are a useful addition to
geological field work. After digital processing, LANDSAT imagery
is suitable for geologic mapping. Programs applied here range from
standards such as grey level stretching and contrast enhancement to
the more sophisticated and time-consuming ones like unsupervised
and supervised multispectral classification.

In the beginning of the cooperation, simple contrast manipulation and grey level addition were tried on LANDSAT scenes of the
central Massif of Morocco, an area from where adequate ground truth
was obtained during geological field work of a recent BGR expedition. Testing area (not shown here) covers only part of LANDSAT
scene, being situated on the northern border of central Massif,
south of Meknes. It was chosen because of its characteristic
geomorphological appearance, thus guaranteeing easy recognition
after processing: long narrow hills striking from southwest to
northeast, indicating folded rocks of carboniferous age. These are
covered with younger flat layers of mostly Triassic-Jurassic and
Tertiary aging rocks, forming the significant heart-shaped plateau
in the north. Distance from western to eastern border line of
plateau is about 15 km.

Applying density-slicing modules to this test area results in
pointing out clearly the structural difference between the flat
plateau and the folded parallel hills (Figure 3). The three grey
levels decided on rather deliberately, illustrate with their two
forms the morphological units: the plateau with its mostly rounded
areas is situated in the central and upper part of the image, while
the lower part of the picture is the hill region with elongated
structures.

Applying regular photographic or digital enlargement to a
scene multiplies the picture elements of that scene by the enlargement factor. Photographic enlargement too big blurs the pixel
borders, while enlargement of pixels can disturb the visibility.
To avoid this the DIBIAS program separates the pixels by the
enlargement factor, while the size of the pixels remains equal.
The space between the pixels is then filled with linearly inter-

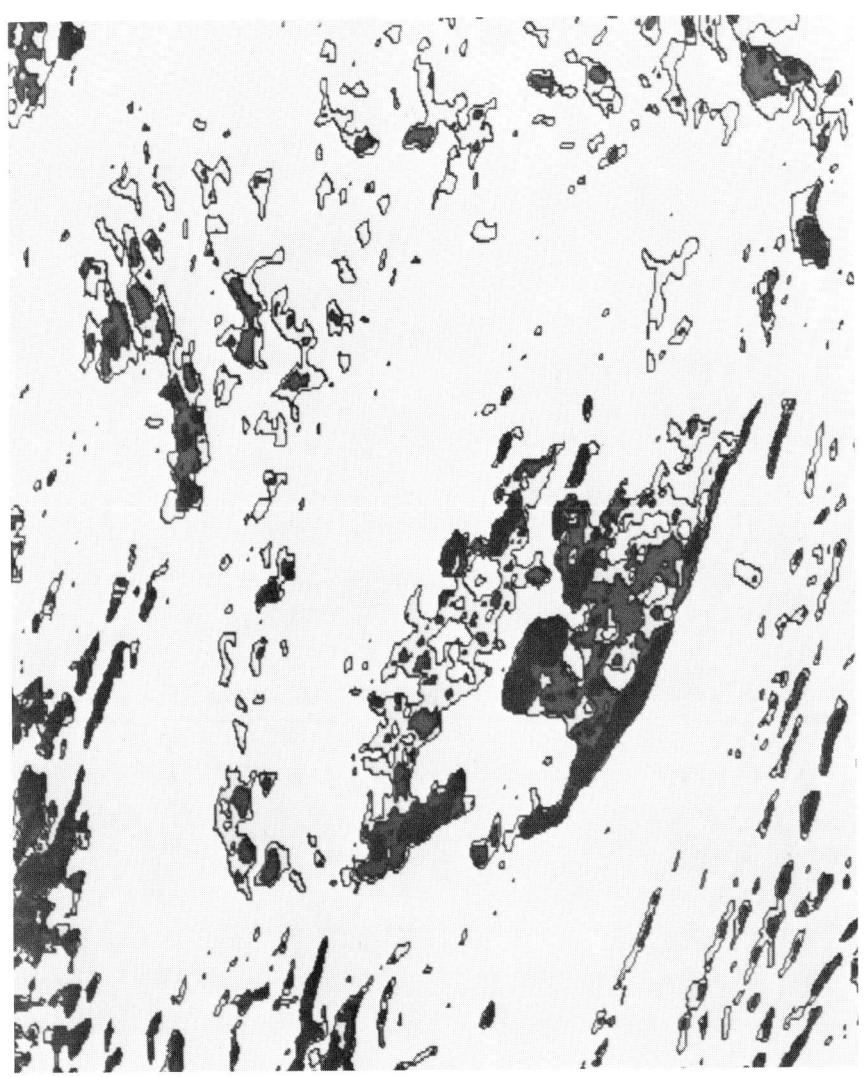

FIGURE 3:   Density-sliced test area, pointing out structural differences.

polated picture elements. Thus the enlargement results in an
enlarged image with the original resolution. This relative
improvement of image resolution provides a better base for topographical maps and therefore easier tracing during field work. The
problem in this case, however, is that all contrast borders become

weaker with pixel interpolation, image appears to be washed out (in
a black/white picture). Since the original pixels aren't real
ground values in the first place, but just represent the spectral
average radiation of a 60 times 80 m rectangle, the image gathered
by pixel interpolation can be fully used for morphological purposes
and investigations, while a pixel classification has to be applied
with caution.

The next step in testing was to apply additive and subtractive
modules to spectral bands, or to use ratios of spectral channels.
The additive module, when applied, resulted in stronger contrast,
while the application of subtractive modules resulted in weaker
relief. On the other hand, contrasts, only weak or shallow before
application of the subtractive module, became distinct and clear:
for instance different crops in the same field with slightly different spectral signature in one channel and strong difference in
another.

High level image processing in geoscientific application
includes programs for classification of spectral reflectance of the
ground. DIBIAS has both unsupervised minimum distance classification and supervised maximum likelihood classification. Classification programs were used both in Morocco and Sudan. In Morocco,
supervised and unsupervised classification both produced unsatis-
factory results. The existing sparse vegetation obviously led to
misinterpretation, by disturbing the spectral reflectance of the
ground. In connection with narrow lithological series and quickly
changing relief, the vegetation then would hide the geological
borders.

The Sudan testing area (Figure 4), situated approximately 300
km north of Khartoum in the Bayuda desert, didn't have this problem. The complete lack of vegetation resulted in direct reflection
of the radiation of the sun by the different rock types. Distance
across the granitic pear-shaped body in the middle of Figure 4 is
about 5 km.

Figure 5 shows maximum likelihood supervised classification.
The classified areas could be verified by ground truth data
obtained by another BGR expedition. Figure 5 is dominated by three

FIGURE 4: Bayuda desert in Sudan, MSS 4, 5, 6,-version, coded
blue-green, and red respectively.

large structures of the old Pre-Cambrian history of the Bayuda
desert: the big oval of flasergneisses in the northern half of the
figure (purple), a circular feature built up by gneisses and
syntectic granite in the southern half (green), and third, the
pear-like module of granitic intrusion between these two units,
near the center of the figure (red). The network all over the
figure represents relics of the youngest developments: dry beds of
the Wadi system, filled with sand and gravel. The rest of the
figure consists mostly of a base gneiss complex and a cover of
desert sand (Ulbricht, 1980).

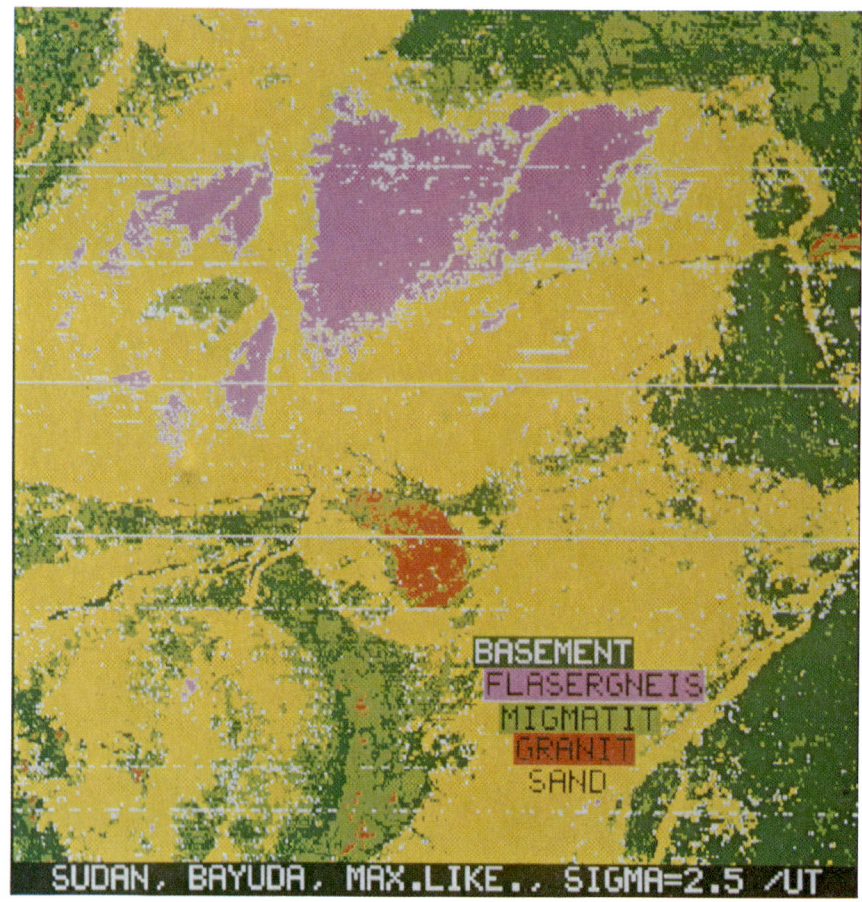

FIGURE 5: Maximum likelihood classification of Bayuda desert, rejection class parameter sigma=2.5.

LAKE OF CONSTANCE, REMOTELY SENSED ANOMALIES

Processing several LANDSAT scenes of the Lake of Constance, taken between 1972 and 1976, provides the chance to gather information on seasonal and other changes throughout the year and lake (Ulbricht, 1977). Water is a rather rewarding substance to cover by remote sensing, because any substance, dissolved or disposed of in the water, can comparatively easily be detected, if it reflects in one of the detection bands used in remote sensing. Regarding a picture of the Lake of Constance of 1972 in it's MSS 4 version, there are several places of anomalous reflection throughout the lake (Figure 6):

a)   the region in the Untersee between the Reichenau-island and the city of Constance

b)   the lighter reflection in form of chains on the middle left side of the lake, Meersburg to Immenstaad

c)   reflection phenomena on the lower middle side of the lake and

d)   the bight of Bregenz, with the mouth of the Rhine leading into the lake on the right hand side of the picture.

Mainly in these above areas a distinct change of the reflectance of the water could be detected within different spectral regions MSS 4 to MSS 7.

Together with limnological scientists, controlling the status of the lake, and together with authorities and interested parties, a supervision of the lake by means of remote sensing for the decrease of contamination and the study of synoptic relations is being discussed.

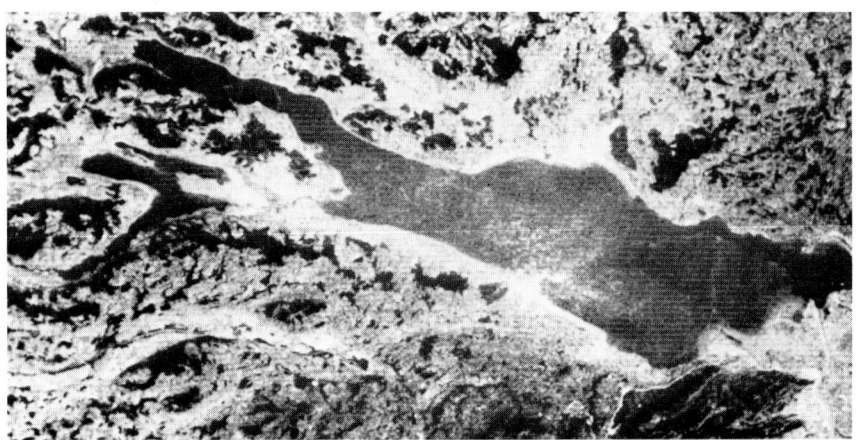

FIGURE 6: MSS 4 version of the Lake of Constance on October 7, 1972.

The above reflectance anomalies show up in all four spectral channels MSS 4 to MSS 7. Discussions with limnologists so far have resulted only in possible explanations: on the western lake (Untersee), flat waters with the possible input of sewage; at Meersburg (middle of the lake), probably chains of algae coming up from the bottom of the lake with a depth of up to 250 m, because in the fall, when this satellite picture was taken, the lake is mixed vertically, due to winds and temperature; at Rohrschach (southern middle lake), a contamination or input of strongly reflecting particles does not seem unlikely either, and at the Rhine estuary, the lightly curved white line going from the mouth of the Rhine to the north side of the lake is obviously due to the Rhine River.

Processing of parts of the lake at different spectral regions leads to Figure 7, the mouth of the Rhine River. Similar grey values have been interpolated to "zones" of equal reflectivity, with a higher variety of zones at shorter wavelengths. This seems to be reasonable because of the higher penetration depth of the shorter wavelength of MSS 4 compared to MSS 5. Obviously there are different currents or streams of the Rhine along the shore. Different reflection within the same bandwith is probably due to different velocity of water along the shore, i.e. reflecting zones are indicative of different depths and matter dissolved.

A processed MSS 4 version (which is not shown here) of the Rhine estuary flowing into the Lake on Constance of August 8, 1975, displays a different curvature of the reflecting zones. They do not seem to reach out into the lake as far as in the LANDSAT scene of February 27, 1975, the beginning of the same year. The size of the reflecting zone probably is a function of the time of the year. An explanation or theory pointing to a definite reason seems rather difficult. A possible explanation might be the fact that the water of the Rhine stays on top longer in wintertime, when it is warmer than the lake, while during the summer the water subsides to a depth of 15 to 25 meters in streaming into the lake and the reflection certainly is caused to a certain extent by sedimentation of the Rhine water.

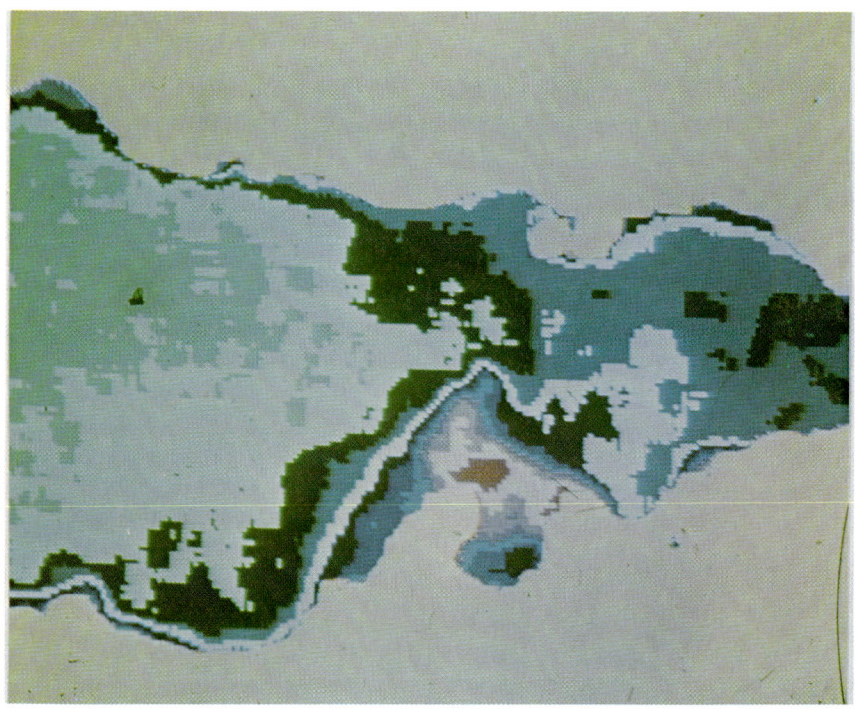

FIGURE 7: Eastern Lake of Constance with Rhine estuary, different grey values are processed and colour-coded, to point out different MSS 4 reflecting zones or currents.

A measurement of the surface temperature of the lake would be valuable for the understanding of limnological processes. Points of interest to be covered by remote sensing studies, could be temperature, chlorophyll contents of the water, sedimentation and streaming, up-welling phenomena and phosphorus contents as well as synoptic relations of these questions.

THEMATIC MAPPING OF PART OF THE RHINE-GRABEN

Two LANDSAT scenes from August 9, and August 28, 1975 have been selected for cartographic mapping.

Due to the technology of the LANDSAT scanner and to special circumstances during the recording time (e.g. rotation of the

earth), LANDSAT images are geometrically distorted, therefore the
original data have to be geometrically rectified. For the recti-
fication, a sample test of corresponding control points has been
measured in the distorted LANDSAT scene and in the reference map.
In the distorted LANDSAT image the coordinates of the control
points have been expressed in image coordinates (line and row) and
in the reference map in Gauss-Krüger coordinates. The control
points have been used to compute the parameters of the mapping
functions between the correct reference map and the distorted
LANDSAT image. The mapping functions were then used to rectify the
distorted image data (Haberacker et al., 1979). The r.m.s. error
of the applied least squares method was about 60 meters.

The training areas have been defined using aerophotographic
material. The boundaries of the training areas have been entered
in maps of the area under investigation. Then the polygons describing the limits of the training areas have been digitized and
by means of DIBIAS software added to the rectified LANDSAT data so
that the positional information was fixed. After the training, the
image data have been classified pixel by pixel, which means that
the classification algorithm tried to relate each pixel to one of
the defined classes or to reject it. The classified image, in
which each class is coded with a distinct colour, has been recorded
with the film recorder and enlarged to the scale 1:200 000 (Figure
8).

The investigation of the quality of the classification results
showed that the accuracy, depending on the class and the position
within the image, is between 60% and 85%. Misclassifications may
be explained with the mixture of spectral radiation in areas with
dispersed land-use regions which are predominant in the Federal
Republic of Germany. Due to the limited visual field of 79 x 79 m
of the LANDSAT scanner, this fine dispersed structure is lost.

Finally the digitized boundaries of communities and counties
have been added to the rectified and classified image data. With
suitable software it was possible to compute the land-use inventory
for the communities and counties. The result can be expressed in
percent, related to the whole area or absolute in hectare.

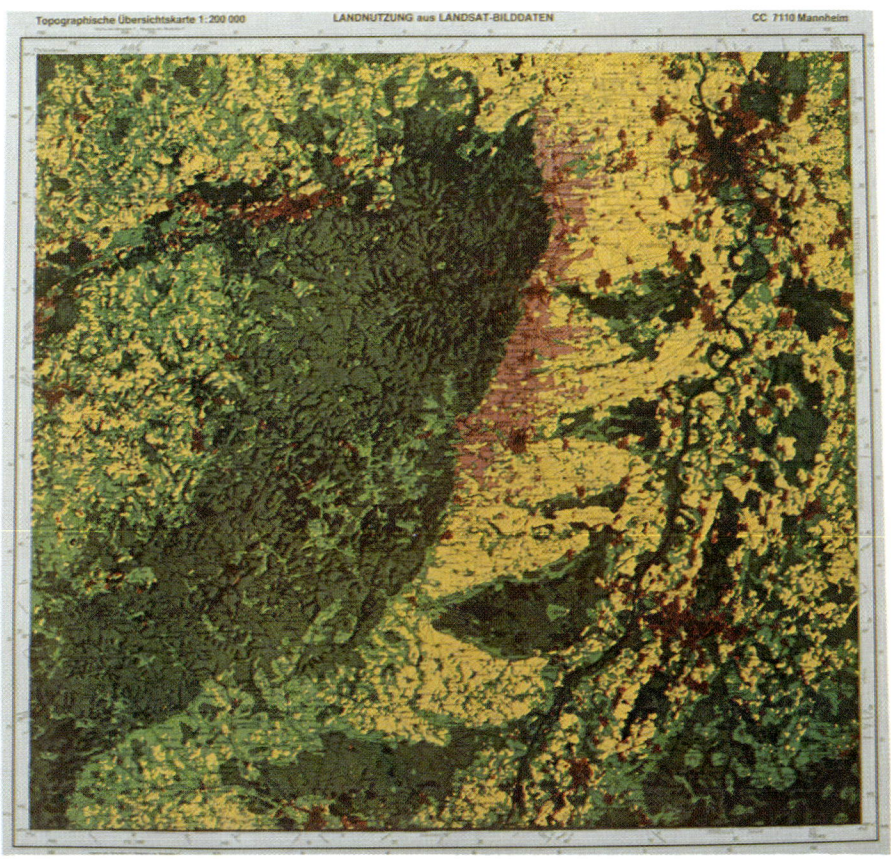

FIGURE 8: Geometrically rectified thematic map of part of the Rhinegraben derived from LANDSAT imagery, verified by ground truth.

MASS APPEARANCE OF BLUE-GREEN ALGAE IN THE BALTIC SEA: EVALUATION OF MULTISPECTRAL LANDSAT SCENES

Scanning through LANDSAT scenes of August 1975, we noticed irregular white traces in the western and southern Baltic Sea, differing in intensity, partially curved, sometimes straight, appearing at different grey levels in each of the four multi-spectral LANDSAT channels. Except for the MSS 4 region, contrast enhancing modules had to be applied, to make the phenomenon clearly visible. Figure 9, courtesy of the GSOC, German Space Operations Center of DFVLR, shows the MSS 4 appearance of the Baltic between

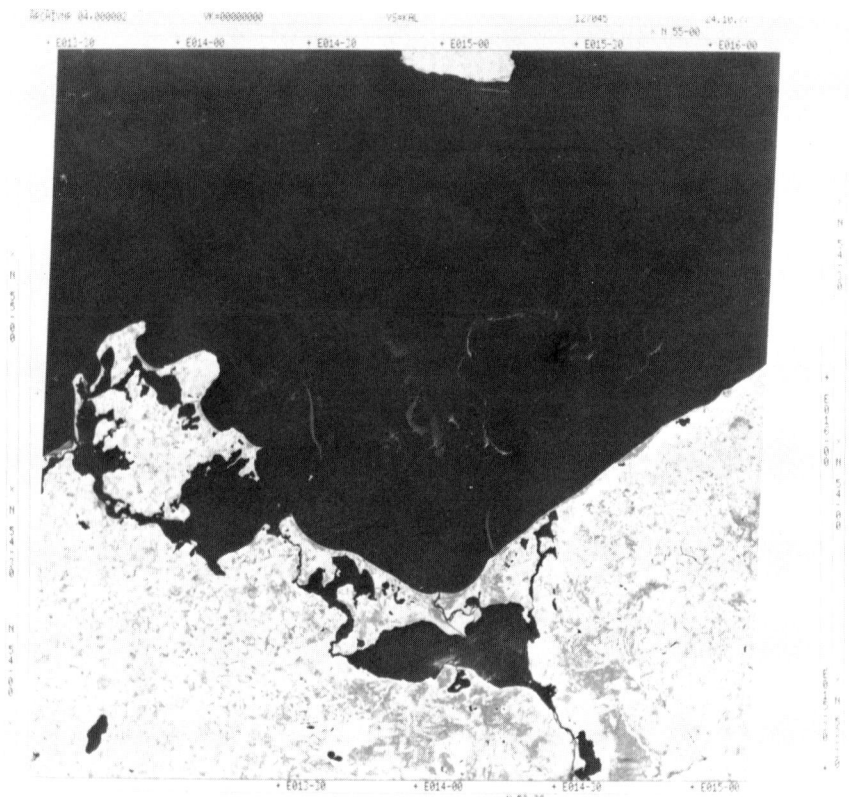

FIGURE 9: MSS 4 appearance of the Baltic northeast of Rügen, showing mass appearance of blue-green algae.

Bornholm and Rügen (GDR) on August 7, 1975, after contrast enhancing modules have been applied. White lines seem to display surface currents or probable wind directions. Figure 10 shows the MSS 7 appearance of same phenomenon. In the near IR-region of MSS 7 water can be considered to react like a black body, only substances above the surface reflect the incoming radiation, the other radiation is absorbed by the water. From different spectral appearance then, it can be assumed that phenomena investigated are situated inside the upper layer of the Baltic, the euphotic layer.

Picture 11 then shows part of the LANDSAT scene north of the island of Rügen, a false colour presentation of the MSS 4, 5 and 6 channels of the LANDSAT satellite, channels being coded with

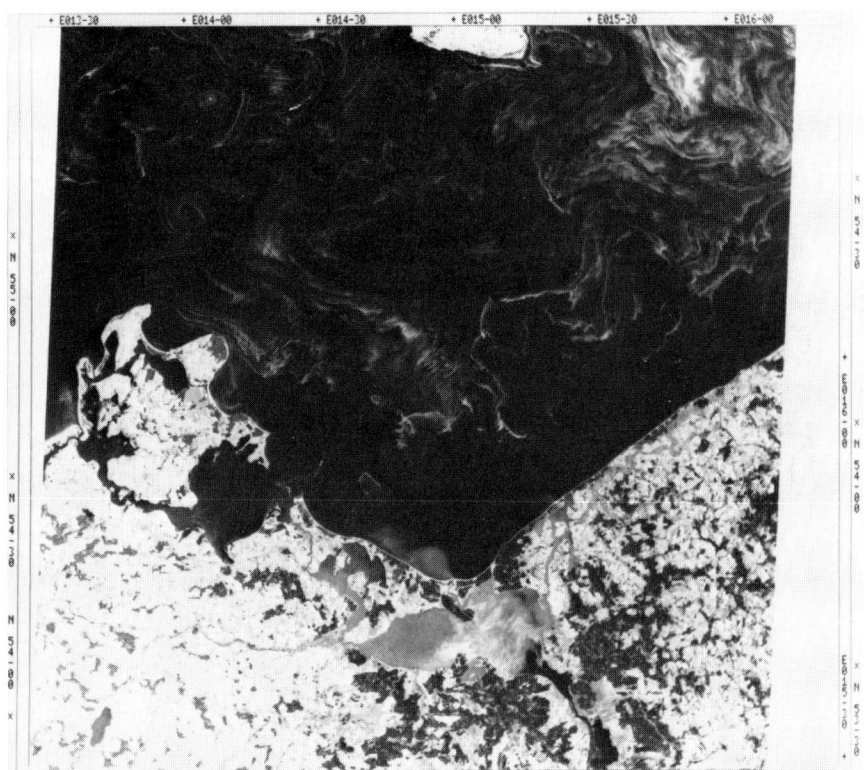

FIGURE 10: MSS 7 version of the above region (Figure 9).

colours blue, green and red respectively, contrast enhanced.
(White appearances show reflecting substance to be at the surface
of the water). Eddies here are among the first to be shown in
land-locked seas, and have been measured by ground truth measurements in the Baltic Sea Experiment Bosex 77 (Kielmann, 1978).

    For an explanation of phenomena we contacted the German
Hydrographic Institute, Hamburg and the Institute for Marine
Research at the University of Kiel. Discussing the phenomenon, and
starting with a much smaller picture than shown beforehand, we
excluded waste dumping as improbable. (According to the so called
"Helsinki convention", an agreement between states bordering the
Baltic, dumping of waste etc. is prohibited. Though not yet rati-
fied the convention laws are applied by the Federal Republic of

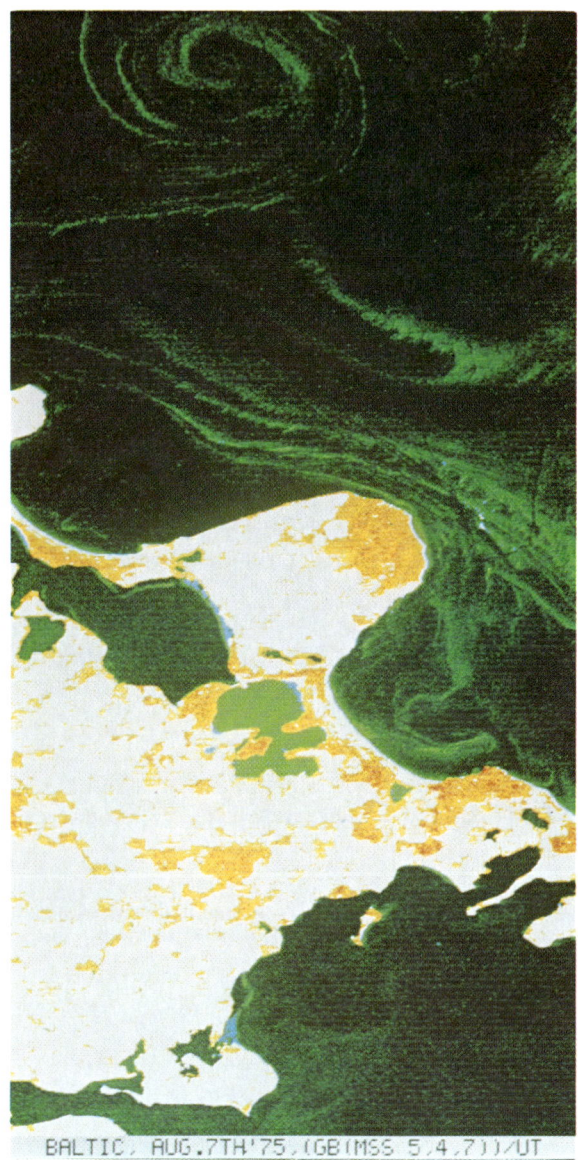

FIGURE 11: False colour presentation of Baltic north of Rügen on
August 7, 1975, MSS 4, 5, and 6 colour-coded.

Germany). In any case, waste dumping of this order would have been
noticed. Oil slicks, on the other hand, were excluded because of

chronological appearance of phenomenon: as to be seen in several
pictures (not shown here), an unstructurized white "cloud" in the
water changed to distinct white lines the next day - oil would
behave the opposite way.

Further discussions and a review of literature showed, that
traces refer to the so-called "blue-green" algae, already mentioned
in 1884, when Prince Albert of Monaco was sailing through the Gulf
of Finland, where he noticed "une grande quantité de petites
algues".

There are two dominant species of blue-green algae that
develop blooms in the western and southern part of the Baltic sea:

a) Aphanizomenon flos aquae (Figure 12), shown in bundles
    here, the filaments of Aphanizomenon flos aquae sticking
    together to bundles, the single cells being 5 to 15 μm
    long and 4 to 6 μm wide. Aphanizomenon has heterocysts,
    cells fixing nitrogen from the air, which are scarcely
    larger than the vegetation cells and lack pigmentation,
    and

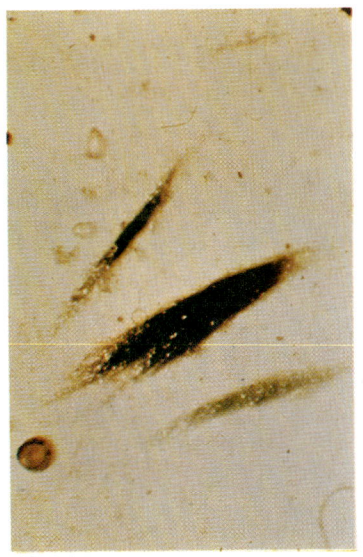 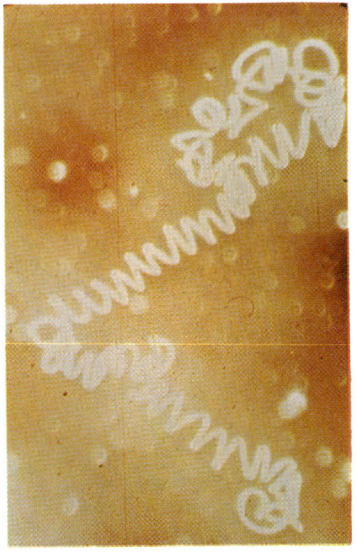

FIGURE 12: Left and FIGURE 13: showing Aphanizomenon flos aquae
in bundles, and Nodularia spumigena.

b)   Nodularia spumigena (Figure 13), 8 - 14 μm in length,
      width being 10 - 16 μm.

Figure 14 is a map of the Baltic with the approximate location
of where we found the algae (Øresund, east of Møn, between Bornholm
and GDR, Danish coast).

FIGURE 14: Map of southwestern Baltic, where blue-greens appeared
east of Kopenhagen, north of the island of Rügen, east of Møn
island, and at Alsen island (far left).

False colour presentation of channels MSS 7, 5, and 4, coded
with blue, green and red and contrast enhancing, results in Figure
15 showing the Baltic east of the Danish island of Møn. False
colour presentation serves to distinguish between phenomena considered to be blue-green algae, and sand banks near shore, here

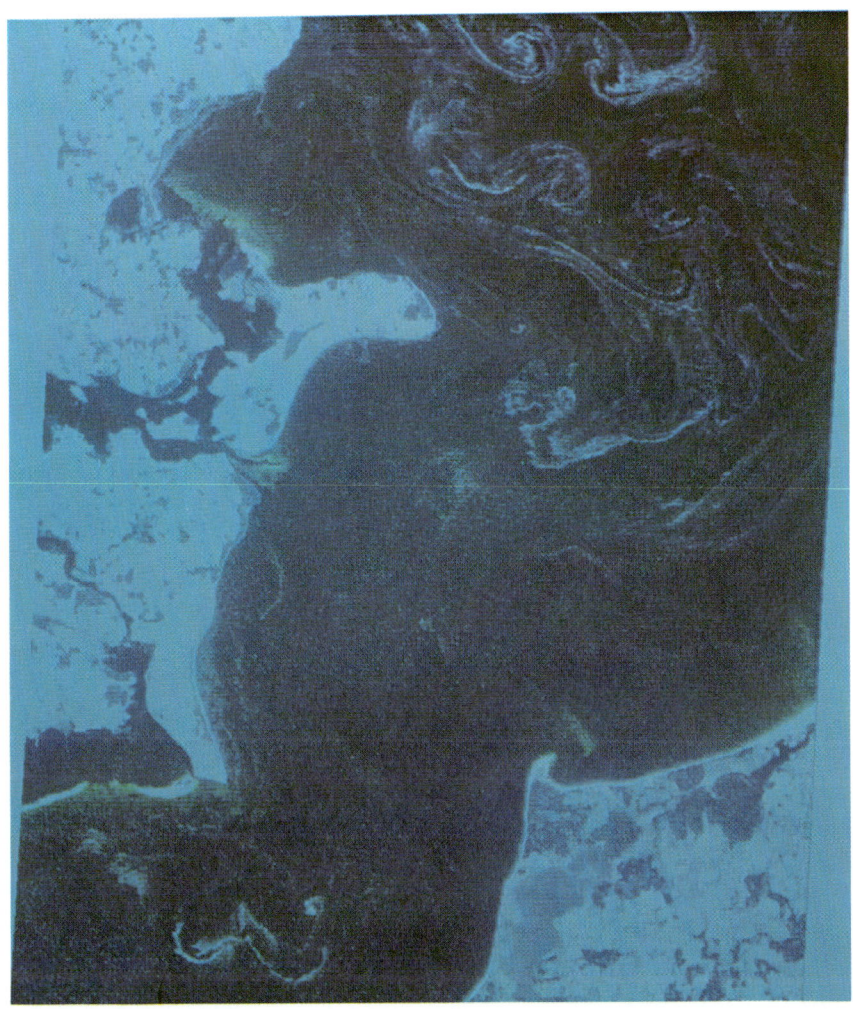

FIGURE 15: Blue-green algae appearance east of Møn, appearing as white traces in the waters of the Baltic (contrast enhanced MSS 4, 5, and 6 LANDSAT spectral channels.

shown in purple. The spectral channels 5, 6, and 4 have been coded with blue, green, and red respectively.

Blue-green algae can be an indicator of increased phosphate concentration: mass production is thought to originate near the coast because of high phosphate contents: blue-green algae use phosphate for their growth.

Blue-green algae have the capability to fix molecular nitrogen from the air and reduce it to ammonia. Blooms in the Baltic start in early summer, after the phytoplankton bloom has depleted the water of mineral nutrients. Because of their nitrogen fixing capacity, blue-green algae are an important link in the "nutrient supply" of the sea because dissolved nitrogen is a limiting factor for many plankton blooms (Ulbricht et al., 1978). Because of this fact, an estimation of the absolute amount of nitrogen fixed by the blue-green algae would be of considerable interest. For this purpose, however, the amount of blue-green algae in the Baltic, detected by satellites and estimated with the help of in situ measurements has to be known.

Figure 16 then shows a contrast enhanced, false colour close-up of the Alsen island, where "clouds" have changed to distinct white lines, as we would expect it for a blue-green bloom.

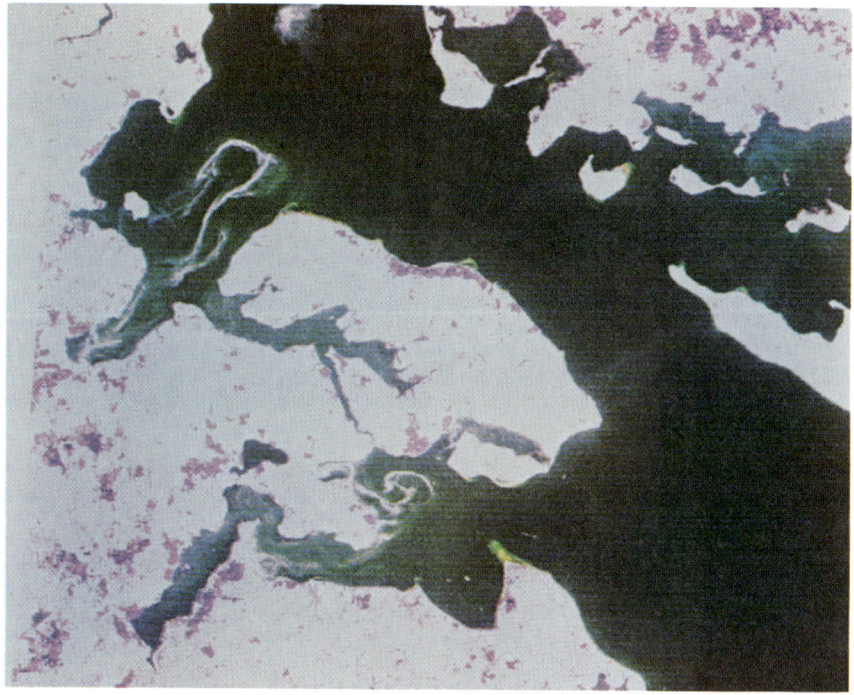

FIGURE 16: MSS 4, 5, 6 contrast enhanced version of Alsen island on August 11, 1975, showing blue-green algae as white traces in the water.

Since the appearance of blue green-algae depends very much on wind
velocity and wave action, and the amount of nitrogen fixed on
biomass and their nitrogen fixing capacity, it would be unrealistic
to calculate the amount of nitrogen fixed by them without in situ
measurements during the algal bloom. Calculations recently given
by the Finnish scientist Rinne et al. (1977), taking into account
vertical distribution as well as differences in daily fixation
rate, arrived at 100,000 tons of nitrogen for 1974 for the Baltic.

Using supervised maximum likelihood classification of the
Baltic of LANDSAT tapes, together with adequate ground truth measurements, arriving at a nitrogen fixing factor of blue-green
algae, it should be possible to arrive at the part of the nitrogen
budget of the Baltic due to blue-green algae. Calculation of the
nitrogen fixing factor should be accompanied by calculation of such
parameters as calculation of the biomass, suspension and salinity,
investigation of eddies and internal waves, influence of the contamination of the Baltic by phosphorus, and last but not least the
investigation of the food web, of which the blue-greens are an
important part.

REFERENCES

Haberäcker, P. 1978. DIBIAS, the digital image processing system
     at DFVLR, system design and applications, proceedings of an
     international conference on earth observation from space and
     management of planetory resources, Toulouse, France, 6 - 11
     March 1978, pp. 233-236.
Haberäcker, P., Kirchhof, W., Krauth, E., Kritikos, G., and Winter,
     R. 1979. Computation of a data structure for a topographical
     map, using multispectral LANDSAT scenes, proceedings, 13th
     International ERIM-Symposium on Remote Sensing, Ann Arbor,
     USA, 1979.
Hoppe, P. Ulbricht, K. A. 1978. Application of digital image
     processing modules on LANDSAT scenes for their improvement and
     geological evaluation, proceedings of an international con-
     ference on earth observation from space and management of
     planetory resources, Toulouse, France, 6 - 11 March 1978, pp.
     237-241.
Horstmann, U. and Ulbricht, K. A. 1978. Detection of eutrophica-
     tion processes from air and space, 12th International ERIM-
     Symposium on Remote Sensing of Environment, Manila,
     Phillippines, 20 - 26 April 1978, proc. pp 1379-1389.

Rinne, I. 1977. Nitrogen fixation by blue-green algae in the
    Baltic Sea, Fifth Symposium of the Baltic Marine Biologists,
    Kiel, Germany, November 1977.
Ulbricht, K. A. 1977. Application of the digital image processing
    system DIBIAS on LANDSAT pictures of central Morocco and
    Southern Germany, XXIV Congresso Internationale Elettronica et
    Aerospaziale, Rome, Italy, proceedings, pp. 81 - 89, 28 - 30
    March, 1977.
Ulbricht, K. A. and Schmidt, D. 1978. Mass occurrence of blue-
    green algae in the Western Baltic, evaluation of satellite
    imagery and implications on marine chemistry and pollution, XI
    Conference of the Baltic Oceanographers, Rostock, GDR, proc.
    pp. 62-69, (April 24-27, 1978).
Ulbricht, K. A. 1980. Comparative experimental study on the
    employment of original and compressed multispectral LANDSAT
    data for applied research, European Space Agency, ESTEC
    Contract 3569/78/NL/HP(SC), April 1979, DFVLR - FB 80-03.

Outline of Meteosat Data Processing at Darmstadt

C. Tomassini

INTRODUCTION

The first European meteorological space program had its origin as a national French project in the late sixties. In 1971, as the French government offered 'Meteosat', such a program became "Europeanized" under the aegis of the European Space Agency (ESA).

The mission was defined by a group composed of the Directors of European Met. Services along with some experts and observers.

THE METEOSAT SYSTEM

The main point of the system is the satellite which in its geostationary position takes radiometric images of the earth and relays meteorological data. The raw images taken by the satellite arrive every half hour at the Central Ground Station (DATTS)

located near Darmstadt (West Germany). Then such images are routed
to ESOC (European Space Operations Centre) some 50 Km away for
processing by the MGCS (Meteosat Ground Computer Centre). The
processing is mainly automatic and consists of removing distortions
from raw images and preparing meteorological products to be disseminated via GTS, or via satellite itself to primary and secondary
user stations (PDUS, SDUS) located in Europe, Africa, Arabia and S.
America.

The quasi-automatic processing is assisted also by human
intervention for purposes of quality control and improvement of the
procedures. The operators work continuously with the automatic
part using television-type displays and keyboards available in MIEC
and DRCC (Meteorological Information Extraction Centre, Data
Referencing and Conditioning Centre). Furthermore the operators in
the MOCC (METEOSAT Operators Control Centre) perform the normal
function of telemetry monitoring and telecommanding.

The satellite can relay processed images and also meteorological data received from sensors installed in various terrestrial
regions. Such sensors are mounted on the so-called DCP (Data
Collection Platforms), which can be fixed or mobile, land-based or
seaborne, or on aircraft. The data from DCP are processed by the
MGCS and relayed to the users via landline, generally using the GTS
but also by telex etc. The RTH in Offenbach is an entry point of
the Meteosat onto the GTS. Finally the ESA ground station in
Belgium (REDU) acts as a back-up ground station for the satellite
control, while a transponder, located in French Guyana, is used for
the accurate determination of the satellite orbit, very important
for the correct earth imaging.

The Satellite

The satellite is a complex of concentric drum-shaped bodies.
The principal body is covered with a skin of solar cells to provide
energy for the vehicle. The main payload, the radiometer, is
mounted inside the main body and looks at the earth through a
cut-out solar panel. The cylindrical structures on the main body
top are antennae. The satellite is spin-stablized at 100 rpm, with
spin axis parallel to earth axis South-North; the radiometer scans

the earth disc every 600 m sec from East to West and after each
revolution is rotated a step in South-North direction. Therefore,
a complete image consists of an array East-West and South-North of
image elements (pixels). The radiometer consists of a complex
system of mirrors with the sensors kept at constant temperature
measuring radiances in the infrared and visible bands. The images
are generated simultaneously in the different bands. Furthermore,
the visible sensors are duplicated to produce a double resolution
respect to the infrared (~5km) that is of about 2.5 km.

The earth is sampled in 30 m sec out of 600 m sec necessary
for a complete revolution line by line whereby the entire image
needs 25 m. Five more minutes are then spent to bring back the
radiometer in the original position and to calibrate the sensors.
The raw image, as it arrives at the DATTS, has a number of imper-
fections, as geometrical distortions caused by radiometric scanning
and by the precession of the satellite, though small, about the
rotation axis. Moreover, other errors are introduced by the electronics, but their contribution can be easily computed, and a dedicated computer at the MGCS is devoted to remove such defects.

Dissemination products

The Meteosat system disseminates data to the users in two
different modes through two dedicated channels. These modes are:

- conventional analogue transmissions (WEFAX) and
- special digital transmissions (High Resolution - HR)

The WEFAX transmissions are in a format compatible with that of
other meteorological satellites (including the APT). The HR
digital data are in a format specific to Meteosat and are designed
for the user requiring pre-processed, full resolution data.

Archiving

The users can obtain from the Meteosat Image Archive many
products including the following:

- digital recording of all VIS, IR and WV data on conventional magnetic tapes;
- positive transparencies or paper copies;
- plots or digital tapes of meteorological products;
- etc.

METEOROLOGICAL PRODUCTS

Study groups of potential users of meteorological products defined the products to be extracted and the resolution of these data, both in space and time. The products defined include:

- Winds. Cloud displacement measurements using a sequence of images are performed twice each day. This product is now operational and generates some 1200 quality controlled cloud motion vectors each day.
- Sea-surface temperature. These are computed for every segment (an array of 32 x 32 IR pixels) twice a day. The objective is to attain an accuracy of better than 1°C.
- Water content of the upper atmosphere. The measure consists of determining for each segment the average relative humidity in the upper atmosphere twice a day with a desired accuracy of 10%.
- Cloud coverage. This product is computer-coded nephanalysis for each segment. For each layer identified, the partial coverage and the top temperature are computed.
- Radiation balance. The measure is obtained for each segment summing up the 24 values computed at every hour of the radiation balance at the top of the atmosphere. The result is very sensitive to calibration errors on VIS and IR bands, whereby two values, respectively for short and long waves, are produced.
- Cloud Top Height Maps. This product presents in the form of image the height of clouds at a scale of 25 km. The height is represented in the image by using levels, black, six grey and white. Every grey value represents a layer of 1500 m between 3,000 and 12,000 m., black indicates absence of clouds or clouds with a height less than 3,000 m and finally the white indicates clouds with a height greater than 12,000 m. The accuracy requested is of 750 m.

Data for the Meteorological Information Extraction Centre.

The MIEC uses the preprocessed images, that is images which under-
went processes of:

- linear deconvolution
- bidimensional deconvolution
- calibration

They are put into suitable format, that is segmented in squares of
32 x 32 infrared pixels (segments). This segmentation is such as
to provide geographical locations of the segment centres independent upon the geometrical deformation present in the raw image.

MIEC requires, in real time, classical meteorological data in
several sectors, on one hand to transform in height or pressure
level information which is purely in form of temperature, on the
other, to evaluate in a rational manner the various corrections to
be applied to the raw measurements in order to have the products,
as for example the correction applied to the measured sea radiance
to take into account the atmospheric absorption.

These necessities are difficult to satisfy using raw meteorological data. Meteorological information, coming from regions
covered by Meteosat, is generally available after a great delay, if
at all. Therefore, it was decided to utilize forecast data coming
from the NOAA Global Numerical forecast model, i.e. humidity and
temperature profiles for a specified geographical grid. These data
are received twice a day.

- Treatment Structure. The MIEC tasks belong to the set of
  the general Meteosat treatment tasks. This structure is
  based on a sub-division into two different categories.
  One contains the operational tasks which run automatically every day, the other contains the so-called support
  tasks which must be executed from time to time. The
  first group represents the essential logic of the system
  and is executed under the supervision of a set of programs called Data Processing Monitor (DPM). The second
  contains all the tasks which run using data independent
  from the real-time image data and which create all the
  support meteorological information.

PRODUCT EXTRACTION

The product extraction implies in every case the following scheme:
-   support data availability
-   image data reception
-   intermediate product extraction in terms of image data (radiance measurements, pixels displacement)
-   transformation from intermediate to final product. A brief description of wind and SST extraction is given in what follows. Wind extraction is based upon IR image for localization of the information and on the remaining channels for identification of the layers. Four points should be noted:

1.   Layer identification

The analysis proceeds in several steps. Computation of bidimensional IR/VIS and IR/WV histograms and monodimensional histograms for all channels available and for all segments satisfying a given quality threshold. Definition of all the peaks present in the histograms considering the histograms as a superposition of several 'Gaussion' peaks. To do that, one determines a peak C on the monodimensional IR histogram, then computes the corresponding variance CIR and in the band C±3CIR of the bidimensional IR/VIS histogram searches for the peaks CVVIS in the VIS channel. Then if CVIS is the relative variance, one subtracts from the bidimensional IR/VIS histogram the 'gaussion' distribution C, CVVIS, CIR, CVIS and so on for all the peaks found in the IR histograms. Same treatment is done for the WV, that is, using the bidimensional histogram IR/WV, but no subtraction is applied to it. Afterwards, each cluster extracted is identified by comparison against a preset interpretation table containing the radiance characteristics of several objects (sea, land, clouds). Peaks not corresponding to any reflector in this table are considered as contamination and merged into the remaining ones according to certain rule. The result of this analysis is that each reflector present in a segment

is described in terms of mean radiance and variance of it in all
the available channels.

2.   Displacement extraction

To determine it the first step is to use the IR images.
Individual segments from the central image of a sequence of three
are treated as target windows and a search is made over a 3 x 3
segment window in each of the adjacent images for a good match,
where the matching is ·achieved by the calculation of cross-
correlation coefficients for each possible displacement. This step
is done by a dedicated minicomputer with an attached array processor calculating the cross products by performing a Fast Fourier
Transform of the data. The two cross matrices are examined and
peaks of value greater than a given threshold (0.4) are selected.
Because of similarity in the images, spurious peaks may occur, so
only those representing displacements, which are within given
limits, symmetrically reversed in the two matrices are accepted as
representing a true wind, and a quality index depending on several
factors is assigned to every one of them.

3.   Attribution of the displacement to a layer

If there is only one layer, the symmetrical peak with the
highest quality index is assigned to the layer. If there is more
than one layer, a peak is assigned to a layer as follows. For all
the peaks and for all the layers present one recomputes, around
peaks positions, cross-correlation values after grey scale enhancement of one layer at a time supressing the others. The displacement and the layer producing the peak within the highest quality
index in such a matrix, are representative of the wind.

4.   Height assignment and conversion of displacement
     into a vector

After a displacement is determined, conversion to a vector is
achieved using a deformation model which computes the displacement
in the case of an ideal image. Then by means of a forecast profile, a table radiance-temperature-pressure is determined. This
automatic system is supplemented by a manual extraction method,
using the MIEC interactive system. Both types of wind determination

are followed by human action and finally by dissemination via GTS in SATOB code.

- Sea-Surface Temperature. If a reflector is identified as sea, its IR radiance value is corrected (if that is the case) according to its visible radiance for the sunglint effect, then the detection of atmospheric contribution takes place. The radiance, so corrected, undergoes quality control and then is converted to temperature according to a calibration table, giving for each radiance measured the equivalent temperature. After human quality control, this value is disseminated via GTS in SATOB code.

ACKNOWLEDGEMENTS

The author wishes to thank all those associated with this project for their past and continuing support.

Dundee University Meteorological Satellite Ground Receiving and Data Archiving Facility

P.E. Baylis

ABSTRACT

The University of Dundee VHRR/AVHRR data reception and archiving facility is described. Details of the normal routine operation of the station and the archived data products available are given.

INTRODUCTION

The Dundee facility is not an offical government funded station. It was created by a small group within the Department of Electrical Engineering and Electronics, University of Dundee, Scotland. The original station became operational in 1966 with APT transmissions from NIMBUS-2 and ESSA-2.

In April 1975 construction of an ITOS/NOAA VHRR receiving station was completed and images were produced from NOAA-4. Later that year the U.K. Natural Environment Research Council awarded the group enough funds to staff the station in order to archive VHRR data on magnetic tape. More recently, the grant has been extended to up-grade the station to receive and archive TIROS-N/NOAA-6 AVHRR data.

OPERATIONAL ROUTINE

NOAA-5 VHRR

NOAA-5 was launched in July 1976 into a 0835 LST southbound equator crossing sun synchronous polar orbit with a height of 1520 Km. Recording at Dundee commenced on August 23rd, 1976, and continued until termination of the spacecraft on February 28th, 1979. Basically, recordings were taken from the closest to overhead U.K. pass each morning and evening.

On occasion, by prior arrangement, extra recordings were taken from passes further to the east or west in support of special projects. Typical examples were the British North Pole Expedition in 1977, JASIN in 1978 and a U.S. Navy project in the eastern Mediterranean in 1978.

There are a total of 2400 recordings in the NOAA-5 archive.

TIROS-N/NOAA-6 AVHRR

TIROS-N was launched on October 13th, 1978, into a 1500 LST northbound equator crossing sun synchronous polar orbit, with a height of 850 Km. Recordings from afternoon close to overhead U.K. passes only were commenced soon after launch. Due to the reduced swath width caused by the lower orbit, two successive passes are always taken. The early morning passes are not recorded due to inadequate staffing.

NOAA-6 (AVHRR) was launched into a similar orbit with 0730 LST southbound equator crossing on June 27th, 1979. Recording from the morning passes commenced after switch-on of the thermal channel about 20 days after launch. Extra recordings from TIROS-N/NOAA-6 passes not normally taken, can be archived by prior arrangement.

COVERAGE

Dundee is at latitude 56.458°N longitude 2.97°W. Figure 1 shows satellite horizon for NOAA-5 and TIROS-N/NOAA-6. Diameter of the NOAA-5 circle is approximately 8000 Km. and TIROS-N/NOAA-6, approximately 6000 Km. VHRR scans from horizon to horizon, but earth curvature limits useful swath width to about 4800 Km. AVHRR scan width is truncated at 3000 Km.

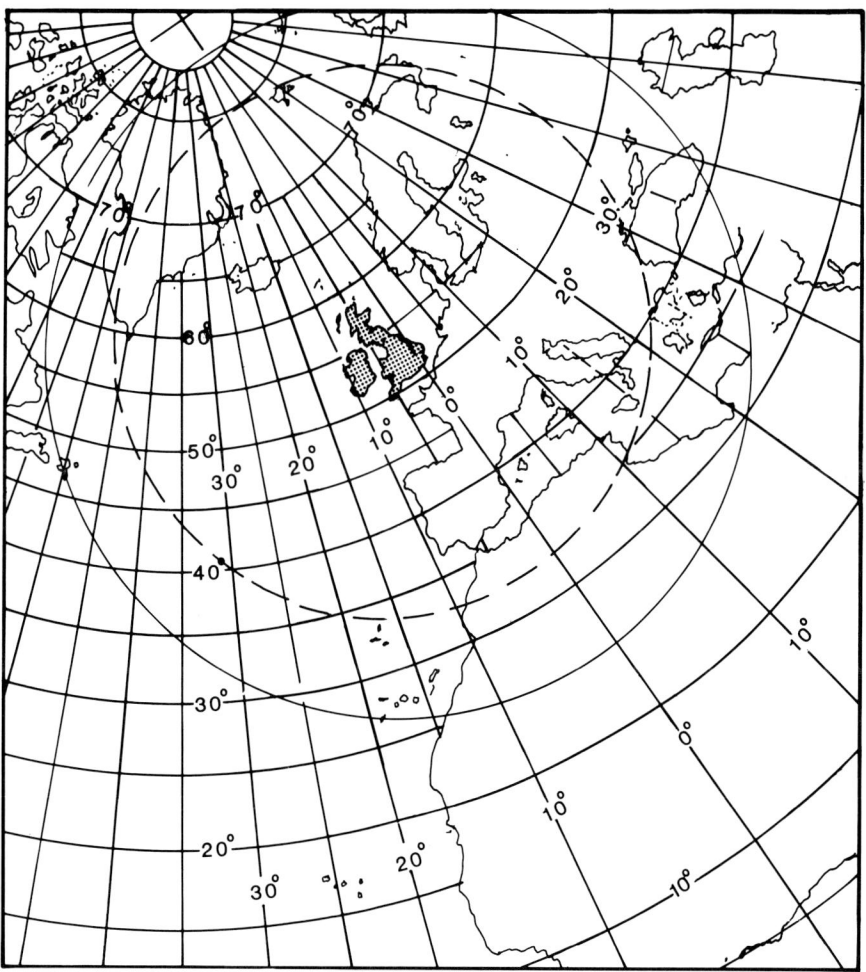

FIGURE 1: Satellite horizons as seen at Dundee, Lat. 56.458°N long. 2.97°W. Outer circle NOAA-5, inner circle, TIROS-N/NOAA-6.

SCANNER CHARACTERISTICS

VHRR

VHRR is a two channel radiometer with spectral responses 0.6-0.7 μm and 10.5-12.5 μm. Scan rate is 400 lines/minute and the nadir resolution is 0.9 Km. An overhead pass takes 23 minutes horizon to horizon, yielding about 9000 scan lines. The spacecraft/ground link is analogous and such that about 100 grey scale steps over the radiometer dynamic range can be resolved.

AVHRR

AVHRR is a four channel radiometer with spectral responses (1) .55-.68 mm, (2) .725-1.10 mm, (3) 3.55 - 3.93 mm, (4) 10.5-11.5 mm. Later instruments will be fitted with a fifth channel, 11.5-12.5 mm. Scan rate is 360 lines/minute and the nadir resolution is 1.1 Km. An overhead pass takes 15 minutes horizon to horizon, yielding about 5400 scan lines. The spacecraft/ground line is digital with 10 bit binary words so that the grey scale resolution is about an order of magnitude better than for VHRR.

DATA HANDLING

NOAA-5 VHRR

VHRR data were recorded unprocessed on an S.E. 7000A 14 track machine running at 30 i.p.s. The spacecraft FM 99 KHz subcarrier was placed directly on the tape along with a Doppler corrected 12 KHz sync. tone, frequency multiplexed on to the same track. One track was sufficient for one pass so that each 4600' x 1" reel of tape contains 14 passes.

Demodulation of the spacecraft 99 KHz FM subcarrier yields a time multiplexed video waveform with a video bandwidth of 35 KHz. At Dundee the video is digitized into 8 bit words in order to aid processing. Data from one of the two VHRR channels are clocked into a buffer memory, line by line. Read out rate is varied in a manner which corrects scale compression at the swath edge caused by earth curvature. The linearization algorithm is stored in a PROM which can be changed to suit a spacecraft launched into an orbit of different height.

One channel was printed on a large negative (10" x 8½") in a photofacsimile machine in real-time during the pass. A grid was added during contact printing and a positive print placed in the station browse file. Extra copies were distributed to users on the mailing list. The visible channel was similarly printed from the tape after day-time passes.

The linearizer is also arranged to read out selected sectors of each scan line in order to produce sectorized electronic enlargement. Three enlargement factors are available which match the available line densities in another photofacsimile machine. They are x2 (300 lines/inch) x4½ (150 lines/inch) and x6.25 (100 lines/ inch). The image size is 10" x 10" and contains 3000, 1500 or 1000 scan lines. Any part of the full pass image, vis. or IR, may be selected for sectorized enlargement. Selection is aided by a transparent overlay placed over the full pass image in the browse file. A vertical scale is calibrated in tape metres and horizontal scale, in numbers which are dialed up on an input decade switch on the linearizer/sectorizer front panel.

TIROS-N/NOAA-6 AVHRR

AVHRR data are recorded unprocessed on an S.E. 7000M 14 track machine running at 29.5 i.p.s. The spacecraft 665.4 kb/s serial bit stream is converted from NRZ to DM (Miller code) and placed directly on to a single track with a packing density of 22.5 kb/ inch. A 4600' x 1" reel of tape can accommodate 28 passes.

Any of the five AVHRR channels may be selected from the serial bit stream and passed through a linearizer/sectorizer similar to the one used for VHRR. Due to the lower height of TIROS-N/NOAA-6, the enlargement factors are reduced to x1½, x3 and x4½.

After linearization/sectorization, the 10 bit pixel words are passed through a digital processor and D/A converter before hard-copying in a photofacsimile machine. The processor can select any

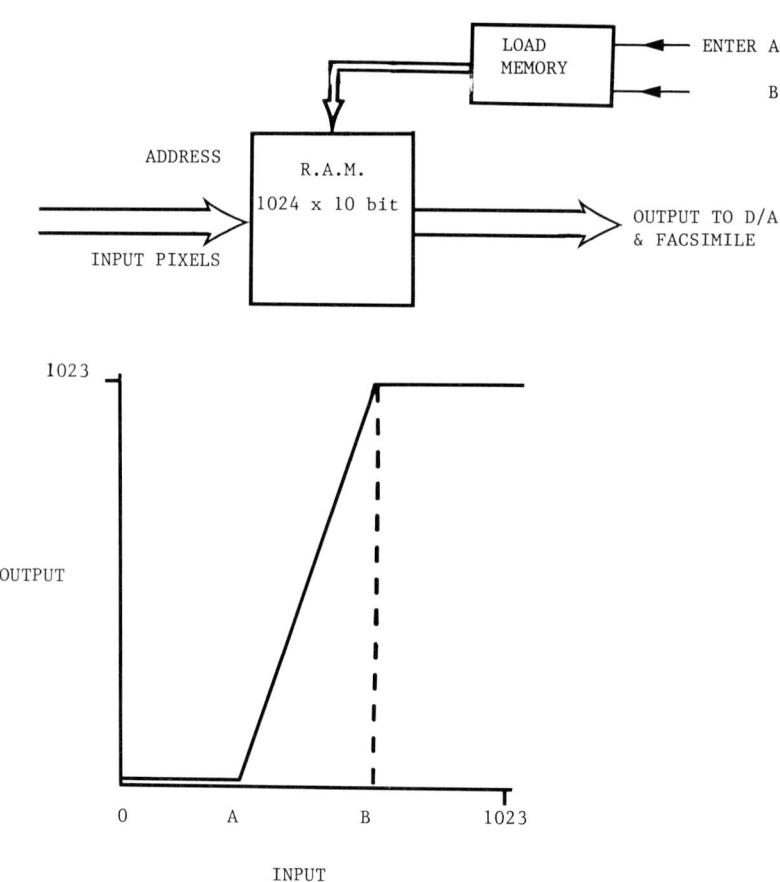

FIGURE 2: TIROS-N/NOAA-6 digital video processor.

part of the 0-1023 (10 bit) dynamic range of the video and match it
to the photofacsimile machine. Selection is by two analogue ten
turn potentiometers and the actual lower and upper pixel values
selected appear on a front panel LED digital readout. The processor consists of a RAM addressed by the incoming pixel numbers.
Numbers at the addressed memory locations form the output. Loading
of the RAM by appropriate numbers is under the control of the two
potentiometers, see Figure 2. The processor is particularly useful
for producing sea-surface temperature images. Typically, such

images are generated using less than 40 grey scale steps out of the
available 1024, see Figure 3.

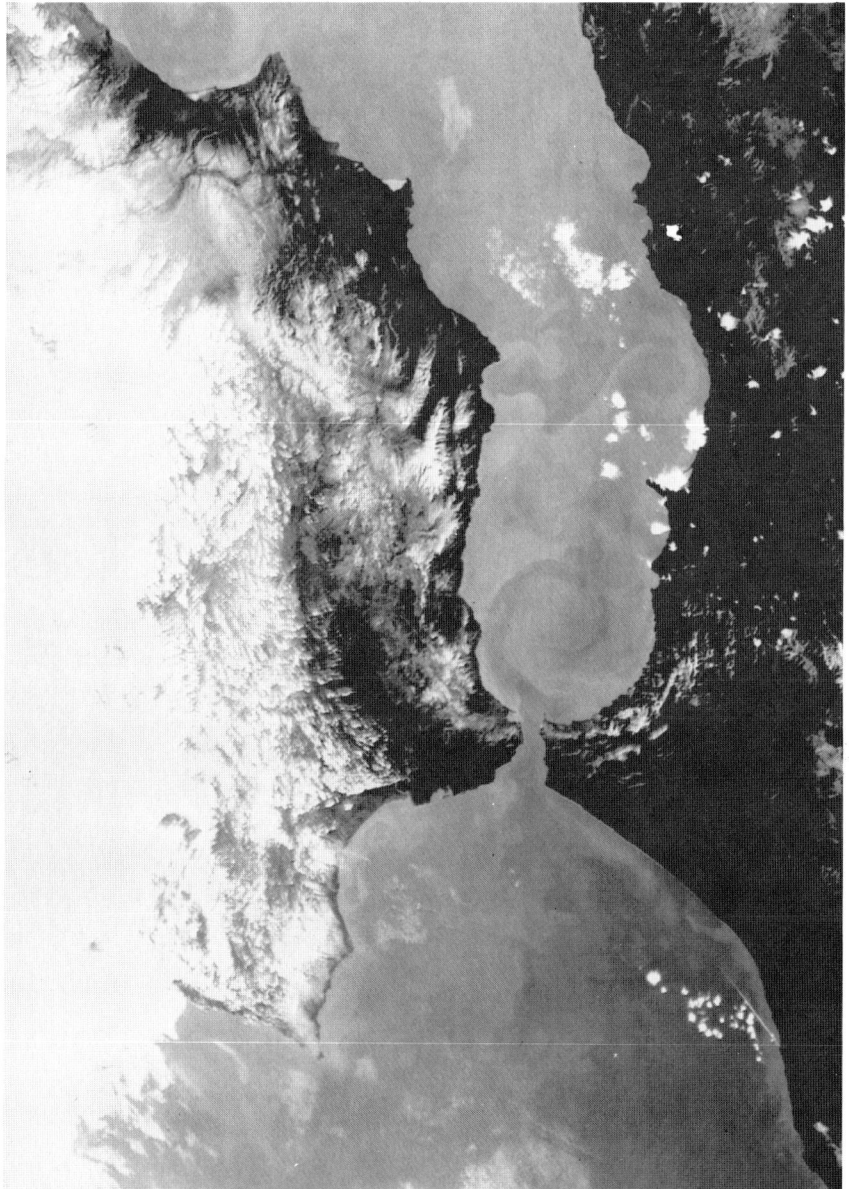

FIGURE 3: Grey scale stretched channel 4 (IR) AVHRR image showing
sea surface temperature patterns in the Straits of Gibraltar at
1440 GMT on December 12, 1978. The video processor was adjusted so
that grey scale level 478 corresponds to black (warm) and 510 to
white (cold).

Channel 1 (visible) and channel 4 (thermal IR) full pass
gridded images are produced routinely from each pass recorded, for
the station browse file and distribution.

COMPUTER COMPATIBLE TAPE RECORDER

An interface between the high bit density AVHRR archiving
recorder and a CCT transport, reassembles 3 successive 10-bit words
into 4 8-bit words. Blocks of 8-bit words are then recorded
PE/1600 BPI on the transport. By this means AVHRR data archived at
Dundee can be interfaced with any computer or image processing
facility that can accept standard PE/1600 BPI CCTs. Another interface between the transport and a photofacsimile machine, allows the
return of processed CCTs for hard copying into black and white
images with line densities of 100, 150 or 300 lines/inch. At
present there is no interface available between data in the VHRR
archive and the CCT transport.

SUMMARY OF AVAILABLE PRODUCTS FROM ACRHIVE

Hard Copy Photofacsimile Positive Images (all linearized):

MAGNIFI-CATION	LINES/INCH	SIZE	TOTAL LINES	GRID	AREA COVERAGE H x W
VHRR x1	625	10" x 8½	6250	OPTIONAL	5625 x 4400 Km
VHRR x2	300	10" x 10"	3000	NO	2700 x 2700 Km
VHRR x4½	150	10" x 10"	1500	NO	1350 x 1350 Km
VHRR x6¼	100	10" x 10"	1000	NO	900 x 900 Km
AVHRR x1	450	9½" x 6¼"	4275	OPTIONAL	4700 x 3000 km
AVHRR x1½	300	10" x 10"	3000	NO	3300 x 3300 Km
AVHRR x3	150	10" x 10"	1500	NO	1650 x 1650 Km
AVHRR x4½	100	10" x 10"	1000	NO	1100 x 1100 Km

Browse File

At least one channel image from every recording in the VHRR/
AVHRR archive has been hard copied and placed in the station browse
file. In addition, 35 mm contact prints, 36 to a page, are distributed regularly to users on the mailing list. Back numbers can be
supplied. Browse file sheets are either positive prints or positive transparencies.

CCTs

Selected scenes from the AVHRR high bit density tape archive may be transcribed on to CCT PE/1600 BPI.

METEOSAT RECEIVER

The Dundee facility is also equipped with an in-house constructed digital Primary Data User Station and an analogue Secondary Data User Station. Any of the PDUS/SDUS formats can be hard copied in photofacsimile machines. Data are not archived routinely, recordings are only taken as required by prior arrangement.

The station antennas are shown in Figure 4:

APT	-	Automatic picture transmission
ITOS	-	Improved TIROS operational satellite
NOAA	-	National Oceanic and Atmospheric Administration
VHRR	-	Very high resolution radiometer
AVHRR	-	Advanced very high resolution radiometer
JASIN	-	Joint air sea interactive network
PROM	-	Programmable read only memory
D/A	-	Digital to analogue
RAM	-	Random access memory
CCT	-	Computer compatible tape
PDUS	-	Primary data user station
SDUS	-	Secondary data user station

Directions in Remote Sensing, Past Present and Future

Dag G. Gjessing

Progress in radio science innovates the environmental sciences.

The last decade has been characterized by substantial progress in the field of radio science. This progress has largely been stimulated by radio communication interest and by military requirements. In searching for better and more reliable communication channels, one has been obliged to consider a wide spectrum of geophysical/biological factors affecting the propagation of electromagnetic waves. Atmospheric waves and turbulence are responsible for long range, over-the-horizon, scatter communication in the troposphere. The earth topography and vegetation have a dominating influence on ground wave communication, and the irregular structure of the sea surface influences the propagation of radio waves which relies on reflection between the ionosphere and the earth surface. The propagation of optical electromagnetic waves (laser radiation) is strongly dependent on the molecular

403

structure of the propagation medium. In using electromagnetic waves (microwaves and optical waves) for detection/indentification of military objects, an improved understanding of the interaction between electromagnetic waves and environmental objects was achieved.

In parallel with, and decoupled from this progress in radio science, an accelerated interest in the field of life sciences has taken place. The progress in the field of environmental sciences is very largely a result of an upsurge of interests in problems concerned with the earth's resources, pollution, and conservation, leading to a general need for a more detailed understanding of the governing principles in environmental systems and in living organisms. We have now reached the point where the two fields are merging, resulting in a proliferating progress in radio science.

The great challenge for the radio scientists now is one of matching new technological achievements to our particular environmental application. To achieve this, to bridge the scientific gap between the radio scientist and the broad field of life sciences, one shall need a detailed understanding of the mechanisms of interaction between electromagnetic waves (the tool of the physicist) and a wide spectrum of parameters describing the environmental state variables.

New components and methods in the field of microwave technique, electro-optics, computer technology, statistics, and electromagnetics open new and very interesting possibilities with regard to detection and identification problems in relation to surveillance of environmental factors, and in relation to detection and indentification of objects and substances in general. On the basis of new concepts (interaction of electromagnetic waves with surfaces and influence of propagation medium on the electromagnetic waves) and on the basis of new technology (surface acoustic waves, computer controlled solid-state frequency generators, charge-coupled devices for pattern matching, opto-acoustic systems, high-speed micropressors, etc.) environmental surveillance systems can be designed so as to give optimum information about the object of interest.

The basic principle of such a "target adaptive surveillance system" is the following: Most of the existing detection/identification systems do not, as a result of limited system flexibility, make optimum use of all the a priori information that one generally is in possession of with regard to the object of interest. Knowing the geometrical shape of the object of interest, and its molecular surface structure, an illumination function can be structured which gives optimum system sensitivity (minimum receiver bandwidth) with respect to the object of interest at the expense of the sensitivity for background objects.

In order to construct such an optimal detection system, we shall need information about the propagation medium, and in the limiting case, we shall also need information about the properties of the background. Figure 1 illustrates this statement. Thus, when viewing the object (or the chemical substance) of particular interest through an atmosphere which absorbs and distorts the rays from our illuminator, we shall select the detection and identification criteria of the environmental parameters in such a way, so as to obtain maximum information and minimum "noise contribution".

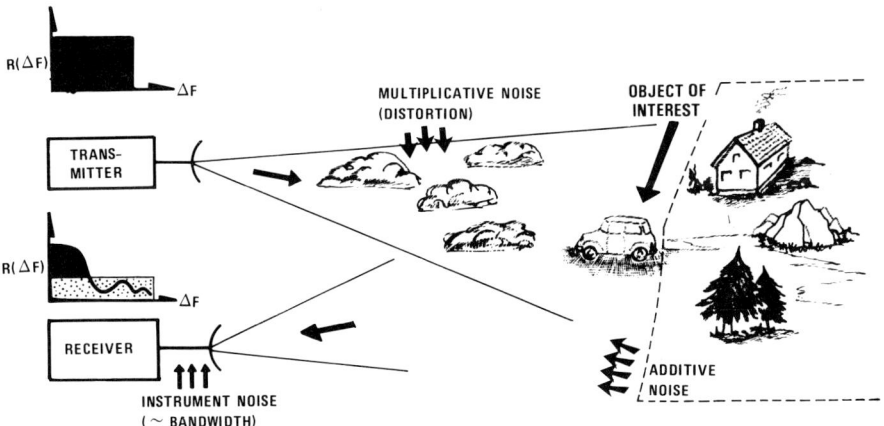

FIGURE 1: Symbolic presentation of the general detection/identification problem. Noise, additive and multiplicative, suppress the signal from the object of interest. The task is to structure the illuminating wave in such a way, so as to obtain optimal signal to noise ratio from the object of interest.

<u>Potential criteria for detection and identification of</u>

<u>environmental parameters when use is made of new technology</u>

New technology and new methods open very interesting possibilities in regard to the detection/identification problem. In exploring new detection criteria for our environmental remote sensing problems in relation to potential techniques, the relative merits of these criteria should be investigated. Potentially, there are numerous such criteria. However, employing a phenomenological classification rather than an application oriented one, a limited number founded on basic physical principles emerge. A set of these are as follows:

- Detection/identification on the basis of shape and size of object (macrostructure of the environment).

- Detection/identification on the basis of the chemical composition of the object (molecular microstructure).

- Detection/identification on the basis of the gas which the environmental feature may emit (scents from plants, odours from organic substances, gases emitted from minerals, etc).

- Detection/identification on the basis of the motion pattern of the object (motion pattern of vegetation, wave motion of the ocean).

- Classification of the environment or of a specific environmental object on the basis of the distribution of temperature over the scene of interest using high resolution radiometry (heat capacity and thermal conductivity mapping).

The environmental object will in general possess useful detection criteria (fingerprints) in various spectral ranges. Some of these are weak, and some are pronounced. For the purpose of emphasizing the contrast between objects of interest against a background and through an atmosphere causing absorption and blurring, we shall have to consider the fingerprints of a wide spectrum of objects and of backgrounds, and also the properties of the intervening medium, the atmosphere, through which the electromagnetic waves will have to pass in order to reach the scene of interest.

Figure 2 illustrates in the way of an artist's conception a set of detection/indentification methods based on molecular absorption processes and tunable laser systems and also what information can be obtained on the basis of scattering/diffraction mechanisms.

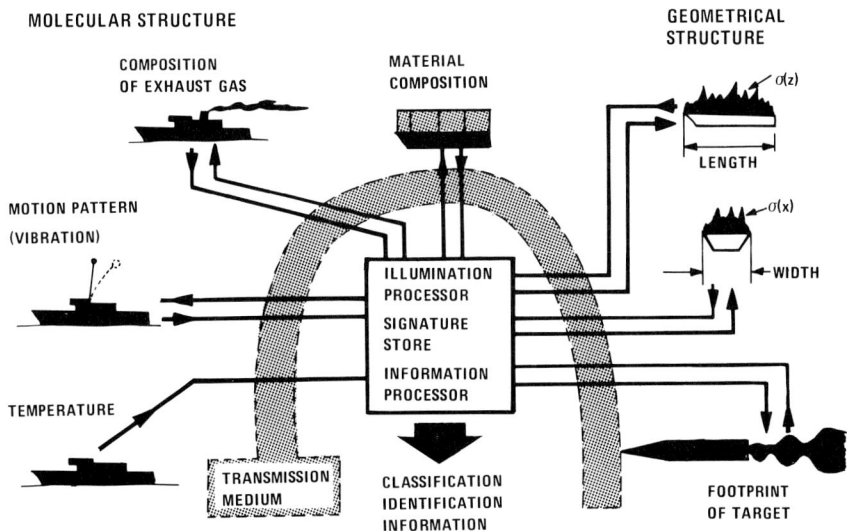

FIGURE 2: Environmental surveillance methods based on molecular absorption mechanisms making use of electro-optical systems with frequency tuning (tunable laser systems) and of scattering/diffraction of radio waves.

Fundamental radio science principles in the study of the geometrical properties of the environment (shape and size of object).

Under this heading there are two types of cases. The first concerns the case when we have at our disposal an electromagnetic radiator capable of illuminating the object of interest in a pre-described manner (composite spectrum of radio wavelength). The second deals with the situation when we have at our disposal one illuminating frequency and an array of spatially distributed receiving elements capable of determining the spatial wave pattern (correlation properties in space of the scattered wave).

We first consider detection/identification on basis of an electromagnetic radiator capable of illuminating the object of interest in a predescribed manner (composite spectrum of wavelengths). Consider first the simplest of all targets consisting of two reflection points spaced Δz longitudinally along the direction of wave propagation. We shine a wide spectrum of frequencies on this object, and analyze the set of interferring waves received. We note that as the difference in frequency is changed, the situation arises when the waves reflected from the two reflection points arrive in antiphase at the receiving site causing a minimum in the frequency spectrum. If the distance in depth between the two reflecting elements is Δz, the phase angle between the waves reflected from the two reflectors is given by

$$\phi = 2\pi \frac{2\Delta z}{\lambda_1} = 2\pi \frac{2\Delta z \, F_1}{c} \tag{1}$$

for the frequency F_1. To change the phase angle ϕ through 180° so as to move from constructive interference to destructive interference requires a frequency change to F_2 such that

$$2\pi = \phi = 2\pi \frac{\Delta z \, F_2}{c} \tag{2}$$

The distance in frequency between two successive inteference minima, is then given by

$$F_2 - F_1 = \Delta F = \frac{c}{2\Delta z} \tag{3}$$

Then consider the more general situation with an object in the form of a periodic structure (a tree with a set of equally spaced branches). We express the size of the object as Δz, and the fine scale structure as δz (exponential damped sinusoidal oscillation). The object is a distance z_o from the transmitter. We illuminate this object with a set of correlated frequencies. The source is a large number of frequency generators which are derived from a common source. When this set of electromagnetic waves is coupled to the general object shown, the reflected wave is decorrelated frequency-wise as shown in Figure 3: The maximum of the correlation envelope is shifted a distance $\frac{c}{2\delta z}$ (where δz is the finest

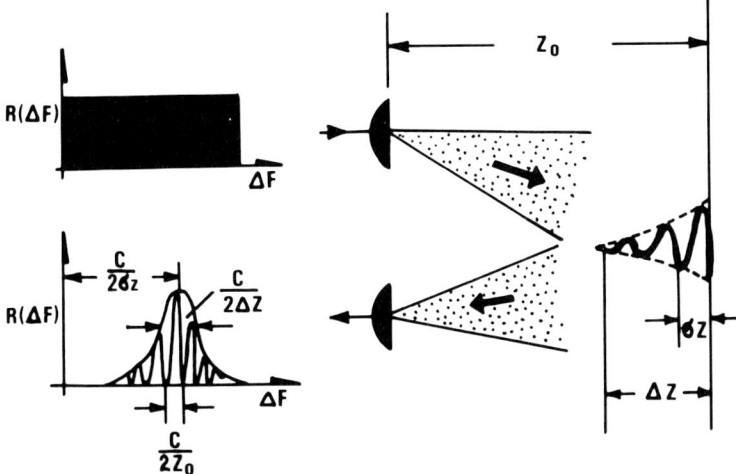

FIGURE 3: Illuminating the object with a set of electromagnetic waves with different frequencies, the correlation function in the frequency domain of the reflected wave gives information about the distance to the object, the size of the object, and the shape.

structure of the object under investigation). The width of the correlation envelope is determined by $\frac{c}{2\Delta z}$ where Δz is the size of the object. Finally, under the envelope of the correlation function there are periodic oscillations stemming from the fact that the system delay-function is non-symmetrical, resulting in a complex autocorrelation function in the frequency domain. In analyzing the fine scale oscillation of the correlation function, we find a period which is given by

$$\frac{c}{2z_o}$$

where z_o is the distance to the target.

To emphasize these very important properties of the scattered/ diffracted field, Figure 4 is presented.

In conclusion, the autocorrelation in the frequency domain of a scattered/diffracted electromagnetic wave is determined directly from the Fourier transform of the objects' delay function. Knowing the position, the size and the shape of the object, the correlation function in the frequency domain is determined. Figure 5 shows

DISTANCE TO OBJECT

SIZE OF OBJECT

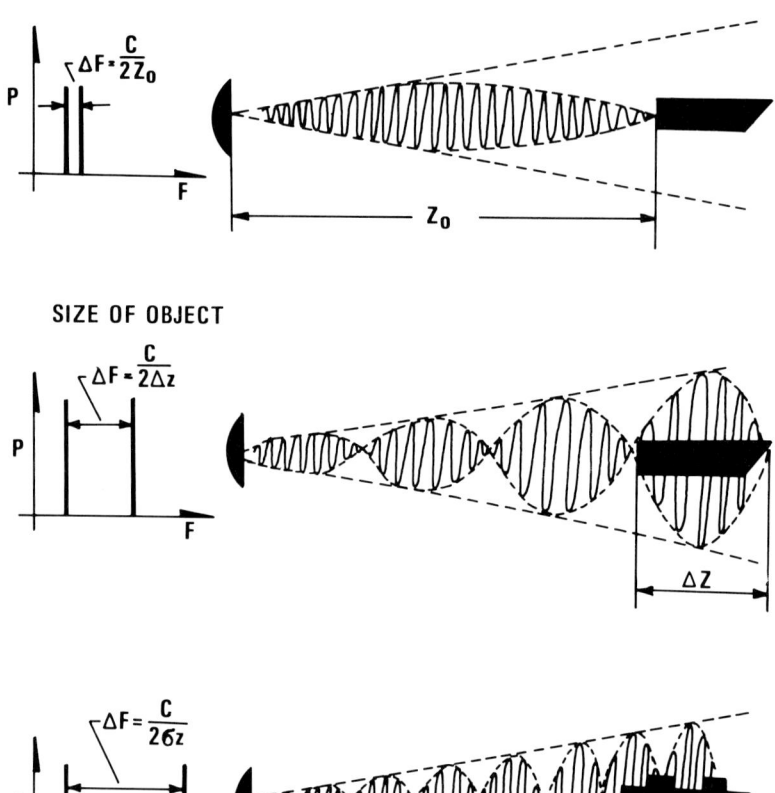

FIGURE 4: The distance to the object z_o is determined by transmitting two microwave frequencies with spacing

$$\frac{c}{2z_o}$$

the size of the object is obtained in the same manner by transmitting two frequencies with spacing $\frac{c}{2\Delta z}$ Finally, the shape is obtained by transmitting waves with the larger frequency separation $\frac{c}{2\delta z}$.

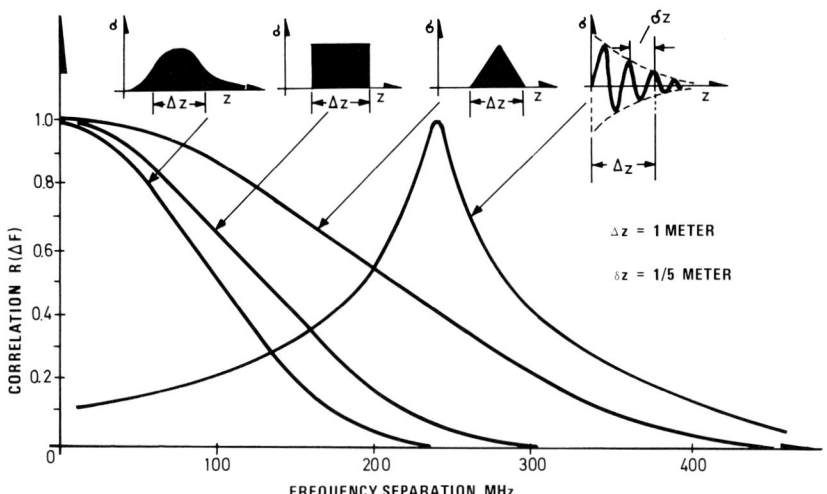

FIGURE 5:   Theoretical results illustrating the identification potential of a multifrequency illumination system.

such correlation functions for a set of different objects whereas Figure 6 shows the signature of ships against a background of sea clutter.

Measurement of object shape and size based on spaced receiving antennas

In the same way as above where the distribution in depth of an object is determined by the autocorrelation function of the scattered field in the frequency domain, the transverse distribution of the object is determined by measuring the spatial correlation properties transversely to the scattered wave. This property of the scattered field is depicted in Figure 7. Two scattering elements a) and b), spaced Δx apart along a direction normal to the propagation direction of the scattered field, give rise to a spatially periodic interference pattern at the receiver. Measuring the spatial period of this interference pattern, gives information about the transverse size Δx of the object. It can be shown (see reference above) that if $\sigma(x)$ is the spatial distribution of the object in the x-direction, then the spatial autocorrelation function of the field along the x-direction at the receiving site is

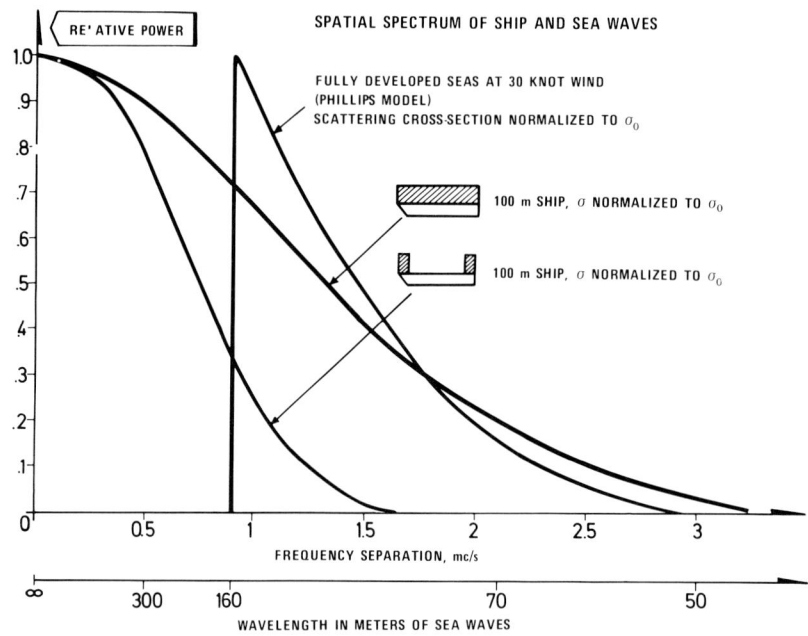

FIGURE 6: Theoretical results, based on the Phillips model of fully developed seas at 30 knots wind showing the "signature" of ships relative to that of the ocean surface.

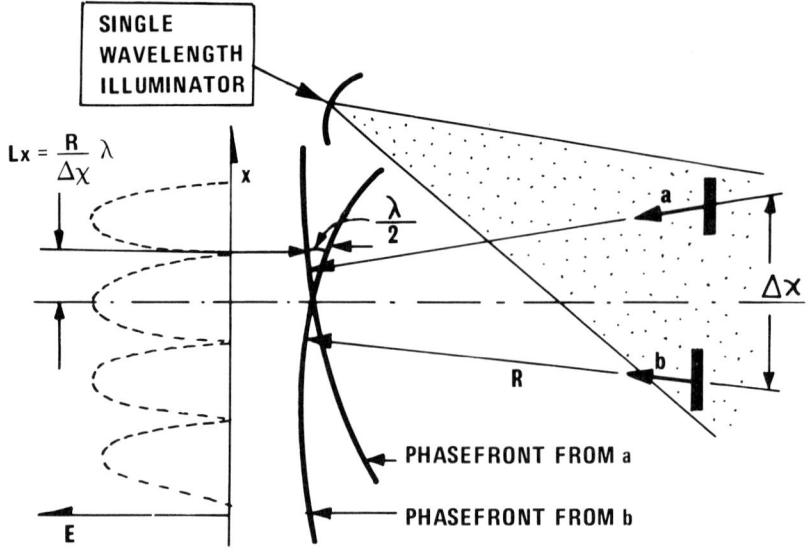

FIGURE 7: Transverse distribution of field strength. The size (width) Δx of the object manifests itself as a "transverse inter-ferogram".

given as the Fourier transform of the σ(x) function normalized with
the distance R to the object.

Thus, in order to measure the "transverse shape" of an object,
we shall need a set of spaced antennas (a broadside array). Figure
8 shows such an array system coupled to a microprocessor.

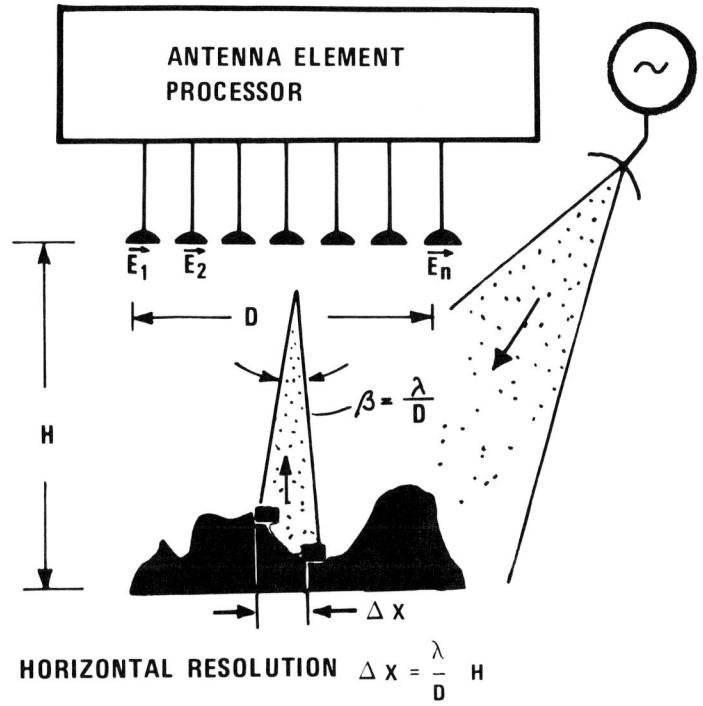

FIGURE 8: Wavefront analysis by array matrix. Spatial correlation
of field strength $R_E(x)$ = Fourier transform of angular distribution
of scattered power. From this the transverse shape of the object
is directly obtained.

If one wants to resolve the fine scale structure of the
object, the phase relationship between distant antennas should be
observed. On the other hand, if the interest is focussed on the
large scale properties of the object, this information is obtained
from antenna elements which are closely spaced. This is in direct
analogy with the case discussed above with spaced frequencies.
Note that the principal of the side looking radar relies on the
spatial property of a scattered electromagnetic wave. An example

from a recent accomplishment is the NASA SEASAT A satellite
launched in 1978, where a ground resolution of some 10 m was
achieved from satellite height.

Sketch of a target adaptive radar system (frequency/time)

The discussion above, dealing with well established physical
principles, forms the basis for a target adaptive radar system.
Figure 9 shows a specific target adaptive radar system, which is
under development by the author's organization for the purpose of
studying ocean waves and also ship signatures. A microprocessor
controls a set of frequency synthesizers, so as to illuminate the
object of interest with a suitable set of frequencies. The
receiver consists of a set of narrowband filters capable of receiving each of the transmitted frequencies. The signal from these
receivers is converted from analogue to digital in the same microprocessor, which in turn performs the appropriate computations so
as to display the autocorrelation functions in the frequency domain
(distribution in depth of the scattering object) within each time
frame of interest. Thus, if the object is at rest and if the
transmission medium is frozen, then one would expect to have a set
of identical correlation functions along the time axis. If, however, the object changes aspect angle (e.g. yaw, roll and pitch
motion of a ship) then we would also see a structure in the time
domain. Referring now to Figure 3 above, it is self-evident that
the microprocessor can be programmed so as to investigate each of
the target properties sketched in Figure 3: range, size of object,
shape of object. Referring again to the receiving end of Figure 9,
we see, that is we know the range to the object, the filter F_1 and
F_n can be infinitesimally narrow, and we obtain a system with
extreme sensitivity. Similarly if we know the size of the target
of interest, and we are interested in knowing whether there is such
a target at distance z_o, then we need two narrow banded receiving
filters spaced $\Delta f = \dfrac{c}{2\Delta z}$ apart. This would give us a system with
extreme sensitivity related to objects of size Δz. A similar
system based on two transmitter frequencies is now in use by Plant
et al. at the Naval Research Lab in Washington. The system is
designed specifically for sea wave investigations.

TARGET ADAPTIVE RADAR SYSTEM
FREQUENCY/TIME

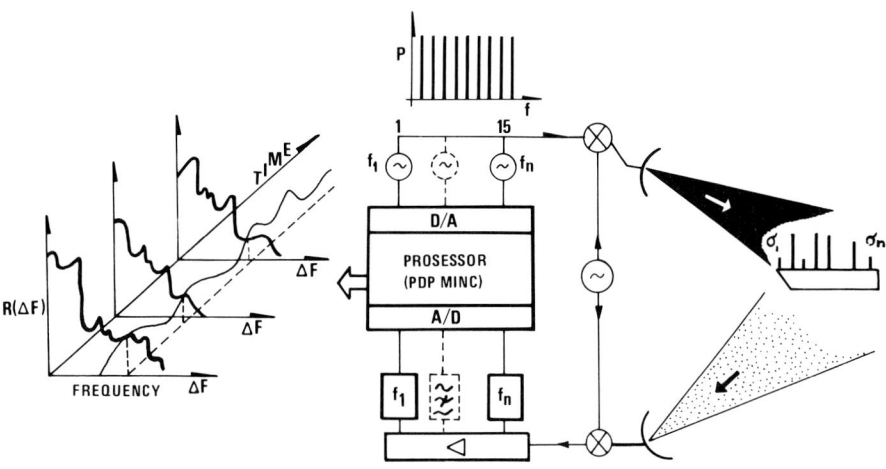

FIGURE 9: A schematic diagram of a target adaptive radar system which in its simplest form is being developed by the author's organization. A microprocessor system measures distance to object, size and shape in turn. By keeping track of the time history of the relevant descriptive parameters $(z_o, \Delta z, \delta z)$ the motion pattern is obtained.

In general, if we know the distribution in depth $\sigma(z)$ of the object of interest (i.e. we know the delay function of the object), then we should illuminate the object with a set of frequencies whose mutual distribution is some inverse function of the delay function characterizing the object (see Ann Arbor reference). The result of this illumination would be a reflected wave whose correlation properties in the frequency domain is reduced to a delta function requiring an infinitesimally narrow bandwidth for detection. This is, not surprisingly, in accord with the classical concepts from information theory: It requires zero bandwidth to convey a known message, it takes zero bandwidth to detect a sinusoidally varying signal when its period is known.

Methods based on molecular absorption phenomena

One characteristic property of an object is, as we have seen, its geometrical shape and size. Now let us consider what information can be obtained about the molecular composition of the object.

We are specifically interested in the possibility of identifying certain objects by their material composition, such as minerals, vegetation, and ecological objects in general. We are also interested in determining the surface composition of certain planets, so as to form an opinion as to the influence of pollution on these organisms. Pollution in general can either be assessed by measuring the pollution burden (e.g. the spatial concentration of a given polluting gas) or we can measure the effect of this pollution on living organisms. Since the degree to which a substance should be classified as a pollutant depends on its adverse effects on living organisms, it will become increasingly important to measure the state of life and chemical surface composition of the organism itself.

When we discussed methods by which the geometrical shape of an object can be determined remotely, we relied on the backscattering capability of the object. In order to determine the material composition, we shall rely on the capability of the object to absorb electromagnetic waves. Specifically we select an illuminating wavelength, which stimulates resonance phenomena in the molecular surface structure of the object. Hitting one of these resonance frequencies causes the material to absorb electromagnetic waves. Noting the frequency at which electromagnetic energy is absorbed, and also the amount of absorbed energy, gives us information about the absorption spectrum which again reveals the molecular composition of the material. Having at our disposal electromagnetic waves in the range from microwaves through infrared waves and visible light to the ultraviolet region, gives us the capability of determining a large number of material compositions.

Consider as an example a di-atomic molecule within which potential forces are acting; the atoms are bound together by elastic forces. If these atoms, having finite mass, are excited by an alternating electromagnetic field, resonance will occur when the stimulating frequency equals that corresponding to the difference between two quantized vibrational energy levels. We can consider these two atoms with masses m_1 and m_2, respectively, as being tied together by forces that can be represented as a ball and spring

system where the restoring force obeys Hooke's law. The frequency
of this simple harmonic motion, oscillating ball and spring system,
is given by

$$v_o = \frac{1}{2\pi} \text{ x } (\frac{K}{\mu})^{\frac{1}{2}} \qquad\qquad (4)$$

where

K = spring constant in Hooke's law

$$\mu = \text{reduced mass } = \frac{m_1 \cdot m_2}{m_1 + m_2}$$

By considering the energy which is being absorbed, we obtain another equation:

$$E_v = (v + \tfrac{1}{2}) \text{ x } \frac{h}{2\pi} \text{ x } (\frac{K}{\mu})^{\frac{1}{2}} \qquad\qquad (5)$$

where

v = 0,1,2, 3 .... is the vibrational quantum number.

The above example applies for two atoms. A complex molecule
will have a large number of oscillating modes leading to numerous
energy levels. Still, if we know the structural details of the
molecule of interest, we can in principle calculate the absorption
spectrum. Also, if we have access to the material of interest,
which generally is the case, then the absorption spectrum can be
obtained experimentally. In any case, we have information about
the "fingerprint", so as to know what to look for. We shall now
discuss methods by the application of which chemical substances of
known molecular structure (absorption spectrum) can be detected and
identified selectively.

One can think of many such systems and, indeed, many have
already been developed. As the flexibility of laser systems
improves, so does the sensitivity and the agent specificity of the
detection systems. Many systems are based on lasers which can be
tuned between two specific wavelengths only. One wavelength is
selected so as to correspond with the absorption line of the
material of interest, whereas the other line is selected, so as to
give minimum absorption. This differential absorption system has

typically a sensitivity of one part per billion in relation to
gases, when the gas of interest is distributed over typically 1 km
and the measuring path is correspondingly 1 km long. If attention
is to be focussed on another gas, we shall have to select different
pairs of frequencies. The progress in the field of laser physics,
however, has now reached the stage where a large number of radiation wavelengths can be produced in rapid succession. Using a CO_2
laser tuned by means of a rotating grid system, a periodic sequence
of something like one hundred 20 watt lines can be produced within
a period of milliseconds. With such a powerful laser tool, a
number of chemical compounds can be detected simultaneously and
down to very minute concentrations. Furthermore, systems will soon
be designed, where use is made of lasers with "tailored" illumination. The wavelength distribution is tailored in an optimum manner
in relation to the specific chemical compound of interest.

Two such multifrequency laser systems will now be discussed.
The first is being developed by the author's organization, whereas
the second exists in the form of a computer verified concept only.
Use is being made of a tunable CO_2 laser in the study of space and
time distributions of a set of atmospheric gases (Figure 10). The
CO_2 laser is tuned from one frequency to the other in a pre-
described manner. The laser beam is directed to a reflector which
upon reflection diverts the beam back to the receiver. Measuring
here the received power as a function of radiation wavelength, the
absorption spectrum of the composite gas mixture over the path
between the transmitter/receiver and the reflector is obtained. In
order to measure the concentration of a specific gas of interest,
this composite absorption spectrum is compared with the absorption
of the particular agent of interest. Basically there are many such
"pattern recognition methods" at hand. The relative merits of
these schemes of pattern identification is determined by the nature
of the composite spectrum, by the mixture of individual patterns.
If the composite spectrum is a sum of individual absorption spectra, the agent of interest is efficiently filtered out by a cross-
correlation process as shown in Figure 10 where the composite
spectrum is cross-correlated with the absorption spectrum of the

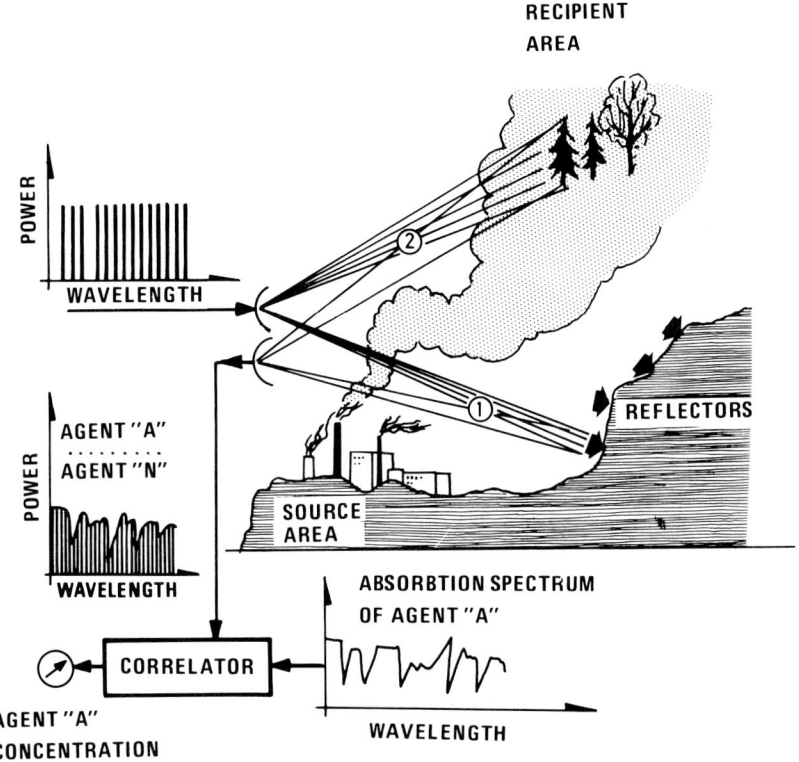

FIGURE 10: A multiwavelength laser system transmits a succession of wavelengths through the atmosphere to be investigated and back to the receiver employing a passive reflector. By comparing the multi-agent absorption spectrum appearing at the receiver with a stored signature of the agent of interest, information about the space averaged concentration of this agent is obtained. The gas spectrometer can be converted to a surface reflection spectrometer (configuration 2 in the diagram) by shining the laser beam onto the surface to be analyzed.

agent of interest. New applications of classical concepts from information theory have led to many powerful information retrieval schemes. One group of such is known as Kalman filtering methods; another more recent one, is based on maximum entropy considerations.

Returning again to Figure 10, note that such a long path laser absorption spectrometer gives the spatial integral of the gas concentration between the position of the transmitter/receiver and that of the reflector virtually instantaneously. As depicted in

Figure 10, spatial profiling is achieved by a suitable positioning of the reflectors. Note also that the system can be converted from a gas absorption spectrometer to a surface reflection spectrometer if the surface to be investigated is interchanged with the mirror reflector. Note that this reflection spectroscopy scheme requires a very sensitive receiver since the reflection properties of a topographic reflector are very much inferior to those of the mirror reflector. Thus, in order to obtain a sufficient signal to noise ratio a superheterodyne receiving system is generally required. These, however, are now about to become available offering an improvement in sensitivity in the order of 60 dB over the conventional direct detection system. Note also that in order to distinguish between the absorption caused by the intervening gas and the absorption caused by the reflecting surface to be analyzed, the spectrum of the intervening gas has to be different from that of the surface.

Distortion effects in the intervening tranmission medium (multiplicative noise contribution) can be minimized if use is made of "matched illumination". The wavelength distribution of the illuminator is tailored in an optimum manner in relation to the specific chemical compound of interest. This matched illumination scheme is shown in Figure 11 (for details see the reference list). We have at our disposal a laser illuminator, which can be amplitude- and frequency-modulated in such a way so as to produce a waveform, which, upon reflection from the absorbing surface gives a waveform which requires minimum detection bandwidth. This follows directly from basic concepts of information theory: a very small receiver bandwidth is required if we know the frequency of the signal to be detected and sufficient integration time is provided. Figure 11 shows the infrared absorption spectrum for vinyl. This is used to structure the waveform of the laser illuminator (through an amplitude and frequency modulation process) so as to produce the narrow-banded power spectrum shown in the upper right hand corner of the figure. Note that in the case of a conventional illumination (no tailoring) where the laser frequency is tuned in a linear saw-tooth manner, the power spectrum referred to would be obtained

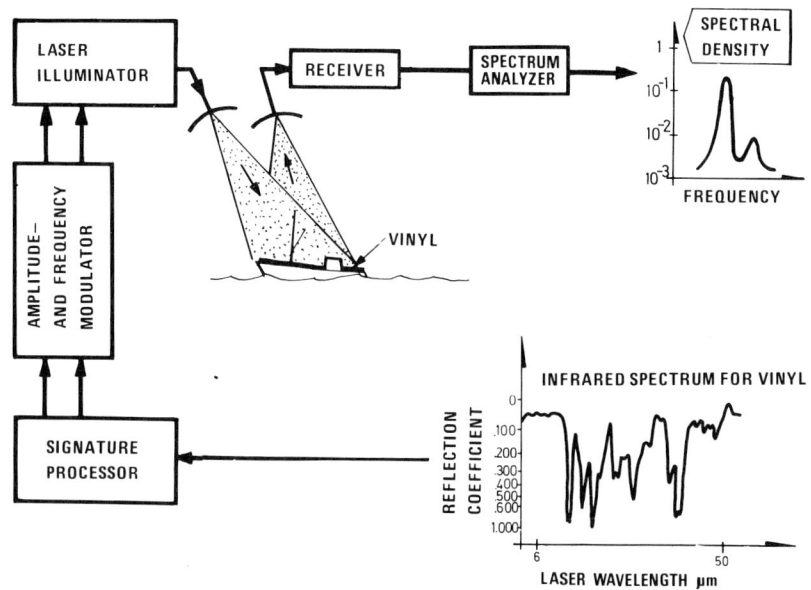

FIGURE 11: If we have detailed information about the absorption spectrum of the chemical compound of interest, the laser illumination can be tailored so as to achieve optimum system sensitivity and minimum influence of interferants.

by a direct spectral analysis (Fourier transformation) of the infrared absorption spectrum of vinyl since this would appear as an amplitude modulation on the laser beam. The merit of the "tailored illumination" scheme is to preprocess the illumination in such a way as to produce a power spectrum of the returned signal envelope with minimum bandwidth.

Before we leave this section on identification methods based on molecular absorption processes, we shall present some recent experimental results. Figure 12 shows an optical fluorosensor developed by the author's organization (T. Lund in press) for the purpose of studying the effect of pollution processes in inland lakes. Water pollution can lead to an enhancement of growth of algal populations. By studying the distribution in space and time of the various algal species, information is obtained about the pollution burden. A set of illuminators with different wavelengths (colours) is switched on in a cyclic manner and coupled to the

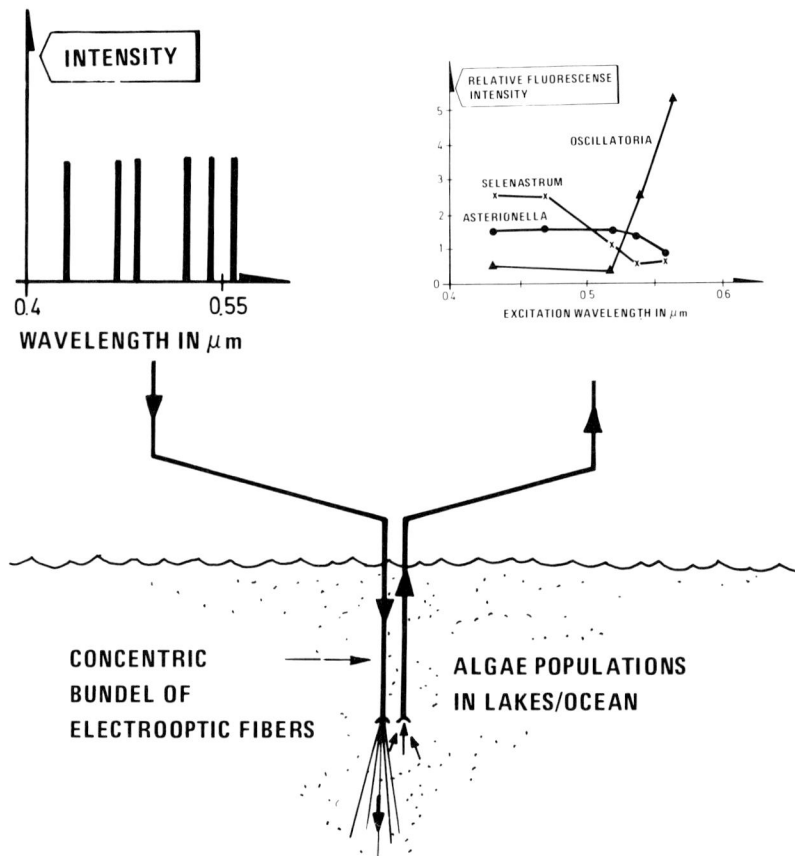

FIGURE 12: Based on electro-optical fibres, the various algal species are excited using light of different wavelengths. The fluorescent light is picked up by another set of fibres thus giving information about the relationship between fluorescent intensity and excitation wavelength. (T. Lund)

TABLE 1: Minimum detectable algal concentrations based on a fibre-optic fluorosensor (T. Lund, in press).

	MINIMUM DETECTABLE CONCENTRATION	
	mm^3/litre	µg/litre chlorophyll-a
OSCILLATORIA SELENASTRUM	0.1 $\quad$ 10^{-2} 0.5	0.2 $\quad$ 10^{-1} 0.3

water mass to be analyzed through a set of electro-optical fibres. This light excites the various fluorescence mechanisms in the algal populations. Studying the intensity of the fluorescent light as a function of excitation wavelength (thus obtaining the excitation spectrum) provides a means of identifying the algal species. Note that a fluorescent material can be characterized by both its excitation spectrum and its fluorescence spectrum. As shown in Figure 12 we can distinguish between three important algal species by studying the excitation spectrum. Including also an analysis of the fluorescence spectrum, an improved specificity is achieved. In addition to providing a method by which the distribution of the various algal species in polluted water can be investigated, the algal fluoresence spectrometer provides a very straightforward and sensitive method for biomass investigations. The table below shows that very minute concentrations of the various algal species can be detected using the simple inexpensive fluorosensor shown in Figure 12.

Finally, Figure 14 shows some experimental results obtained with the algal sensor. The left hand set of curves shows the results obtained when the fluorosensor was moved over a distance of approximately 1 km in lake Mjosa during fall 1978. Note the patchy nature of the algal concentration. The spatial distribution of this patchy structure is quantified by computing the spatial (wave-number) spectrum shown. Note that this spectrum is in the form $E(k) \propto k^{-9/3}$ which is not in accordance with what would be expected if the distribution was caused by turbulent motion in the water. Homogeneous isotropic turbulence would lead to a $k^{-5/3}$ distribution. This steep spatial spectrum is indicative of either internal mechanical waves in the water strata or dominating biological processes. Two ostensible explanations of the form of the spatial spectrum can be offered: If the water motion is not governed by more or less homogeneous isotropic turbulence superimposed on a laminar flow, but rather is dominated by internal waves, strata of high algal concentrations may periodically be brought to the surface by the wave motion causing a patchy structure. Energy is then

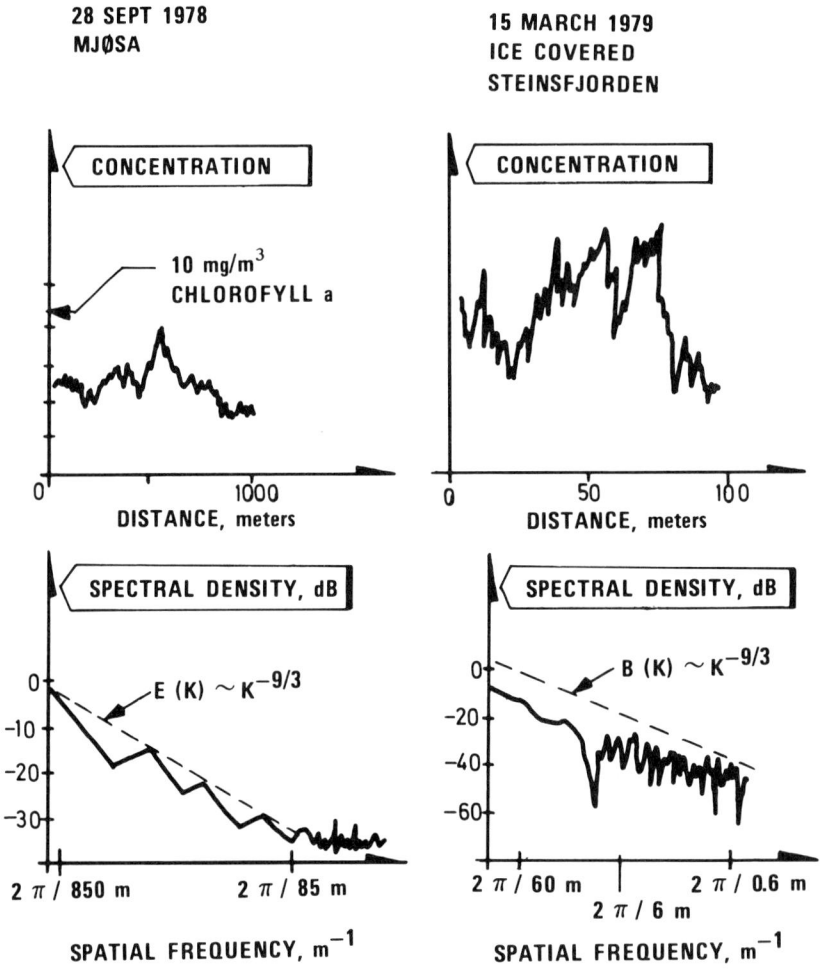

FIGURE 14: Spatial distribution of algae intensity obtained with the algal sensor based on fibre optics. Note that one set of results is obtained just below the water surface in October, whereas the second is obtained underneath the ice-cover in March.

added to the low wavenumber region of the spatial spectrum of algal concentration resulting in a steep slope.

A second explanation is based on biological consideration: if the nutrient conditions under which the algal populations are growing are favorable, then the algae may form a closely clustered pattern. Conversely, if the survival conditions are poor, if the dissipation rate is large in comparison with the rate at which the populations grow, then the mean distance between algal patches may

be large. The former situation leads to a flat spectrum whereas
the latter leads to a steep spectrum slope. To exclude the possibility of wave motion, the experiment was repeated during the month
of March, 1979, where the probe was moved underneath the ice-cover
of an inland lake by means of a torpedo arrangement. Note that
also under such conditions a steep spatial spectrum of the form
$k^{-9/3}$ was obtained. In passing note that in connection with the
ice-covered lake experiment, the spatial distribution of light, as
well as of temperature and of light showed no similarity with that
of algal concentration.

So much about molecular structures. Then let us focus the
attention on another potential detection/identification parameter.

Classification of environmental state parameters by motion pattern mapping

In order to identify an environmental object on the basis of
its motion pattern, we shall have to rely on the Doppler principle.
Illuminating a body which is in motion by electromagnetic waves,
the wave upon reflection is subjected to a Doppler shift f. This
Doppler shift is characterized by the well known Doppler equation
stating that the Doppler shift $f = \dfrac{1}{2\pi} \cdot \vec{K} \cdot \vec{V}$ where $\vec{K}$ is the
wave vector of the electromagnetic field ($\vec{K} = \vec{k}_i - \vec{k}_r$, the
vector difference between the incident and the reflected wave-
number), $\vec{V}$ is the velocity of the object. Observing then both the
Doppler shift and the Doppler spread, we obtain information about
the mean systematic velocity of the object, as well as information
about the velocity distribution.

We use this Doppler principle to investigate the motion
pattern of ocean waves as depicted in the left hand side of Figure
9. Studying the time history of the correlation between two radio-
frequency lines, we obtain information about the motion pattern of
a particular ocean wavelength. The wavelength L which the experiment selects is given by:

$$L = \frac{c}{2\Delta f} \tag{6}$$

where Δf is the frequency spacing of the radiowaves. When we, using Doppler spectrum investigations, have obtained information about the motion pattern of the sea surface, we can make certain deductions regarding the wave pattern. Specifically, if we are dealing with gravity waves, then there is a direct relationship between velocity v and ocean wavelength L

$$v = (\frac{g \cdot L}{2\pi})^{\frac{1}{2}}$$

where g is the acceleration of gravity. This property of gravity waves forms the bases for over-the-horizon "sky-wave" investigations of the ocean wave spectrum.

Figure 13 shows a set of theoretical curves illustrating the potential of a detection/identification technique based on the measurement of motion pattern. Here we have calculated the Doppler shift caused by a ship at 20 knots speed. Note that this Doppler shift is calculated for each radar frequency separation (radar frequency pair) within the range of frequencies required to match the radar illumination to the target of interest (see Figure 6). Specifically, the Doppler shift caused by the ship is given by

$$f_{SHIP} = \frac{2V}{c} \Delta F \qquad (7)$$

where V is the speed of the ship and ΔF is the matched radar difference frequency. ($F = \frac{c}{\Delta l}$ where Δl is the target scale to which the radar is matched). The wave velocity of the sea surface, however, is not linearly related to the sea wavelength

$$v = (\frac{gL}{2\pi})^{\frac{1}{2}}$$

This dispersive relationship increases the detection potential of our "Doppler filtering method" as follows:

Velocity of the gravity wave is given by

$$V = (\frac{gL}{2\pi})^{\frac{1}{2}} \quad \text{(Deep water, fully developed seas)} \qquad (8)$$

The Doppler shift from this wave is

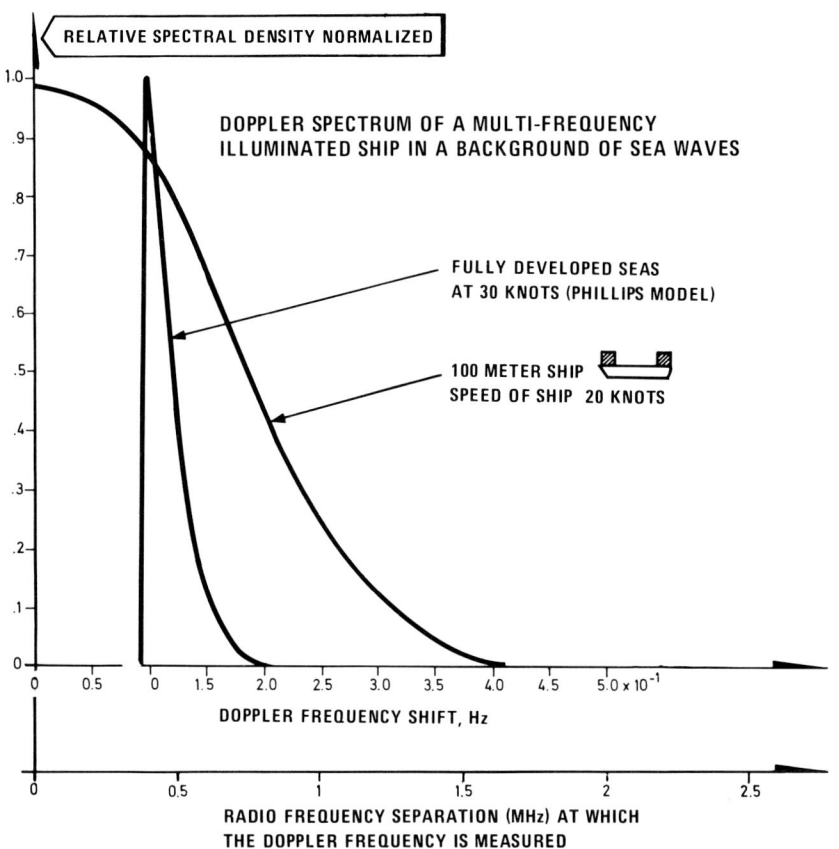

FIGURE 13: The Doppler spectrum of the dispersive sea waves (30 knots wind) is compared with the Doppler shift of the "matched radar illumination" caused by the motion of a ship.

$$f_{SEA} = \frac{2\Delta F}{c} \quad (\frac{gL_{SEA}}{2\pi})^{\frac{1}{2}} \tag{9}$$

and since $\Delta F = \dfrac{c}{L_{SEA}}$ $\tag{10}$

$$f_{SEA} = (\frac{g}{\pi c} \Delta F)^{\frac{1}{2}} \tag{11}$$

We have now studied methods by which the geometrical shape and size of environmental objects can be determined remotely. Furthermore, we have discussed classification schemes for molecular surface structures and we have discussed the question of classification of environmental state parameters by motion pattern mapping.

Measurement of surface temperature

Remote probing of surface temperature distribution is based on the familiar Planck's law of radiation:

$$P = \frac{\varepsilon \nu^{-3}}{e^{h\nu/kT} - 1} \qquad (12)$$

Here

P is radiated power (the unit is watt)
ε is the emission coefficient
ν is the frequency of the radiation field
h is Planck's constant
k is Boltzmann's constant and
T is absolute temperature

The instrument measuring radiated power, known as a radiometer, is capable of measuring the received power P and the radiation frequency ν. Accordingly, we shall have to require information about the emission coefficient ε in order to measure the surface temperature T. If we have no information about ε, the surface temperature T cannot be determined remotely. In general, however, there is a relationship between the emission coefficient ε and the radiation frequency. Accordingly, if we measure radiated power at different values of the radiation frequency, the emission coefficient ε can be determined. Furthermore, the emission coefficient varies with polarization and with the direction of radiation relative to the orientation of the radiating surface. Figure 15 shows the emission coefficient as a function of frequency in the microwave region for various polarization angles and angles of incidence. Furthermore, the figure shows that the emission coefficient increases with increasing frequency reaching values in excess of 90% for wavelengths in the infrared region.

We have seen above that the molecular surface structure had a dominating influence on the reflection properties of a body. Since the emission coefficient is one minus the reflection coefficient, it is obvious that the molecular surface structure will "modulate" the Planck emission relationship. Thus, if the particular form of Planck's law is taken into consideration, information about the molecular absorption spectrum can be determined also from passive

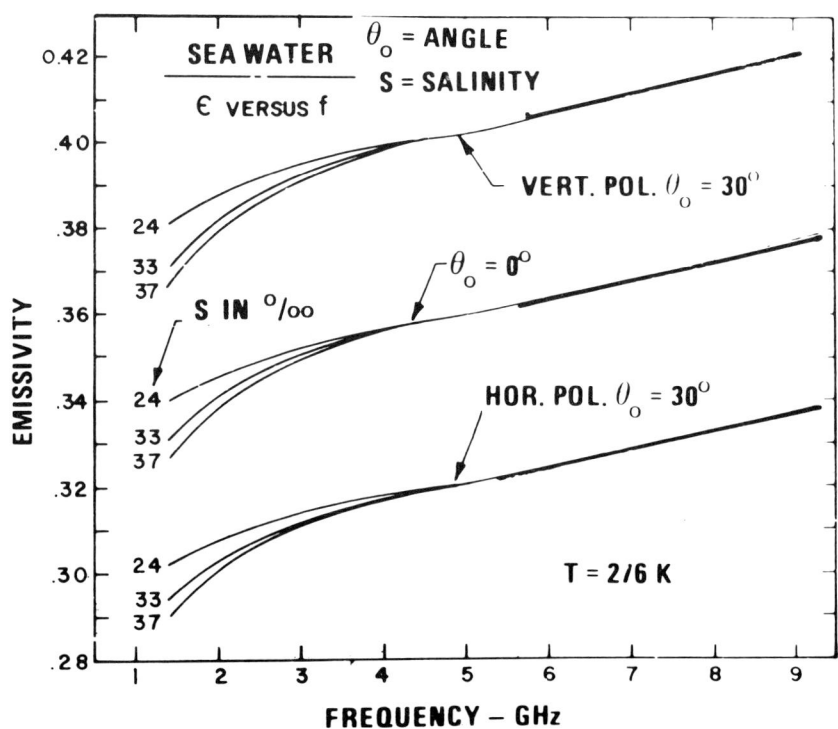

FIGURE 15: Emission coefficient ε is plotted as a function of radiation frequency. Note that the emissivity also varies with angle of incidence, polarization and salinity. Using several radiometer frequencies, information about the surface temperature can be obtained. (From Alishouse)

radiometry measurements. Note also that, by measuring the surface temperature at different conditions of surface heating (day and night), we can draw definite conclusions about the heat capacity and heat conductivity properties of the surface.

Finally, Figure 16 shows the temperature distribution of a 1800 x 1000 m² section of the sea surface. Note that the surface resolution is approximately 10 x 10 m, whereas the temperature resolution is approximately 1/100°K. This picture was obtained using a linescanning infrared radiometer in the 10 micrometer-band. Note the difference in surface structure of the two adjacent water masses and note also the "wavy" structure of the radiation pattern.

SPATIAL RESOLUTION
IS ~ 10m

1800 m

NOISE EQV TEMP
(CALC) 5/1000 K

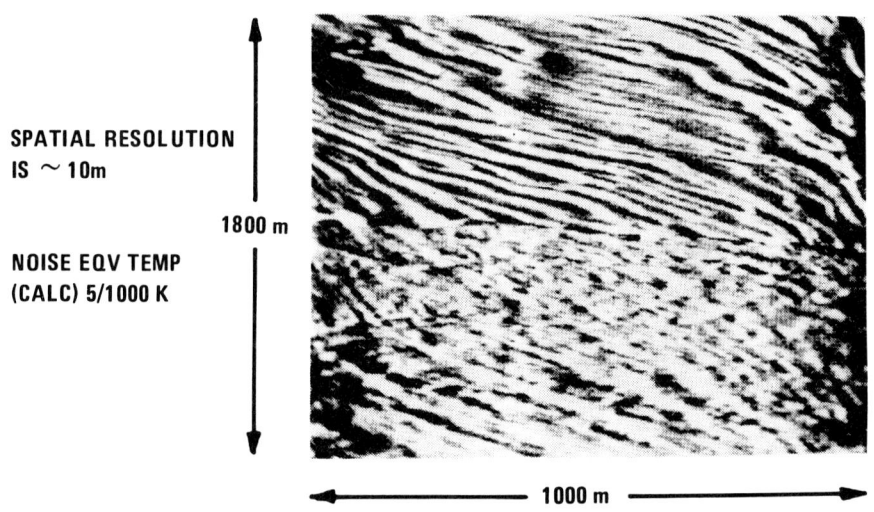

◄──────────── 1000 m ────────────►

FIGURE 16: Using an aircraft mounted IR-Linescanning radiometer, spatial resolution is obtained. The results shown are based on a radiometer developed by NDRE and flown during summer 1978 in Skagerak. Note the detailed structure of surface radiation temperature. (Ref. T. Lund, B Gronlien and O. Wenstop).

ACKNOWLEDGEMENTS

"Acknowledgement is due to The Institute of Physics for permission to include in this paper portions of an article "Environmental Remote Sensing by Radio and Optical Waves: Progress and Perspectives" by the present author, Physics in Technology, Vol 10, 1979".

REFERENCES

Alishouse, J. C., 1975. "A Summary of The Radiometric Technology Model of The Ocean Surface in the Microwave Region". NOAA Technical Memo. n=ss 66.

Gjessing, D. T. 1978 "Remote Surveillance by Electromagnetic Waves for Air-Water-Land", Book on Ann Arbor Science Publishers, Inc., USA.

Gjessing, D. T. 1978 "A generalized method for environmental surveillance by remote probing", J. Radio Sci. March/April.

Lund, T. Ed. 1978 "Surveillance of Environmental Pollution and Resources by Electromagnetic Waves", Proceedings of the NATO Advanced Study Institute held in Spatind, Norway, 9-19 April, 1978.

Lund, T. 1979 "A simple and sensitive in situ algal fluorensence
     sensor based on fibre optics", PFM TN-15/79, and to appear in
     IEEE Journal of Oceanic Engineering.
Plant, W. J., 1977. "Studies of Backscattered Sea Return with a CW
     Dual-Frequency x-Band Radar". IEEE Trans. Ant. Prop. Vol.
     AP-25 No. 1.

Overview of Some Remote Sensing Activities Currently Being Supported by the Office of Naval Research

Peter C. Badgley

INTRODUCTION

There are numerous environmental factors that affect naval systems and operations.

Most of these factors are important for coastal and Arctic marine areas, two of the regions for which the writer has research responsibilities. Some of the phenomena listed (e.g. salinity, temperature, currents, waves, winds, etc.) are of great importance also in open ocean areas, but many of the parameters change more rapidly (distance-wise, horizontally and vertically) and more frequently (time domain) in coastal areas. Also, there are parameters in coastal zones which can be sensed remotely which do not exist in the open sea, e.g. bathymetry, salt water wedges in estuaries, beach characteristics, etc. Thus, the demands upon remote sensing research, in terms of resolutions required and in

433

terms of difficulty of the research involved, are of more severity
in the shallow water coastal regions.

A number of the environmental factors shown in Table I can be
measured remotely, particularly after experience has been gained in
calibrating the remotely sensed data with "ground truth" or "sea
truth" measurements. Some of the environmental factors can be
sensed from satellite platforms while others require the higher
resolution that can be provided only by airborne or surface-based
platforms. Some of the environmental factors can be measured only
by in situ means.

The field of remote sensing is a large and fast moving one and
this paper will cover only those types of remote sensing which are
being actively supported by the writer's office at this time and
among these only those that will not be discussed in detail by
other authors in this volume. These types of research being supported are listed in Table II.

Table II - Types of Remote Sensing of Interest For
Naval Maritime and Coastal Operations

o   Synthetic Aperture Radar - For waves, currents, ocean fronts,
     ice types; the high resolution and all weather capability are
     of significant value.

o   Multispectral Photography - Ratioing capability combined with
     laser sounding, valuable for underwater topography and traf-
     ficability.

o   Acoustic sounding for underwater topography.

o   Raman spectroscopy for underwater temperature and salinity.

o   Infrared and passive microwave sensing, of value for sea
     surface temperature, ocean fronts, coastal currents, etc.

o   HF surface wave and skywave radars coupled with air droppable
     disposable buoys or drogues for measuring waves, currents,
     etc.

TABLE I

SENSITIVITY (IMPORTANCE) AND CURRENT STATE OF KNOWLEDGE OF ENVIRONMENTAL FACTORS FOR VARIOUS NAVAL ACTIVITIES AND SYSTEMS

ENVIRONMENTAL FACTORS THAT AFFECT NAVAL SYSTEMS*

Column group headers:
- *Coastal and Arctic Geography and Oceanography – Includes Air-Sea-Land and Air-Ice-Water Interfaces
- Geophysics – Coastal, Arctic, Coastal and Some Aspects of Earth-Ionosphere Waveguide

TYPES OF NAVAL SYSTEMS	Tides, Storm Surges	Surface Currents	Subsurface Currents	Waves, Surf, Winds	Water Salinity and Temperature	Bio-Luminescence, Fluorescence, Biologics	Vegetation	Water Chemistry	Ice Dist. Horizontal	Ice Dist. Thickness	Ice Ridging and Roughness	Snow Cover	Climatology	Atmospheric Propagation	Bathymetry, Topography	Moisture Content – Soils	Shear Strength in Soils, Sediments	Acoustic Propagation	Seismic Propagation	EM Propagation / EM Properties	Propagation Across Air-Water, Water-Bottom Interfaces	Geomagnetic Field	Gravity Field	Nuclear Radiation
Instrumentation and Devices for Surveillance, Fire Control, Communications and Navigation																								
Electromagnetic/Optical					2/3		2/2							3/3						3/3	3/3			3/3
Acoustic				3/3	3/3	2/3					2/2				3/3			3/3			2/3			
Pressure				2/2																				
Magnetic (MAD)				3/3																3/3		3/3		
Seismic																	3/3		3/3		3/3			
Gravity																							2/3	
Nuclear Detection																				3/3	2/2			3/3
Weapons Dynamics																								
Torpedoes	2/2					3/3			3/2		3/2							3/3						
Mines		3/3	3/3	3/3													3/3	3/2		2/3		3/2		
Ballistics				3/1											3/2								2/3	
Vehicle Operations																								
Ships	3/3	3/2		2/2		1/3			3/2	3/2		2/2	2/2		3/2			3/2						3/1
Landing Craft (Conventional)	3/3	3/3		3/3					3/2	3/2		3/2	3/2		3/3				3/3					3/1
Landing Craft (ACV)				2/2					3/2	2/2	3/2	2/2	2/2											3/1
Submarines			3/3		3/3	2/3		2/2	3/3		3/2				3/3			2/2						
Aircraft											3/2	3/3	3/3	3/3		3/3	3/3			3/3	3/3			3/1
Tanks, Trucks							3/2		3/1	3/1	3/2	3/2	2/3	3/2		3/2						2/2		3/1
Personnel (Effects on)																								
Divers, Swimmers	3/3	3/3	3/3	3/1	3/2		3/2		3/1															3/3
Support Services																								
Coastal Arctic Engineer Facilities	3/1	3/1	3/1	3/2	2/3	3/3	3/1		3/3			3/2	3/2		3/1		3/3	3/2						
Environmental Information Systems	2/3	2/3	2/3	2/3	2/3	2/3	2/3					3/3	3/3	2/3	2/3			3/3	3/3	3/3	2/3			3/2
Prediction Models	3/3	3/3	3/3	3/3	3/3	3/3			3/3			3/2	3/3	3/3		3/3	3/3	3/3	3/3	3/3	3/3			3/2
Cartography	2/3														3/3									

LEGEND

IMPORTANCE (SENSITIVITY) FOR NAVAL SYSTEMS OR OPERATIONS.
- 1 = Minor
- 2 = Major
- 3 = Critical

LEVEL OF KNOWLEDGE
- 1 = Adequate
- 2 = Marginal
- 3 = Inadequate

THOSE BOXES HAVING A 3/3 RATING ARE OF HIGHEST IMPORTANCE AND ARE RECEIVING RESEARCH SUPPORT

*This chart does not indicate what instruments are most important for obtaining information on the environmental factors that are important for various naval systems and operations

SYNTHETIC APERTURE RADAR (SAR)

This type of radar is receiving considerable attention because
its resolution is independent of altitude; it can penetrate through
clouds, and there are good indications from work performed to date
that SAR can detect internal waves, surface wave spectra, wave
heights (maybe), surf, currents, water mass boundaries (fronts),
and sea surface slicks (thin sea-surface biological or chemical
films) that affect the small scale surface waves (capillary waves,
i.e., waves related to surface tension). The detailed work of
documenting and publishing results on such detections is currently
in progress. A number of technical problems are being investigated
including the following:

- o   What is the exact nature of the interaction of electro-
        magnetic energy (EM wave propagated outward from antenna,
        e.g. SAR) with the wave field which exists at the sea
        surface?

- o   What is the nature of the distortion (doppler shift) due
        to movement of the target? How can defocused signals be
        evaluated?

- o   How can the high data rate inherent with SAR's be handled
        more efficiently and how can this information be used in
        a operational sense?

Even though all of these problems have not been solved,
several recent examples illustrate the potential of SAR.

Figure 1 shows the following phenomena:
internal waves, a ship wake, and a surface ship. These data were
acquired by L-Band radar in the HH mode from an aircraft flying
over the Georgia Straits near Vancouver, British Columbia, Canada.
The figure was provided by courtesy of the Environmental Research
Institute of Michigan (ERIM).

Figure 2 is a view from SEASTAT SAR (L-Band) taken over Gray's
Harbor, Washington State, provided by courtesy of the Jet
Propulsion Laboratory of the California Institute of Technology.
The bright line effect noted on the south edge of the harbor inlet
is due to the ebb tide moving out of the inlet. The SAR energy

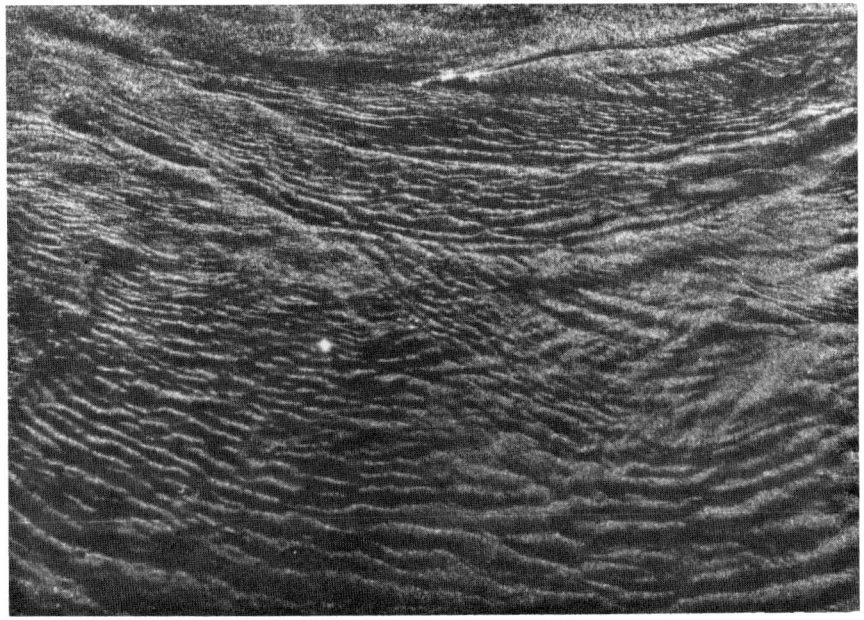

FIGURE 1: Internal waves and ships observed by an airborne Synthetic Aperture Radar (SAR) operating in the L-Band (21 cm) and with dual polarization (HH polarization in this instance). This image is from the Georgia Straits near Vancouver, Canada. The bright spot near the center of the image is the research vessel "Endeavor". Note also a second ship and ship wake near the top of the image. Internal waves are well displayed throughout the image. The distance from top to bottom of this image is 5.6 km. This image was provided by courtesy of the Environmental Research Institute of Michigan.

returning to the spacecraft antenna has been changed slightly in frequency due to the change in range (Doppler effect) of the moving water. The resulting SAR signal history (image) is modified. An algorithm has been developed for calculating the speed of this motion, i.e., the current velocity. Internal waves are also apparent on this SAR image.

ADDITIONAL SAR RESEARCH NEEDED

Additional airborne and satellite flights are needed with SAR over controlled test sites because we still do not fully understand

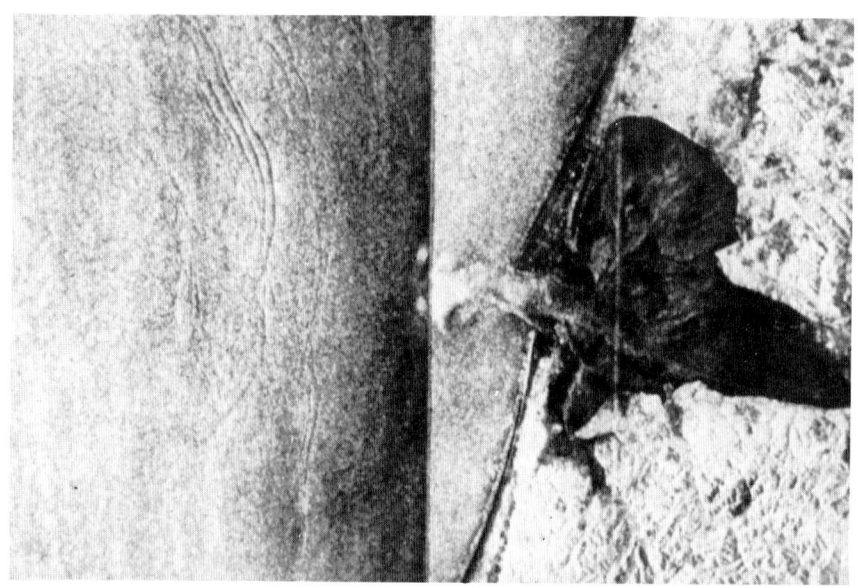

FIGURE 2: Synthetic Aperture Radar (SAR) image (L-Band) taken by
SEASAT I on 7 July 1978 over Grays Harbor, Washington State. At
that time the entrance to the harbor was experiencing an ebb tide
(outflow) with a velocity of 1.5 meters/sec. The moving water
(tidal current) shows up on the SAR image as a bright line along
the south boundary of the tidal jet. The width of the bright line
represents a doppler shift induced by the current and is proportional to the magnitude of this current. An algorithm has been
developed for calculating the velocity of this current motion. The
SAR image also shows presence of internal waves, which are 800 m
long, offshore of the entrance. These waves are sub-surface waves
which form at a depth of maximum density gradient, usually 50-150 m
below the surface. The internal waves, however, generate orbital
velocities which extend to the surface and cause variations in the
ocean surface roughness. The SAR detects the variations in the
ocean surface roughness. Image provided by courtesy of the Jet
Propulsion Laboratory.

the optimum flight trajectories (azimuths), depression angles, wave

lengths (X-, L-Band, etc.), and polarizations (HH, HV, etc.) needed

for extracting the full data content from the signals received.

The exact effect of environmental noise (waves, etc.) on signal

returns from surface ships is not well understood and also needs

more attention. Work is underway to extract only significant

portions of the data return from moving SAR platforms (airborne or

satellite) and to transmit only the relevant information. Addi-

tional work on this and many of the other problems alluded to above
is needed.

MULTISPECTRAL PHOTOGRAPHY AND RATIOING TECHNIQUES COUPLED
WITH LASER BATHYMETRY

Multispectral photography, which is available both from satellites such as LANDSAT as well as from aircraft overflights, is
proving to be useful for determining topographic details on the
seafloor in shallow waters as noted in Figure 3. This is an image
of coastal waters in the Gulf Coast region of the U.S. The reflection of light (visible wave length energy) from the sea surface has
been filtered out in the lower half of the image and the residual
returns have been processed to produce a ratio image of the
$\frac{0.50\text{-}0.54 \ \mu m}{0.48\text{-}0.52 \ \mu m}$ wavelength bands.
The absorption effects of the water column are removed by means of
an algorithm and the residual reflections from the seafloor are
indicative of the surface roughness and composition of the seafloor. Such images are also being calibrated by pulsed lasers.
The laser energy is reflected both by the sea surface and the
seafloor and the time difference is indicative of the water depth.
By a combination of such multispectral ratio images and a scanning
pulsed laser system, it is possible to derive an accurate picture
of seafloor details (channels, offshore bars, obstacles, etc.),
which is important for a variety of coastal operations (both naval
and civilian). Breadboard models of a pulsed scanning laser system
are now being tested by several government agencies in the U.S.and
an operational system will be developed soon. This multispectral
photography and laser approach, although useful in relatively clear
waters, does not work well in turbid waters and as a consequence
another approach is being investigated for such areas.

ACOUSTIC SOUNDING FOR UNDERWATER TOPOGRAPHY

Figure 4 illustrates an acoustic bathymetry approach which is
being investigated currently. A CO_2 laser is the source of the

FIGURE 3: Ratio image $\dfrac{0.50 - 0.54 \text{ um}}{0.48 - 0.52 \text{ um}}$ from the Florida Coast.

Acquired by an airborne multispectral scanner operating in the
visible wavelengths. The reflection from the sea surface has been
filtered out in the lower half of the image and the absorption
effects of the water column have been removed by means of an algorithm. The residual reflections are from the sea floor and are
indicative of surface roughness. Image provided by courtesy of the
Environmental Research Institute of Michigan (under ONR contract).

acoustic energy ($\cong$ 40 KHz). Part of the energy is reflected from
the air-water interface and part of the energy is reflected from
the seafloor. The time difference between the two reflections
indicates the depth to the seafloor. Hydrophones are being used in
the current test program to monitor the level of energy that is
transmitted back to the surface from different depths and to determine the total attenuation at the air-sea interface. If this
approach works, an airborne system could be developed as shown in
Figure 5.

FIGURE 4

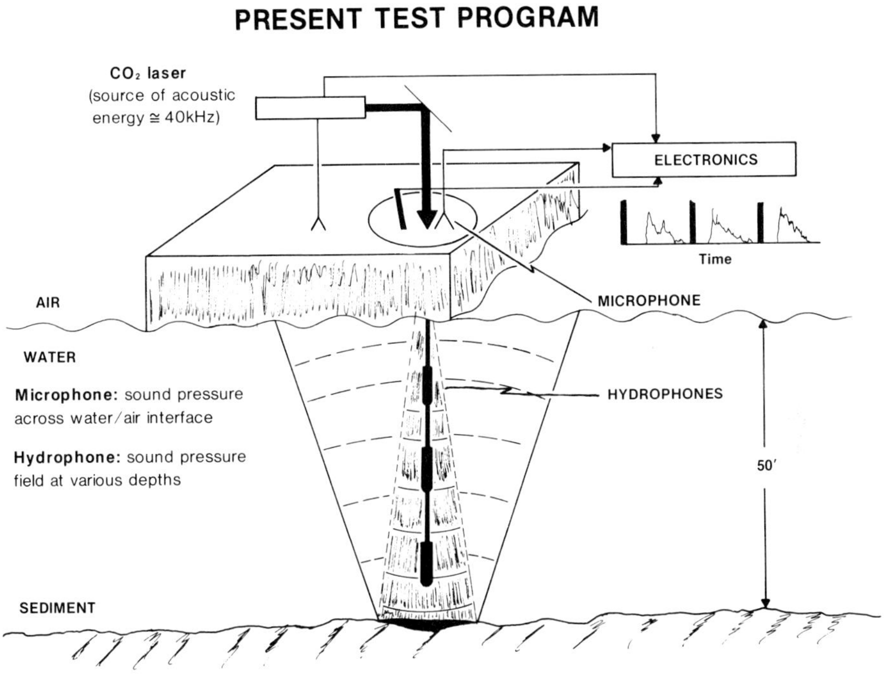

PRESENT TEST PROGRAM

RAMAN SPECTROSCOPY FOR UNDERWATER TEMPERATURE AND SALINITY

A pulsed laser system is being used to measure temperature and salinity below the sea surface (Figure 6). The laser energy is circularly polarized and a two channel radiometer is used to sense the right and left polarization of the returning energy. The intensity and spectral signature of the backscattered radiation is related to the hydrogen bonding of the water molecules which, in turn, is related to the water temperature. The signals are time-gated to permit temperature measurements for specific depths. A catalog of unique spectral responses for different water temperatures is being developed for both left and right polarized back-scattered radiation intensities as shown in Figure 7. Salinity changes also affect the degree of hydrogen bonding of the water molecules and thus the depolarization ratio of the left and right

FIGURE 5

ACOUSTIC BATHYMETRY

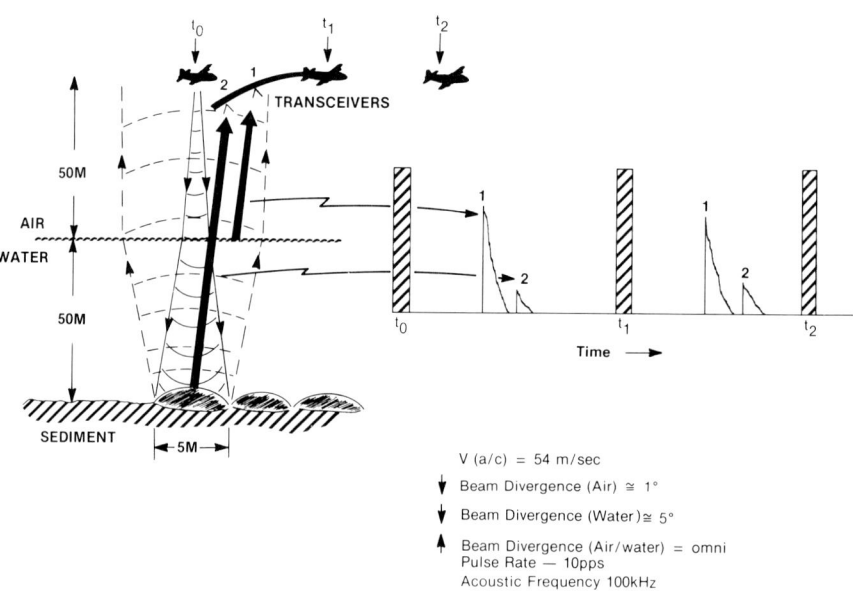

V (a/c) = 54 m/sec

Beam Divergence (Air) ≅ 1°

Beam Divergence (Water) ≅ 5°

Beam Divergence (Air/water) = omni
Pulse Rate — 10pps
Acoustic Frequency 100kHz

polarized backscattered radiation. The possibilities of this Raman
Spectroscopy approach are being investigated at this time by a
ship-based system, which is hoped will lead eventually to an airborne system for both subsurface temperature and salinity profiling.

INFRARED AND PASSIVE MICROWAVE SENSING OF SEA SURFACE TEMPERATURE,
OCEAN FRONTS, COASTAL CURRENTS, ETC.

Satellite sensing of sea surface temperatures has been performed for a number of years but the accuracy of this method is
being currently improved as we gain new insight on the potential
masking effects of atmospheric moisture. Figure 8 is a satellite
infrared image of the Gulf Stream (dark band) off the east coast of
Florida. The dark band of the Gulf Stream is hardly visible toward
the bottom of the image due to the masking effect of a mass of
moist air sitting over the southern end of the Florida peninsula.

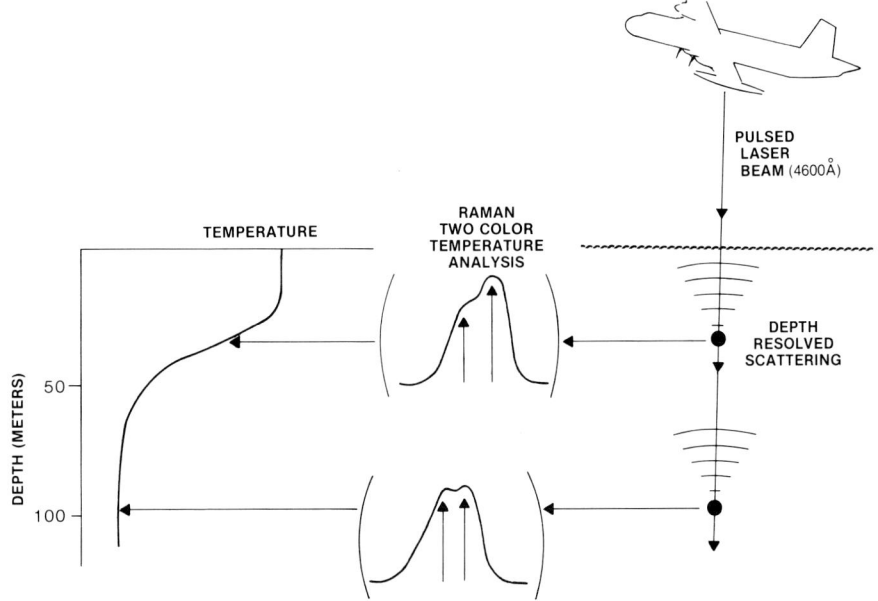

The intensity and spectral signature of the backscattered radiation is
related to the hydrogen bonding of the water molecules which in turn
is related to water temperature. The signals are time gated to permit
temperature measurements for specific depths.

FIGURE 6: Raman Spectroscopy test program for detecting subsurface
water temperature and salinity.

An explanation for this masking effect is shown in Figure 9. At
satellite altitudes, the difference between two spots on the sea
surface having two degrees (Celsius) difference in temperature only
shows up as a difference of 0.8°C where warm moist (absorbent) air
intervenes between the sea surface and the satellite sensors. An
ocean front might thus be masked in such an area. Where the air is
drier (left hand side of Figure 9 where polar continental air
exists), the difference in temperature apparent in the satellite
sensors at altitude is 1.4°C (i.e. 18.6°C -17.2°C) and thus, this
atmospheric situation is more conducive to the detection of water
mass boundaries having different sea surface temperatures. During
the past several years, our understanding of these atmospheric
effects has been increased considerably and we are getting to a
point where we can predict those areas and times of year when it is
possible to locate and monitor sea surface currents and water mass

FIGURE 7

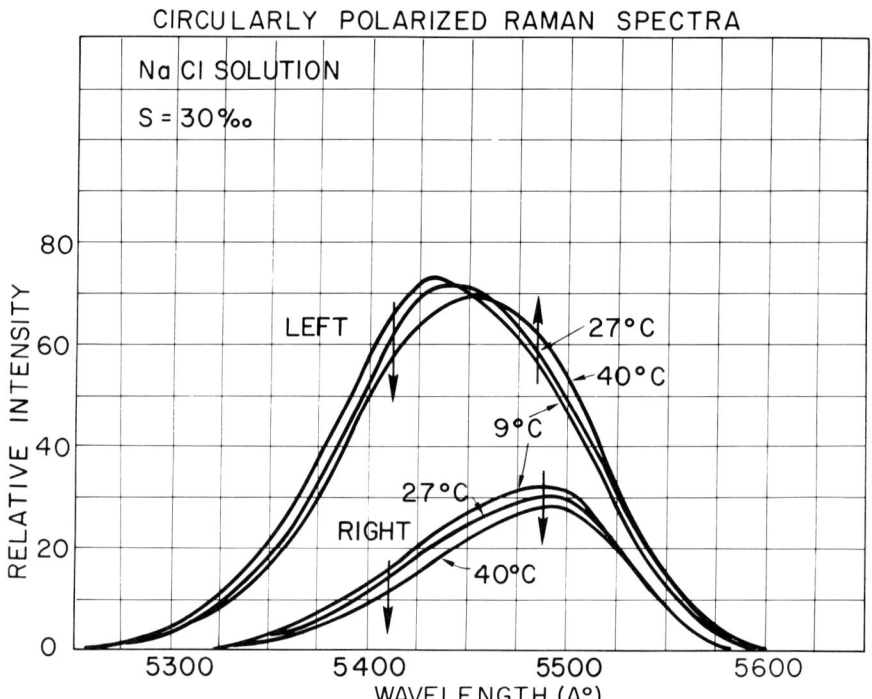

boundaries with greater precision. We are also interested in learning better how to use passive microwave radiation from the ocean for sea surface temperature remote sensing. A particular concern here is the disturbance from man-made microwave transmitters, and it can only be hoped that the international conference for the allocation of radio frequencies will duly consider this.

HF SURFACE WAVE AND SKYWAVE RADARS COUPLED WITH AIR DROPPABLE DISPOSABLE BUOYS OR DROGUES FOR MEASURING WAVES, CURRENTS, ETC.

Considerable research has been conducted by ONR on the use of HF radars for detection of ships, aircraft, and other targets using the skywave mode. In this mode, the HF wave is refracted from the ionosphere and the radar wave then impinges upon the sea surface some 400-1600 nautical miles away from the originating antenna.

FIGURE 8

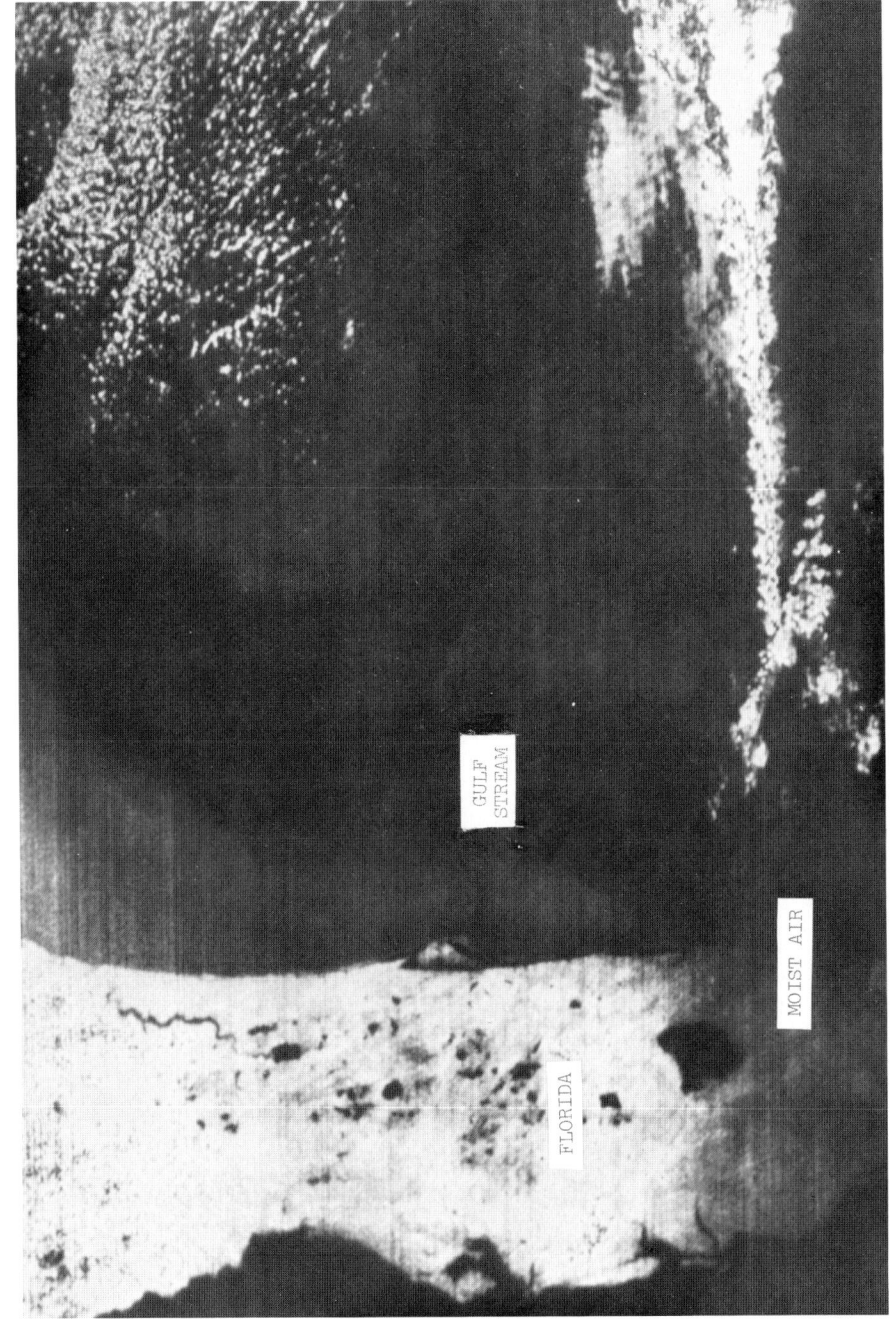

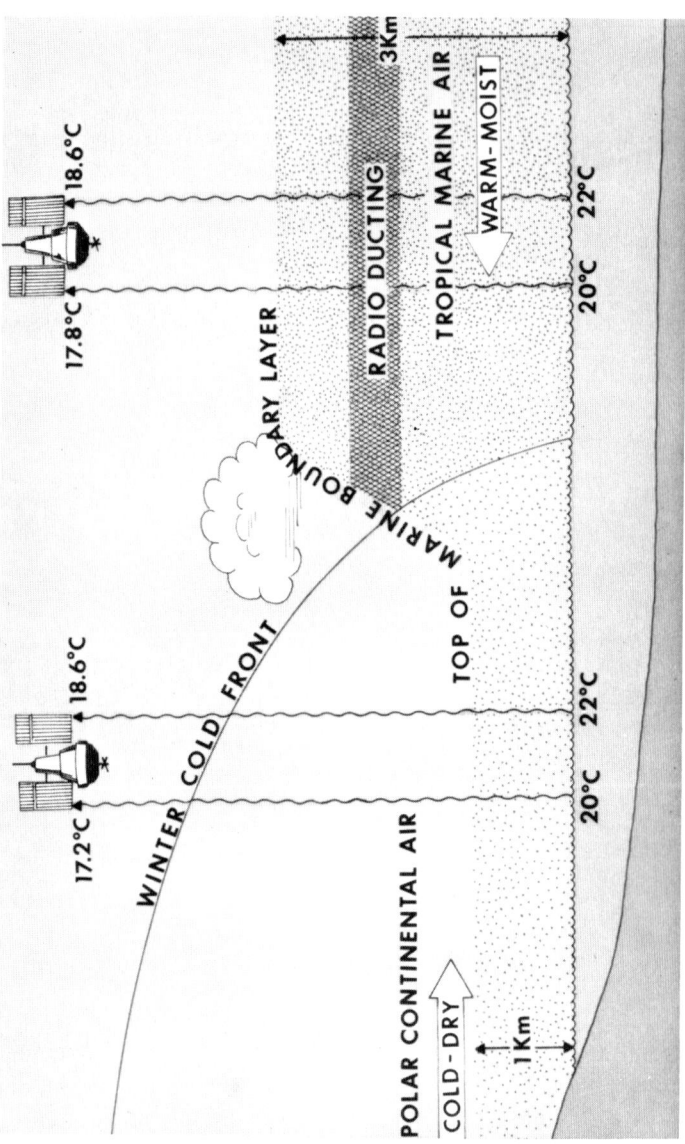

FIGURE 9: Effect of atmospheric moisture column upon the sea surface temperature differences which are actually recorded at the satellite sensors. Small sea surface temperature differences (of 2°C) can be completely missed by the satellite sensors if their detectability is only on the order of 1°C or more.

The HF energy has also been used to measure wave heights using the Bragg resonance effect and the Doppler shift of the reflected signal measures the propagation direction of the waves. Such skywave HF propagation requires large antennas. More recently, several investigators have begun to use portable HF antennas which are capable of determining wave spectra via the surface wave mode out to distances of 400 n miles. ONR researchers have also developed, in the past several years, expendable, low cost (approximately $200.00) air droppable buoys and drogues. These buoys contain sensors that will measure the wave height, wind direction, and many other parameters. By using such sensors in the drifting mode, the current direction can also be determined both at the surface and at depth. Temperature and other sensors can also be added. These buoys and drogues have been used to provide ground truth for both the HF skywave and surface wave radar signals. A direction finding loop antenna at the shore has been used to track the movement of the drogues in the drifting cases. Recently, a more elaborate version of these buoys has been used to support several near shore wave prediction models being developed by ONR (see Figure 10). A deep water wave prediction model which uses synoptic scale weather maps (data derived from weather ships, satellites, etc.) can be checked by the buoy (Figure 10). The buoy contains microprocessors which receive pressure-sensor data for checking the waves predicted from the deep water model. Knowing the bathymetry and topography of the adjoining coastal area, the microprocessor can then predict the wave height and longshore current direction and speed from the buoy into the adjoining shore areas. These inshore predictions are based on a set of algorithms formulated by ONR researchers. The buoy illustrated in Figure 10 is currently being developed. It will be air droppable, self mooring, self skuttling, and for all intents and purposes, its location and existence are known only to those who have air dropped the buoy. It will transmit the processed information in coded bursts only when interrogated, operates in microwatts of power, and will have a long life (months). The range of reception is 400 miles (ground wave mode) and 2000-3000 miles in the skywave mode.

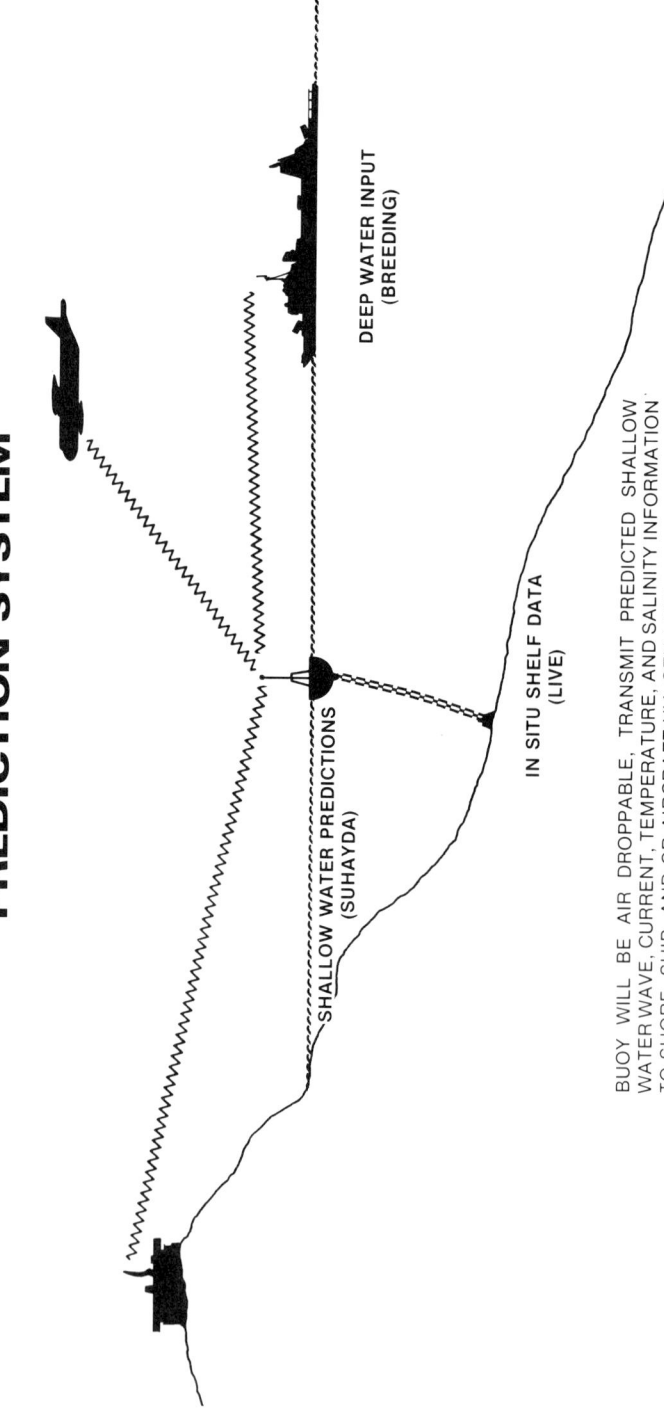

COASTAL DYNAMICS
PREDICTION SYSTEM

DEEP WATER INPUT
(BREEDING)

SHALLOW WATER PREDICTIONS
(SUHAYDA)

IN SITU SHELF DATA
(LIVE)

BUOY WILL BE AIR DROPPABLE, TRANSMIT PREDICTED SHALLOW
WATER WAVE, CURRENT, TEMPERATURE, AND SALINITY INFORMATION
TO SHORE, SHIP, AND OR AIRCRAFT VIA OTH LINK.

FIGURE 10: Air droppable buoy for measuring environmental parameters in coastal regions, coupled to over the horizon data communication link.

SUMMARY

The above examples give an idea of some of the types of remote sensing that are of interest to the U.S. Navy for coastal and deep water areas. The list is by no means complete as several of the other speakers at this conference will touch upon such items as satellite altimetry for mapping geoidal undulations and for monitoring storm surges, both of which are of interest to our office also. In closing, I would like to say that the Office of Naval Research is interested in collaborating with other agencies and scientists in both the U.S. and abroad in tackling some of the more important and urgent problems of remote sensing. I would like to acknowledge the help that I have received from a number of Navy colleagues and ONR supported researchers in compiling this paper.

Remote Sensing-Research Experiences and Problems

Sueno Ueno

ABSTRACT

The apparent radiance of ground materials as measured by a remote sensor differs from the intrinsic surface radiance, because of the presence of the intervening atmosphere. One of the important consequences of the atmospheric effects is the blurring of the observed radiance map. Hence, the removal of the blurring should improve the accuracy of pattern recognition and image interpretation in remote sensing.

In the present paper, the first section deals with an invariant imbedding procedure, which enables us to evaluate the total spectral radiance at the top of an inhomogeneous, anisotropically scattering atmosphere bounded by horizontally nonuniform albedos of underlying surfaces, i.e., an initial-value solution of the diffuse reflection pattern for multidimensional planetary problems, allow-

451

ing for the approximate solution. In the second section, it is
shown how to evaluate the atmospheric blurring for the MSS digital
data from air and space. Assuming the above bounded atmosphere,
the procedures to evaluate the fraction of path radiance to the
total radiance are described. In the case of aircraft data, we
must evaluate the path radiance within the atmosphere and also the
scan angle effect. Due to these complexities, a somewhat simplified average method is formulated. In the case of LANDSAT data, a
convolution-type integral equation, relating the surface radiance
distribution to the observed radiance map, is derived. This equation can be solved efficiently by means of the FFT algorithm.
Furthermore, we describe the AECS, a system of computer software
programs, which has been developed to perform the atmospheric
corrections for the pure atmospheric path radiance effect, the
blurring effect due to the interaction of the photons with surface
of background, and the scan angle effect on both LANDSAT and airborne multi-spectral digital data. In the third section we treat
with inverse problems in remote sensing. In other words, an identification of such optical properties as the atmospheric thickness,
the albedos of the underlying surface in earth-atmosphere or
atmosphere-ocean system is presented with the aid of quasilinearization and invariant imbedding. Finally, the last section deals
with the oilspill detection by the use of LANDSAT MSS data. It
enables us to evaluate an apparent contrast of the total radiance
on oil polluted calm sea surface, making use of MSS data from
space.

FUNDAMENTALS IN RADIATIVE TRANSFER

INTRODUCTION

Since the development of transfer theory by Ambarzumian (1958)
and Chandrasekhar (1950), the emphasis of discussion has been
firstly put on the one-dimensional case (cf. Sobolev 1963,
Busbridge 1960, Bellman, Kalaba, and Prestrud 1963), because of the
tractability from the analytic and computational aspects. On the

other hand, in the field of applied mathematics and astrophysics, the study of multidimensional transfer theory has also gradually been carried out by several authors (cf. Ambarzumian 1945, Elliott 1959, Chandrasekhar 1958, Sobolev 1959, Bellman, Kalaba, and Ueno 1965, Matsumoto 1969, Uesugi and Tsujita 1969, Rybicki 1971, Breig and Crosbie 1974, Ueno 1976a). Furthermore, the scattering matrix approach, order-of-scattering theory, and an integral operator method have been extended by several authors, allowing for an extension to multidimensional case (cf. Poon and Ueno 1973, Ueno, Vasudevan, and Bellman 1973, Ueno and Wang 1973). In order to treat with the atmospheric scattering effects on remotely sensed earth's imagery from space and air, we should refer to the multidimensional transfer theory (cf. Ueno 1977).

MULTIDIMENSIONAL TRANSFER THEORY

In this section we deal with a three-dimensional radiative transfer model consisting of a free atmosphere bounded by a horizontally nonuniform diffuse reflector.

Suppose that the top $z=z_1$ of a plane-parallel, vertically inhomogeneous, anisotropically scattering atmosphere of optical thickness τ is monodirectionally illuminated by parallel rays of a constant net flux πF per unit area normal to the direction of propagation. Furthermore, it is assumed that a horizontally nonuniform albedo of the bounding diffuse reflector gives rise to a three-dimensional optical inhomogeneity of the radiation field. Let the upwelling intensity of radiation emergent in the direction Ω from the level $z(0 \leq z \leq z_1)$ at horizontal rectangular coordinates (x,y) be denoted by $I(z,x,y;-\Omega)$, and furthermore, let the downwelling intensity of radiation in the direction Ω at horizontal coordinates (x,y) be denoted by $I(z,x,y;-\Omega)$. In the above Ω stands for $(\theta=\cos^{-1}v,\phi)$, where θ is a polar angle measured from the outward normal and ϕ is an azimuthal angle. Let the level-dependent phase function be denoted by $p(z;\Omega,\Omega')$, in a manner similar to the level-dependent attenuation and scattering coefficients, i.e., $\alpha(z)$ and $\sigma(z)$. The equation of transfer appropriate to this case takes the form

$$\Omega \cdot \nabla I \ (z,x,y;\Omega) + \alpha(z)I = \frac{\sigma(z)}{4\pi} \int_{4\pi} p(z;\Omega,\Omega')I(z,x,y;\Omega')d\Omega', \qquad (1.1)$$

where $-\infty \leq x, y \leq \infty, d\Omega' = dv'd\phi'$, and $\Omega \cdot \nabla I$ is the directional derivative of I in the direction Ω. Equation 1 should be solved subject to the boundary conditions

$$I(z_1,x,y;-\Omega) = \pi F\delta(\Omega-\Omega_o), \qquad (1.2)$$

$$I(0,x,y;+\Omega) = \frac{1}{v} \int_{2\pi} k(x,y;\Omega,\Omega') \ I(0,x,y;-\Omega')v'd\Omega', \qquad (1.3)$$

where $\delta(\Omega-\Omega o) = \delta(v-u)\delta(\phi-\phi o)$. In Equation (1.3) the bidirectional reflectance $k(x,y;\Omega,\Omega')$ represents the probability that a photon incident on the bottom $(0,x,y)$ in the direction Ω' will be reflected from it in the direction Ω within an elementary solid angle. In the case of isotropic reflection we put

$$k(x,y;\Omega,\Omega) = vA(x,y)/\pi \qquad (1.4)$$

where $A(x,y)$ is the albedo of the reflecting surface in accordance with the Lambert's law. It represents the ratio of the total energy reflected by the surface to the incident energy. In this case Equation (1.3) becomes

$$I(0,x,y;+\Omega) = \frac{A(x,y)}{\pi} \int_{2\pi} I(o,x,y;-\Omega')v'd\Omega'. \qquad (1.5)$$

In Equation (1.5) $I(0,x,y;-\Omega')$ consists of the downwards directly transmitted intensity and the downwards diffusely transmitted intensity as below;

$$I(0,x,y;-\Omega') = \pi F\delta(\Omega'-\Omega o)e^{-\tau/v'} + I^*(0,x,y;-\Omega'), \qquad (1.6)$$

In Equation (1.6) τ denotes the optical thickness, and I^* represents the diffuse radiation field.

In such a case of Chandrasekhar's planetary problem with the horizontally inhomogeneous ground albedo of diffuse reflector, the initial-value solution of Equation (1.1) is expressed in terms of the ground albedo, the scattering and transmission function of the free atmosphere, the scattering and transmission functions of the searchlight problem in three-dimensional slab, whose point of incidence is assumed to be $(0,0,0)$, i.e., $A(x,y)$, $S(z_1;\Omega,\Omega_o)$, $T(z_1;\Omega,\Omega_o)$, $S(z_1,x,y;\Omega,\Omega_o)$, and $T(z_1,x,y;\Omega,\Omega_o)$. Based on the invariance principles in three-dimensional slab, we have the recursive solution as below:

$$I^* (z_1,x,y;+\Omega) = \frac{F}{4v} S(z_1;\Omega,\Omega_o) +$$

$$\exp(-\tau/v)I^*(0,x-z_1\tan\theta\sin\phi,y-z_1\tan\theta\cos\phi;+\Omega) + (1/4\pi v) \times$$

$$\int_{-\infty}^{\infty}\int dx'dy'\int_{2\pi} T(z_1,x-x',y-y';\Omega,\Omega')I^*(0,x',y';+\Omega')d\Omega', \qquad (1.7)$$

$$I^*(0,x,y;+\Omega) = A(x,y)F\exp(-\tau/u)u +$$

$$(A(x,y)/\pi) \int_{2\pi} I^*(0,x,y;-\Omega')v'd\Omega', \qquad (1.8)$$

$$I^*(0,x,y;-\Omega) = (F/4\pi v)T(z_1;\Omega,\Omega_o) + (1/4\pi v) \times$$

$$\int_{-\infty}^{\infty}\int dx'dy'\int_{2\pi}S(z_1,x-x',y-y';\Omega,\Omega')I^*(0,x',y';+\Omega')d\Omega', \qquad (1.9)$$

where τ is the atmospheric optical thickness. The initial-value solutions of the scattering and transmission functions for the three-dimensional searchlight problems are found in our preceding paper (Ueno, 1976).

Numerical computation of Equation (1.7) through Equation (1.9) is difficult even with the aid of high-speed digital computers. Then, in the next section we consider an approximate solution of the total spectral radiance by the use of the so-called adding procedure.

APPROXIMATE ADDING EQUATION FOR MULTIDIMENSIONAL PLANETARY PROBLEM

In this section, extending the scattering matrix method (cf. Ueno 1976) to the adding procedure of solving the generalized Chandrasekhar's planetary problem, the total spectral radiance $I(z_1,x,y;\Omega)$ is expressed in terms of the scattering and transmission functions· of the free atmosphere, allowing for the horizontally nonuniform albedo $A(x,y)$. In the case of free atmosphere the diffusely reflected and transmitted intensities of the radiation field, i.e., $I(z_1,+\Omega)$ and $I(0,-\Omega)$ are expressed in terms of the scattering and transmission functions of the free atmosphere

$$I(z_1,+\Omega) = \frac{F}{4v} S(\tau;\Omega,\Omega_o), \qquad (1.10)$$

$$I(0,-\Omega) = \frac{F}{4v} \, T(\tau;\Omega,\Omega_o). \tag{1.11}$$

The Riccati-type of the integro-differential equations governing the S- and T-functions are well-known (cf. Ueno and Wang 1973). Furthermore, the numerical computation of these global functions is also done with the aid of the adding procedure (cf. Poon and Ueno 1973).

On making use of the generalized invariance principles, Equations (1.7) through (1.9) (Ueno and Terashita 1979), and assuming a single reflection approximation at the bottom, we get the following total spectral radiance at the top:

$$I^*(z_1,x,y;+\Omega) \simeq \frac{F}{4v} \, S(\tau; \, \Omega,\Omega_o)$$

$$+ \, FuA(x-z_1\tan\theta\sin\phi, y-z_1\tan\theta\cos\phi) \, e^{-\tau(1/v \, + \, 1/u)}$$

$$+ \, \frac{e^{-\tau/v}}{4\pi} \, FA(x-z_1\tan\theta\sin\phi, \, y-z_1\tan\theta\cos\phi) \int_{2\pi} T(\tau; \, \Omega',\Omega_o)d\Omega'$$

$$+ \, \frac{F}{4\pi v} \, e^{-\tau/u} u \int_{2\pi} d\Omega' \int_{-\infty}^{\infty} \int T(\tau; \, \Omega,\Omega') \, A(x',y')dx'dy' \, +$$

$$\frac{F}{16\pi^2 v} \int_{2\pi} d\Omega' \int_{-\infty}^{\infty} \int T(\tau; \, \Omega,\Omega') \, A(x',y')dx'dy' \int_{2\pi} T(\tau;\Omega'',\Omega_o)d\Omega''. \tag{1.12}$$

On the right-hand side of Equation (1.12) the first term represents the diffuse reflection of the radiation by the free atmosphere, the second term is due to the direct transmission of upward radiation after a single reflection of the directly transmitted incident radiation by the reflector, the third term is also due to the direct transmission of upward radiation after a single surface reflection of the diffusely transmitted downward radiation, the fourth term denotes the diffuse transmission of upward radiation after a single reflection of directly transmitted incident rays by the reflector, and the last term represents the diffuse transmission of upward radiation after a single reflection of diffusely transmitted downward radiation by the reflector. In other words, the transmitted radiance consists of the second and third components, whereas the path radiance is equal to the sum of the first, fourth and last components. In our present case, assuming a strongly forward-elongated phase function and the narrow view angle from the space-borne sensor, Equation (1.12) is approximately converted into a more simple space integration formula, particularly in the case of checkerboard type of flat background bounded by a diffuse reflector, which is scanned by a sensor in a nadir direction within the view, the approximation may be readily made.

After minor rearrangements of terms, we get the required expression appropriate to the numerical computation

$$I^*(z_1,x,y;+\Omega) \cong \frac{FS}{4v} (\tau;\Omega,\Omega o)\ \bar{I}^*(z_1,x,y;+\Omega) + \frac{Fu}{4\pi v} \{\ ^{-\tau/u} +$$

$$\frac{1}{4\pi u}\int_{2\pi}T(\tau;\Omega'',\Omega o)d\Omega''\}\int_{2\pi}d\Omega'\int_\infty^\infty\int T(\tau;\Omega,\Omega')A(x',y')dx'dy'\ ,\qquad(1.13)$$

where

$$\bar{I}^*(z_1,x,y;+\Omega) = Fe^{-\tau/v}u\ A(x-z_1\tan\theta\sin\phi,y-z_1\tan\theta\cos\phi)\ x$$

$$\{e^{-\tau/u} + \frac{1}{4\pi u}\int_{2\pi}T(\tau;\Omega',\Omega o)d\Omega')\}.\qquad(1.14)$$

In Equation (1.14) $\bar{I}^*(z_1,x,y;+\Omega)$ represents the upward intensity of radiation directly transmitted after a single reflection by the bottom.

Equations (1.13) and (1.14) are approximately converted into the convolution form of the total radiance expression used in our preceding paper (cf. Kawata, Haba, Kusaka, Terashita, and Ueno 1970). Finally, it should be mentioned that the evaluation of internal radiation field is made with the aid of adding procedure on the assumption of the mean ground albedo. The procedure is available for the average method used in the AECS system.

In the above single reflection approximation of the multiply scattered radiation we used the S- and T-functions of the free atmosphere.

SUPERPOSITION APPROACH TO THE SPECULAR REFLECTION PROBLEM

It has been verified that the planetary problem with a diffuse reflector can be reduced to the standard problem (cf. Chandrasekhar

1950, Kagiwada and Kalaba 1971, Bellman and Ueno 1972). On the other hand, the specularly reflecting planetary problem could be also reduced to the standard problem with the aid of superposition approach (cf. Ueno and Mukai 1973). This procedure is convenient for the case of rapid convergence of the required scattering function in terms of the scattering and transmission functions of the standard problem for a free atmosphere, particularly, with the calm sea surface. In what follows, we show some of the analytical results of this problem useful for the total radiance by an atmosphere-ocean system.

Consider a plane-parallel, homogeneous, and anisotropically scattering atmosphere of the optical thickness x bounded by a specular reflector at the bottom t=0. Let parallel rays of net flux πF per unit area normal to the direction of propagation be incident on the top t=x of the atmosphere in a direction $(-u,\phi_0)$, where $u(0<u\leqq1)$ represents the cosine of inclination to the inward normal and $\phi_0(0\leq\phi_0<2\pi)$ is the azimuth referred to suitably chosen axis on the bottom. On deriving the relationship between the source function of the planetary problem and the source function of the standard problem, after some complicated manipulations, we get the required results, i.e., the $S(x,A;\Omega,\Omega_0)$-function of the planetary problem in terms of the $S(x;\Omega,\Omega_0)$- and $T(x;\Omega,\Omega_0)$-functions of the standard problem

$$S(x,A;\Omega,\Omega_0) = S(x;\Omega,\Omega_0) + A(u)\exp[-x(1/v+1/u)] \ x$$

$$\sum_{n=1}^{\infty} B(n;\bar{\Omega},\tilde{\Omega})+\exp(-x/u)A(u)T(x;\Omega,\tilde{\Omega})+\exp(-x/u) \ x$$

$$\sum_{n=1}^{\infty} C(n;\Omega,\tilde{\Omega})+\exp(-x/v) \sum_{n=1}^{\infty} D(n;\bar{\Omega},\Omega_0)$$

$$+ \sum_{n=1}^{\infty} E(n;\Omega,\Omega_0), \qquad\qquad (1.15)$$

where

$$B(1;\bar{\Omega},\tilde{\Omega}) = A(v)S(x;\bar{\Omega},\tilde{\Omega}),$$

$$B(n;\bar{\Omega},\tilde{\Omega}) = [A(v)/4\pi]\int_{2\pi}S(x;\bar{\Omega},\Omega')B(n-1;\bar{\Omega}',\tilde{\Omega})d\Omega'/v',$$

$$C(n;\Omega,\tilde{\Omega}) = [A(u)/4\pi] \int_{2\pi} T(x;\Omega,\Omega')B(n;\bar{\Omega}',\tilde{\Omega})d\Omega'/v',$$

$$D(1;\bar{\Omega},\Omega o) = A(v)T(x;\bar{\Omega},\Omega o),$$

$$D(n;\bar{\Omega},\Omega o) = [A(v)/4\pi] \int_{2\pi} S(x;\bar{\Omega},\Omega')D(n-1;\bar{\Omega}',\Omega o)d\Omega'/v',$$

$$E(n;\Omega,\Omega o) = (1/4\pi) \int_{2\pi} T(x;\Omega,\Omega')D(n;\bar{\Omega}',\Omega o)d\Omega'/v'. \qquad (1.16)$$

In Equations (1.15) $\tilde{\Omega}$ and $\bar{\Omega}$ stand for $(u,\phi_0-\pi)$ and $(v,\phi+\pi)$, respectively, and A denotes the specularly reflecting albedo.

Once when the scattering and transmission functions of the standard atmosphere without the reflector have been numerically computed, Equation (1.15) permits us to evaluate the scattering function of the specularly bounded atmosphere with the aid of the high speed digital computer. It is of interest to mention that the quantity n in Equation (1.15) represents the number of reflections by the specular reflector. Furthermore, it should be mentioned that the comparison of the numerical values of S-function by Equation (1.15) with those of S-function by solving the Riccati-type integro-differential equation (cf. Casti, Kalaba and Ueno, 1969), yields a satisfactory degree of accuracy in the isotropically scattering case of $\tau=0.2$, $A=1$, $\lambda=0.1$ (cf. Ueno and Mukai, 1973).

REFERENCES

Ambarzumian, A. 1945. Bull. Erevan Astr. Obs., No. 6.
Ambarzumian, V. A., (Ed). 1958. Theoretical Astrophysics, Pergamon Press.
Bellman, R., R. Kalaba, and M. C. Prestrud. 1963. Invariant Imbedding & Radiative Transfer in Slabs of Finite Thickness, American Elsevier Publ. Co., New York.
Bellman, R., R. Kalaba, and S. Ueno. 1965. ICARUS, 4, 119.
Bellman, R., and S. Ueno. 1972. Astrophys. Space Sci., 16, 241.
Breig, W. F., and A. L. Crosbie. 1974. J. Math. Analy. Appl., 46, 104.
Busbridge, Ida W. 1960. The Mathematics of Radiative Transfer, Cambridge University Press.
Casti, J., R. Kalaba, and S. Ueno. 1969. J. Quant. Spectrosc. Radiat. Transfer, 9, 537.
Chandrasekhar, S. 1950. Radiative Transfer, Oxford University Press.
Chandrasekhar, S. 1958. Proc. Nat. Acad. Sci., USA, 44, 933.

Elliott, J. P. 1959. Proc. Roy. Soc. A. 228, 424.

Kagiwada, H., and R. Kalaba. 1971. J. Quant. Spectrosc. Radiat. Transfer, 11, 1101.

Kawata, Y., Y. Haba, T. Kusaka, Y. Terashita, and S. Ueno. 1978. Proc. Twelfth Intern. Symp. on Remote Sensing of Environment, by J. Cook (Edit), 1241.

Matsumoto, M. 1969. Publ. Astr. Soc. Japan, 21, 1.

Poon, P. T.Y., and S. Ueno 1973. J. Atmosph. Sci., 30, 954.

Rybicki, G. B. 1971. J. Quant. Spectrosc. Radiat. Transfer, 11, 827.

Sobolev, V. V. 1959. Dokl. Nauk SSSR, 129, 1265.

Sobolev, V. V. 1963. A Treatise on Radiative Transfer, Van Nostrand, Princeton, N.J.

Ueno, S., and S. Mukai. 1972. J. Quant. Spectrosc. Radiat. Transfer, 13, 1261.

Ueno, S., R. Vasudevan, and R. Bellman. 1973. Math. Biosci., 18, 55.

Ueno, S. and A. P. Wang. 1973. Astrophys. Space Sci., 23, 205.

Ueno, S. 1976. Mem. Kanazawa Inst. Techn., A-6, 1.

Ueno, S. 1977. Proc. 10th Lunar and Planetary Symposium, by M. Shimizu (Editor), Tokyo, 166.

Ueno, S., and Y. Terashita. 1979. "Evaluation and Control of Environmental Systems", Y. Sawaragi (Editor), Kyoto, 75.

Uesugi, A., and J. Tsujita. 1969. Publ. Astr. Soc. Japan, 21, 370.

CORRECTION OF THE ATMOSPHERIC
SCATTERING EFFECTS ON EARTH'S IMAGERY

Sueo Ueno

INTRODUCTION

In recent years, with the advent of LANDSAT, SKYLAB, and other
advanced earth monitoring space- and air-crafts, it has become
increasingly important to evaluate the extent of the atmospheric
effects on remotely sensed data, because the presence of the terrestrial atmosphere tends to diminish the ability to discriminate
between target and surroundings. In other words, apparent ground
radiance distribution obtained by space- or air-borne sensors
differs from the intrinsic ground radiance distribution, because of
the multiple scattering of radiation in the atmosphere-ground
system. The atmosphere can selectively scatter, absorb, re-emit,
and refract radiation that traverses through it. In this context,
the atmosphere has such a filtering function that changes spatially, spectrally, and temporally.

Hence, an allowance for the atmospheric effects should improve the accuracy of pattern recognition and image interpretation in global monitoring. To insure the usefulness of remote sensing, it is necessary to determine the atmospheric effects due to scattering, absorption and reflection on the radiation field of the atmosphere-ground system. Evaluations of the atmospheric effect based on horizontally uniform radiation field have been attempted by several authors (cf. Kattawar and Plass, 1968; Turner and Spencer, 1972; Potter and Shelton, 1974; Griggs, 1974; Potter and Mendlowitz, 1975; Takashima, 1975; Fraser, 1975; Ueno and Mukai, 1977). In the case of horizontally non-uniform atmosphere, the scattering effect has been analytically studied, in connection with the search-light problem (cf. Ueno, 1976).

In the real atmosphere-ground system, the assumption of the horizontally uniform radiation field is seldom realized. Even if the atmosphere were spatially uniform over an extended area, non-uniform atmospheric effects must be present because of the non-uniform ground albedo (what we may call the 'blurring effect'). Several attempts have been made to take into account the diffuse radiation field due to the horizontally non-uniform ground albedo, with the aid of the adding-and-doubling procedure, the Fourier transform method, the invariant imbedding technique and others (cf. Odell and Weinman, 1975; Tunner, 1975; Haba et al., 1977; Ueno, 1977; Ueno et al., 1978; Kawata et al., 1978a; Kusaka et al., 1978; Kawata et al., 1978b; Ueno and Terashita, 1979; Haba et al., 1979).

In the present paper, based on the three-dimensional transfer model consisting of a one-dimensional free atmosphere bounded by a horizontally non-uniform diffuse reflector, we express the required total spectral radiance at the top in terms of the scattering and transmission functions for the free atmosphere, and the horizontally inhomogeneous ground albedos.

In the second section it is shown that the total spectral radiance recorded by the remote sensor can be expressed in terms of the free-atmospheric path radiance, the directly and diffusely transmitted components of the surface radiance. Among them the diffusely transmitted component of the surface radiance gives rise to the blurring of the ground image. Two approaches for correcting the atmospheric effects are outlined. It should be noted that the correction procedure for aircraft data is much complicated in comparison with that for satellite data, because the computation of the internal radiation field requires a large amount of numerical works, and in addition we have to evaluate the scan angle effect properly. Furthermore, realistic optical models of the atmosphere are established, allowing for the scattering and absorption due to the molecular gas and the stratospheric and coastal types of aerosols. In the third subsection the procedure of correcting the blurring effect on satellite data is described, and the procedure is applied to a LANDSAT data set, for a series of atmospheric models with different haze levels. In the fourth subsection, we describe a somewhat simplified correction procedure that is appropriate for aircraft data. This procedure is applied, using the standard atmosphere, to a data set obtained by Japan Research Committee of Environment Remote Sensing (JRCERS). In the last subsection the atmospheric effects correcting system is outlined.

RADIATIVE PROCESSES IN THE ATMOSPHERE

Diffuse Radiance in the Atmosphere

The radiative processes in the atmosphere bounded by a non-uniform reflector may be described schematically by Figure 1. The diagram represents the case of satellite observation. The detector above the atmosphere measures the upward radiance in the direction of its line of sight. Since the range of the view angle by a space-borne sensor is small, we may consider only the upward normal radiance. In the case of aircraft observation there are two significant differences; 1) the detector is within the atmosphere, 2) the scan angle effect must be considered because of the large range of the view angle. When the space-borne or air-borne sensor sees a target on the background, the following three types of photons reach the detector:

    A.   Photons reflected by the target and then directly
           transmitted by the free atmosphere.

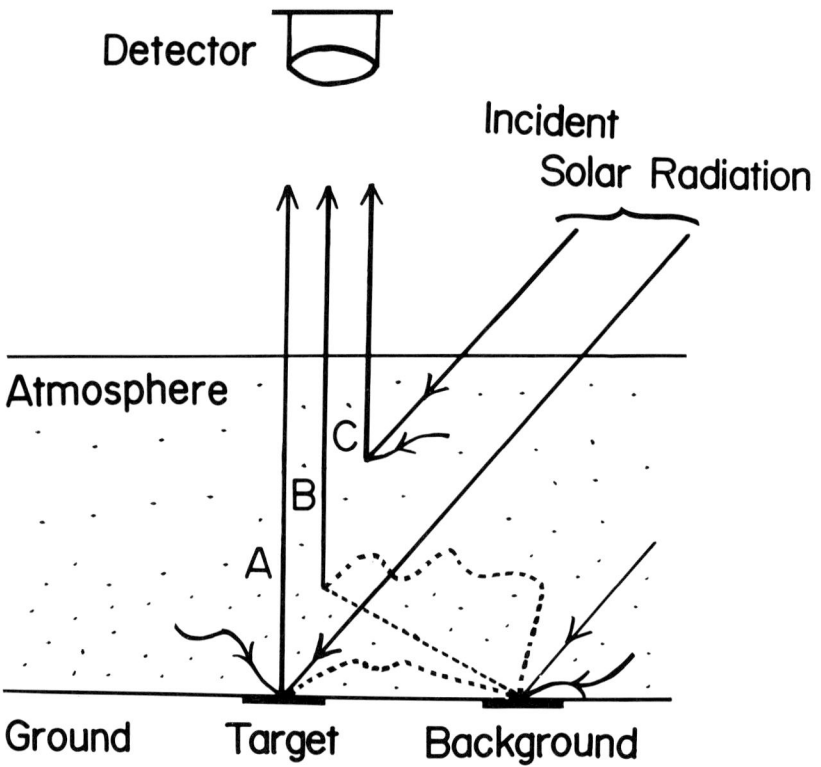

FIGURE 1:   The radiative processes in the atmosphere with a non-uniform reflector.

B.   Photons transmitted diffusely by the free atmosphere, after having interacted by the background.

C.   Photons reflected diffusely by the free atmosphere.

It may be said that Type-A photons carry the direct information on the target, Type-B photons represent the combined effects by both the background and atmosphere, and Type-C photons represent the free-atmospheric effects. The radiance due to Type-B and Type-C photons is usually called the path radiance. In particular, Type-B photons produce the blurring of the ground patterns in much the same way as the instrumental blurring. The radiative processes those photons undergo, before reaching the detector, are as follows: 1) direct transmission through the atmosphere (shows by solid straight lines in Figure 1), 2) pure atmospheric diffusion (solid wavy lines), and 3) atmospheric diffusion involving the ground reflection (dotted wavy lines). A dotted straight line is used to represent direct transmission after the ground reflection.

The presence of the path radiance obscures the intrinsic surface image. Therefore, it is very important to remove these types of radiance and deal only with the radiance by Type-A photons. In order to eliminate the contributions by both Type-B and Type-C photons exactly from the observed radiance, we must solve a three-dimensional radiative transfer problem in the atmosphere-ground system, and it would be an extremely difficult task.

Instead of solving it exactly, we formulate two approximate correction methods for the atmospheric effect, which are applicable to the real three-dimensional environment. Both approximation methods utilized the results from the one-dimensional radiative transfer models.

In the first correction method (the convolution method) we obtain a convolution-type integral equation relating the albedo distribution to the observed radiance map, while the kernel of the equation is given by the properties of the atmosphere. Although this method is considered to be more exact than the second one, the numerical procedures for applying it to aircraft data would be much more complicated. The second correction method (the average method) is found to be applicable to the multi-spectral scanner (MSS) data obtained by the space-borne sensor, as well as to those obtained by the air-borne sensor. This method was used to remove the atmospheric effect on our previous work (cf. Ueno et al., 1978).

Atmospheric Models

In order to solve the radiative transfer equation, we must have necessary atmospheric parameters. In this investigation we shall use the atmosphere of Elterman (1968) as a standard model of the earth's atmosphere. This is a reasonable choice at least for our test imagery, because it is apparently free from cloud. Since it is known that the characteristics and the size distribution of the aerosols are different in the lower atmosphere (altitude < 15 km) from those in the upper atmosphere (> 15km), we divide the atmosphere into two layers. The ground is assumed to be a

Lambertian surface. The necessary atmospheric parameters, such as
the optical thickness of the molecular gas, that of the aerosols,
the turbidity factor of the molecular gas, and the absorption by
ozone in each layers of the atmosphere, are provided in the
Elterman's table.

We use the Haze M and Haze H for the types of the aerosols in
the lower and the upper layers of the atmosphere, respectively.
The Haze M represents maritime- and coastal-type aerosols. Since
our study site is in the coastal area, it is justified to use the
Haze M in the lower layer. The Haze H is the stratospheric type of
aerosols. Discussions on the aerosol types, including Haze M and
Haze H, are given by Deirmendjian (1969). The single scattering
phase functions for the Hazes M and H show strong forward scattering pattern, and may be computed by means of Mie's theory. The
single scattering phase function for a molecular particle is given
by Rayleigh's scattering formula. In solving the one-dimensional
radiative transfer problem, we treat the multiple scattering
exactly for a three-layer model by using the (one-dimensional)
doubling and adding methods. These methods have been commonly used
for radiative transfer in planetary atmospheres (Hansen and Travis,
1974).

The solid curve in Figure 2 shows the attenuation profile due
to aerosol particles, at the wave-length of 0.55μm, as taken from
Elterman's table. Corresponding profiles for other wavelengths
have similar characteristics. It should be noted that the attenuation coefficient decreases exponentially as the altitude increases
in the first 4 kilometers. It will be a reasonable assumption that
the haze level (or visibility) of the atmosphere is in most cases
determined by the aerosols, which are concentrated heavily in the
lowermost portion of the atmosphere. It is also noted that the
concentration of the aerosols dominates that of the molecules in
this portion of the atmosphere. Therefore, in order to account for
the variation of the haze level of the atmosphere, we extend the
standard aerosol profile by changing the gradient of the straight-
line portion (in the semi-logarithmic diagram), in a manner shown
by the dotted lines. For 0.55μm, this portion of the aerosol

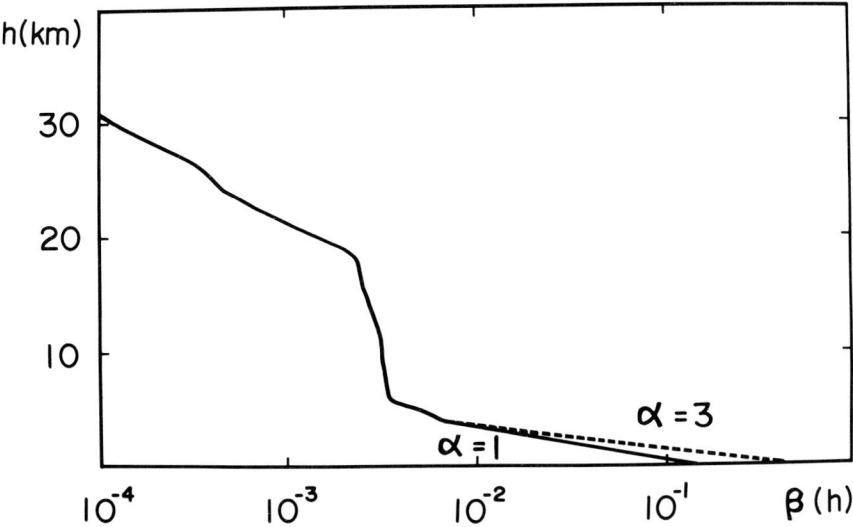

FIGURE 2:   The aerosol attenuation profiles.

attenuation profile, $\beta_A(h)$, is numerically expressed as,

$$\beta_\alpha(h) = 0.158e^{-0.7916h} \cdot \alpha^{1 - \frac{1}{4}h}.$$ (2.1)

Here, h is in units of kilometer. Similar expressions (having different multiplying factors) have been obtained for other wavelengths. In the standard case of Elterman, we have $\alpha = 1$.

ATMOSPHERIC CORRECTION FOR LANDSAT DATA

Single-reflection Approximation

We assume that, for a given wavelength, the atmosphere is horizontally uniform. Let us denote by $\pi F f_G$, the downwards flux at the bottom of the free atmosphere, i.e., the amount of radiation reaching the ground, either directly or diffusely transmitted through the atmosphere, where the input solar radiation is given by πF. When we are dealing with the LANDSAT data, it is only necessary to derive the expression for the vertical component of the radiation emerging from the top of the atmosphere, since in this case the scan angle of observation is sufficiently small.

In the single reflection approximation, the second- (and higher-) order processes for those photons, which have had interactions with the ground, are neglected. The accuracy of this approximation has been checked, using a standard atmosphere with a uniform ground (cf. Ueno et al., 1978). For the ground albedo of less than 0.5 (true for the usual ground materials), the errors are found to be several percents, which are small enough for most applications.

Under the geometry shown in Figure 3, assuming the single reflection approximation, the total radiance detected by the space-borne sensor, $I_o(x,y)$, is expressed in terms of the quantities in observed units such that $I'_o=CI_o$ and $I'_A=CI_A$, with the gain factor C (cf. Equation (1.13); Kawata et al. 1978a).

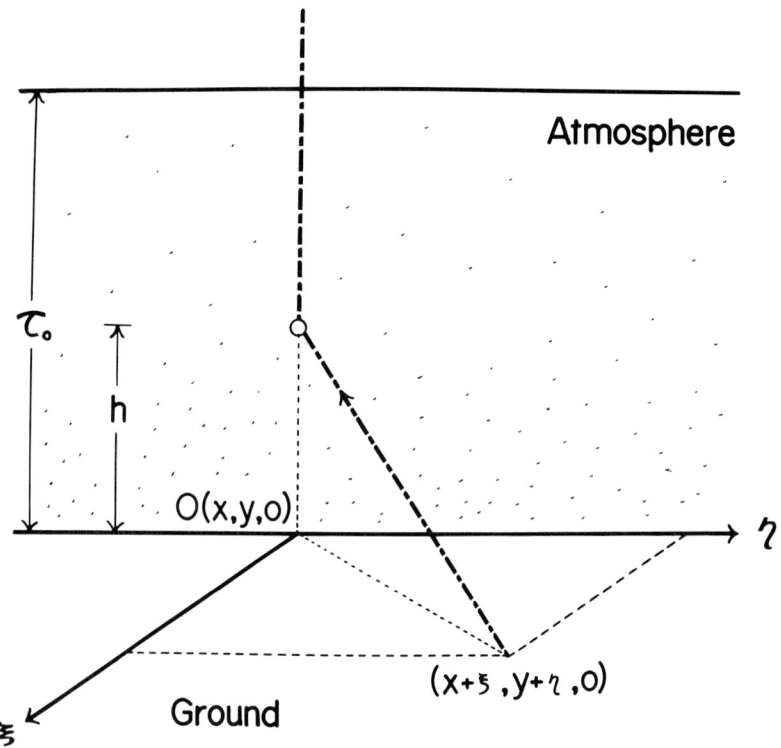

FIGURE 3: The diagram of the signal reflection and re-scattering approximation.

$$I_0'(x,y) = I_A'(x,y) + \int_{-\infty}^{\infty}\int k(\xi,\eta)\, I_A'\,(x+\xi,y+\eta)d\xi d\eta. \qquad (2.2)$$

In Equation (2.2)

$$I_A(x,y)=Ff_G A(x,y)e^{-\tau 0}, \qquad (2.3)$$

where τ_0 is the optical thickness of the atmosphere, and $A(x,y)$ is the ground albedo at the point (x,y). The function, $k(\xi,\eta)$, is defined by,

$$K(\xi,\eta) = \int_0^H \frac{h}{(\xi^2 + \eta^2 + h^2)\,3/2} \frac{P(\theta')}{4\pi}\, e^{-(\frac{1}{\mu};-1)\tau(h)}dh. \qquad (2.4)$$

Here, $P(\theta')$ is the single scattering phase function, μ' the direction cosine $\cos\theta'$ and H the geometrical thickness of the atmosphere. It should be mentioned that Equation (2.2) is found, provided that our concern is only in the removal of the blurring (namely, the Type-B component), thus allowing for an undetermined constant in the solution.

In Equation (2.4), the kernel $k(\xi,\eta)$ can be computed, when the optical properties of the atmosphere are given. Then the distribution $I_A'(x,y)$ can be determined for the observed radiance map, $I'o$ (x,y). This latter quantity gives the blurring-free radiance map, aside from an undetermined additive constant.

Correction Procedure (Convolution Method)

In order to solve Equation (2.2) for $I_A'(x,y)$, we consider the two-dimensional Fourier transform of the equation in discrete form:

$$O_{mn}=A_{mn}+K_{mn}A_{mn}, \qquad (2.5)$$

where A_{mn}, O_{mn}, and K_{mn} are the Fourier coefficients of order (m,n), of the quantities, $I_A'\,(x,y)$, $I_0'(x,y)$ and $K(\xi,\eta)$. From Equation (2.7), we obtain,

$$A_{mn} = \frac{O_{mn}}{1+k_{mn}}. \qquad (2.6)$$

This discrete form of $I_A'(x,y)$ is then given as,

$$a_{k1}= \frac{1}{N^2} \sum_{m=0}^{N-1} \sum_{n=0}^{N-1} (\frac{Omn}{1+K_{mn}})e^{\frac{2\pi j}{N}(km+1n)}. \qquad (2.7)$$

Here we have assumed that the observed image data are given as a square matrix of the N x N pixels. The indices (k,1) give the position of a pixel in the matrix.

Results and Discussions

The convolution method described in the above has been applied to a LANDSAT data set. The data set was obtained by the LANDSAT-II satellite on September 11, 1975, covering the central part of Honshu (the main island) of Japan. The portion of the data set we have studied consists of 256 x 256 pixels, including the City of Kanazawa and its vicinity (roughly corresponding to a 20km x 20km ground area). The original data are displayed by the printouts, Figure 4a (Band 4). The geography consists of the sea (at the upper left corner), a lake (near the upper center), the wooded mountain area (on the right-hand side), and the flat land otherwise. The City of Kanazawa occupies the central portion of the image.

A sequence of atmospheric models, with $\alpha=1$ (standard model), 3, and 10, is used in the correction procedure. The case with $\alpha=10$ is unlikely for this particular observation (according to the weather forecast, the sky was 'clear' on that day). However, the case was included to examine the effect of over-correction. The blurring-corrected images for these models are shown in Figure 4b (Band 4).

In the case of Band 4, in a comparison between Figures 4a (uncorrected) and 4b (corrected with $\alpha=3$) we observe the following improvements: 1) The water channel going parallel to the coast line (near upper left corner) is more clearly seen in the corrected image. 2) The river system on the left hand side near the central region is also better recognized in the corrected image. 3) There is a slight improvement in the ground patterns of the flat region. There does not seem to be significant differences in the wooded region.

ATMOSPHERIC CORRECTION FOR AIRCRAFT DATA

Evaluation of the Internal Radiation Field

In order to evaluate the atmospheric effects for the MSS data

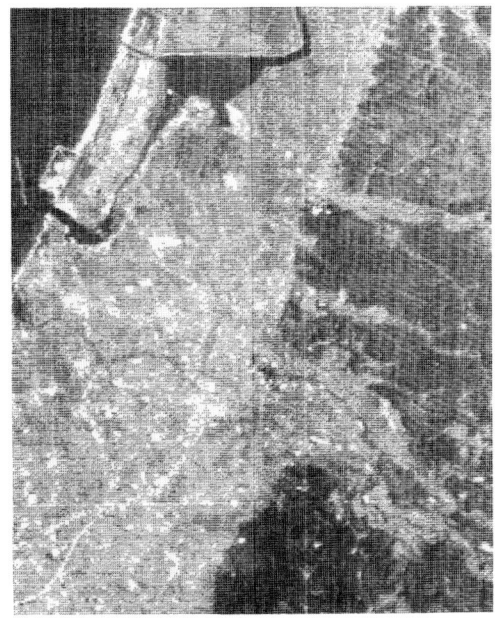

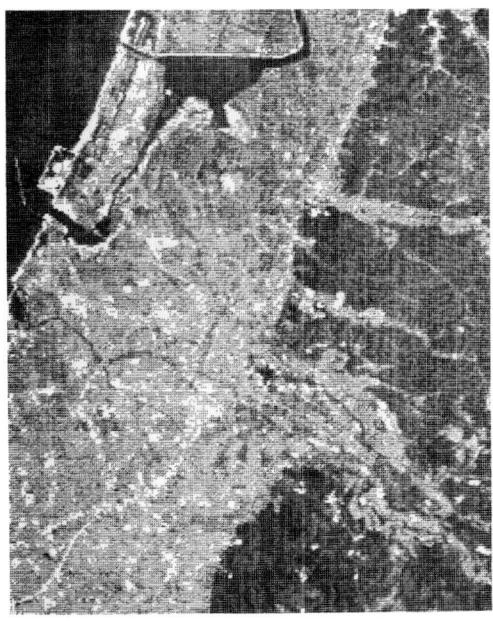

FIGURE 4: The LANDSAT imagery of the Kanazawa in band 4.

obtained by the air-borne sensor, it is necessary to consider an
internal radiation field. We compute the necessary quantities of
internal radiance at three different altitudes of 1km, 3km, and
50km. In the computation the atmospheric model of Elterman (α=1)
is assumed to represent the Earth's atmosphere. We make computations for three different wavelengths (λ=0.475μm, 0.575μm, and
0.670μm). They correspond to the central wavelengths of the particular three bands of aircraft MSS data supplied by JRCERS.

Let us introduce the following important quantities for the
evaluation of the scattering effects within the atmosphere. The
direct upward radiance at a certain altitude due to Type-A photons
is denoted by U_A, the corresponding other kinds of radiances due to
Type-B and -C are denoted by U_B and U_C, respectively. The total
upward radiance is given by the sum of the above three radiances,
U_A, U_B, and U_C. It is denoted by U_T. The sum of the radiances U_B
and U_C is called the upward path radiance, denoted by U_P.

In Figure 5 we present the theoretically computed U_T as a
function of the ground albedo and the scan angle. The linear
dependence of U_T on the ground albedo A holds up to A<0.5. The
slope of the total upward radiance curve is flattened as the scan
angle increases. The ratios U_A/U_T and U_P/U_T are shown as a function of the ground albedo A and the scan angle Θ in Figure 6. The
shapes of the right side portion and the left side portion of the
curves are not symmetric about the line of zero scan angle. This
indicates that the degree of the atmospheric effects depends on
which side the sensor scans with respect to the forward direction
of moving aircraft. In the course of scanning from the right to
the left the transition of the azimuthal angle occurs at zero scan
angle. In Figure 6 we plot the case of the azimuth transition from
50° to 130°. The asymmetric nature of the ratios U_A/U_T and U_P/U_T
about zero scan angle is shown in Figure 6. It is due to atmospheric scatterings. It is, therefore, understood that the above
asymmetry about zero scan angle becomes less with the increase of
the ground albedo, or with the decrease of the altitude. The above
asymmetry about zero scan angle may be simply explained by differences of the azimuthal angle.

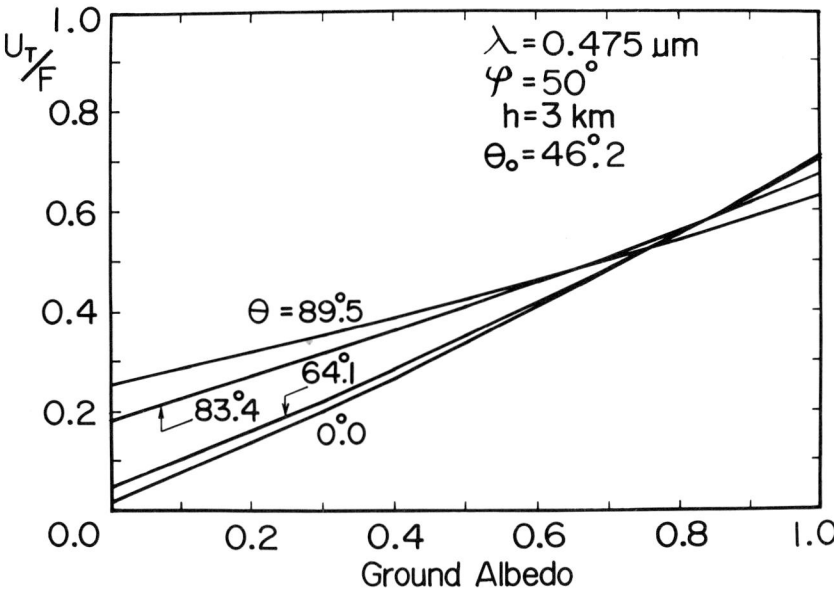

FIGURE 5: The total upward radiance as a function of the ground
albedo and the scan angle.

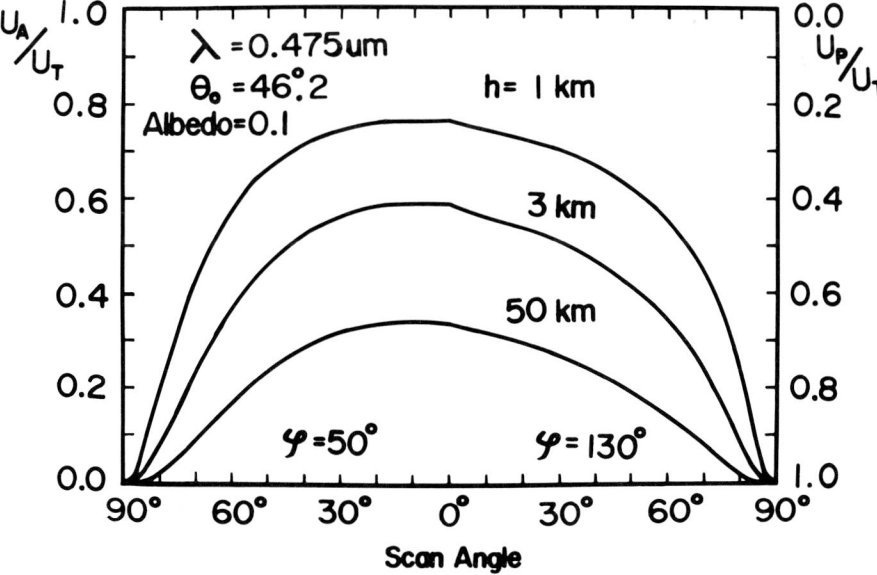

FIGURE 6: The internal radiance ratios as a function of the ground
albedo and the scan angle.

Results and Discussions

In the application of the average method to the real image data, we must estimate the size of the area over which the averaging should be performed. The radius of the area is found to be about 200km, corresponding to 25 pixels in the test image. The actual application is made to the imagery of the Kanazawa port area which was taken by JRCERS aircraft on Oct. 14, 1977. This area is located in the left central portion of the LANDSAT imagery used in Section 3. The studied imagery consists of 768 x 768 pixels, roughly corresponding to a 6km x 6km ground area. We observe the following differences when two grey map images before and after the atmospheric correction are compared. The atmosphere-uncorrected and atmosphere-corrected images are given in Figures 7a and 7b, respectively. In the scattering-corrected, thin surface structures appear more clearly. This is particularly true for a breakwater in the sea and a coastal shore line in the lower left portion of the map. The border between the land and the water is more clearly recognized at the central portion of the map in the corrected image. Some of the roads may be found in the corrected image, while they are hardly recognized in the uncorrected image. The results shown in Figure 7 are for the wavelength $\lambda=0.575\mu m$.

It is shown that the correction of the atmospheric effects for the aircraft data can be accomplished by the removal of both the blurring effect and the scan angle effect. The application of the average method to the real aircraft image data is made for the first time and we obtain the impressive results.

Then, we conclude that the atmospheric correction of the remotely sensed earth's imagery enables us to make a more accurate interpretation of the data in the case of where no such correction is made.

OUTLINE OF THE AECS

The AECS is a system of computer software programs and performs corrections for the effects of the pure atmospheric path radiance, the blurring due to the interaction of photons with

FIGURE 7: The aircraft imagery of the Kanazawa Port area.

surface background materials and the view angle. The block diagram
of the AECS is shown in Figure 8. The AECS consists of two subsystems named the Correction Subsystem and the Inter-Extrapolation
Subsystem. In the Inter-Extrapolation Subsystem we have three
blocks, namely, the Radiance File, the Inter-Extrapolation Program
and the Correction Table. In the Radiance File we store the necessary information on the theoretical radiance quantities for typical
atmospheric and observational conditions. To make this file we
computed the theoretical values on the radiances for combinations
of the following different cases. In the wavelength region we
considered four wavelengths, i.e., 0.47μm, 0.575μm, 0.6μm and
0.845μm which correspond to observed wavelength used by Japan
Research Committee of Environmental Remote Sensing's airborne
sensor. As for the altitude three cases of 1km, 3km, and 50km were
considered. The first two altitude cases correspond to the typical
airborne sensor's altitudes, whereas the case of 50km is applicable
to LANDSAT data. We considered the cases of four atmospheric
models with different optical thickness, corresponding to the cases
from clear standard atmosphere to thick hazy sky. We also considered eleven different values of ground albedo, starting from 0.0
to 1.0 with an increment of 0.1. Thirteen different values of the
incident sun angle and view angle were used in the computation for
each combination of the above cases. A considerable amount of
computer time is required for such computations. Once computations
are carried out and the parameterized radiance coefficients are
stored in the Radiance File, the Inter-Extrapolation Program computes the fraction of the path radiance to the total radiance as a
function of view angle quite efficiently.

Since it is found that most photons received by the sensor
have at most twice interaction with ground materials, both the
theoretical total radiance I_T and the direct radiance I_D may be
expressed in the quadratic function with respect to the ground
albedo A.

$$I_T = P + QA + RA^2, \tag{2.8}$$

$$I_D = SA + TA^2. \tag{2.9}$$

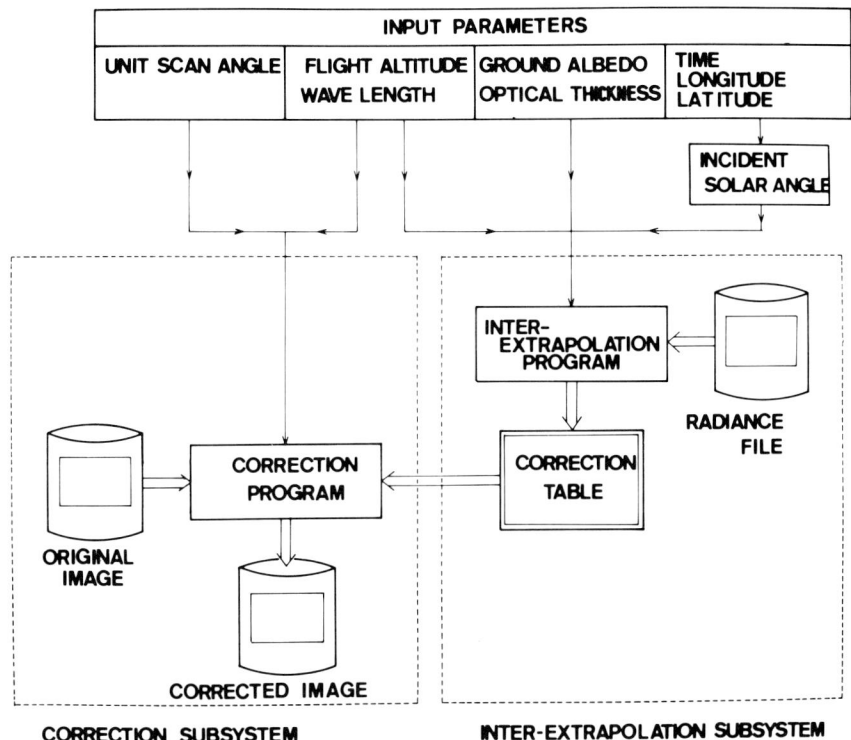

FIGURE 8: The block diagram of the AECS system.

These coefficients P, Q, R for the total radiance and S, T for the
direct radiance are found by applying the least squares method to
the computed radiance results. These parameterized coefficients
for the radiances are stored in the Radiance File.

If input parameters, such as the optical thickness of the
atmosphere, the flight altitude, wavelength and incident sun angle
deduced from the longitude and the latitude of the observed site
and the observed time are given, the fraction of the path radiance
to the total radiance is computed by the Inter-Extrapolation
Program and the results are stored in the Correction Table. Then
the Correction Program accepts the unit scan angle, the flight
altitude and the wavelength as inputs and uses the values stored in
both the Correction Table and the Original Image File to produce
the atmospheric effects free image. The Correction Program adopts

the average method for the blurring and the pure atmospheric path radiance effects, and then performs the multiplication of inverse attenuation factor for the scan angle effect. The information on both unit scan angle and the flight altitude is needed to establish the correspondence between the pixels in the image and the actual ground location.

With regards to the atmospheric corrections one of the most important parameters is the atmospheric condition at the time of the observation. The atmospheric condition may be represented in terms of the atmospheric optical thickness. The block diagram for the estimation of the optical thickness is shown in Figure 9. The optical thickness should be given either by direct measurement or by the iterative estimation from observed data sets themselves. We may obtain the optical thickness of the atmosphere by measuring the horizontal visibility. Thus the ground truth should be done at the same time, when the MSS sensor is collecting the data. Since the simultaneous observations are not always possible, it is preferable to estimate the optical thickness from the MSS data themselves. From this point of view the implementation of the right half flow of the block diagram in Figure 9 is urgently needed and is now being developed by us.

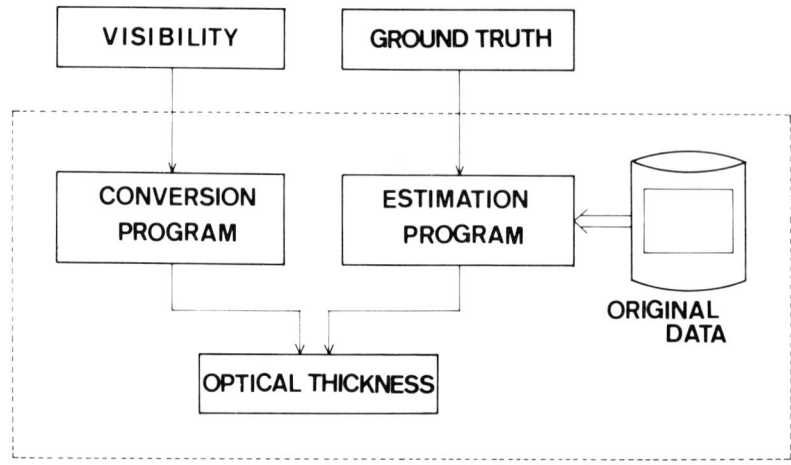

FIGURE 9: The block diagram of the Estimation Subsystem.

RESULTS AND DISCUSSION

As the remote sensing of the earth's surface becomes operational and the volume of data increases, it is very important to have the atmospheric effects correction system. Without correcting the atmospheric effects, the same ground object might be assigned to different signatures in the classification and we will face great difficulties in signature extensions. The correction for the atmospheric effects is particularly important when we study areas with small-scale formation such as those often found in a country like Japan.

The AECS at Kanazawa Institute of Technology is designed to perform the atmospheric correction with a high speed and it is based on the rigorous treatment of a quasi- one dimensional radiative transfer model which can deal with the real three-dimensional environment.

REFERENCES

Deirmendjian, D. 1969. Electromagnetic scattering on spherical polydispersions. American Elsevier, New York.
Elterman, L. 1968. UV, Visible and IR attenuation for altitude to 50km. Rept. AFCRL-68-0153. AFCRL, Bedford, Mass.
Fraser, R. S. 1975. Interaction mechanisms within the atmosphere, in Manual of Remote Sensing, R. G. Reaves, Editor. American Association of Photogrammetry, pp. 181-233.
Griggs, M. 1974. Determination of aerosol content in atmospheres from ERTS 1 data, in Proceedings of Ninth International Symposium Ann Arbor, Michigan. pp. 471-481.
Haba, Y., Y. Kawata, and Y. Terashita. 1977. Atmospheric effects on the classification of remotely sensed earth's imageries, in Proceedings of IFAC Symposium on Environmental Systems Planning, Design and Control, H. Akashi, Editor, Pergamon Press, Oxford. pp. 409-414.
Haba, Y., Y. Kawata, T. Kusaka, and S. Ueno. 1979. The system of correcting remotely sensed earth imagery for atmospheric effects, in Proceedings of 13th International Symposium on Remote Sensing of Environment. April 23-27, Ann Arbor, Michigan, 1979, pp. 1883-1894.
Hansen, J. E., and L. D. Travis. 1974. Light scattering in planetary atmospheres. Space Science Review, Vol. 16, pp. 527-610.
Kattawar, G. W., and G. N. Plass. 1968. Radiance and polarization of multiple scattering light from haze and clouds. Appl. Optics, Vol. 7, pp. 1519-1527.

Kawata, Y., Y. Haba, T. Kusaka, Y. Terashita, and S. Ueno. 1978. Atmospheric effects and their correction in air-borne sensor and LANDSAT MSS data, in Proceedings of the Twelfth International Symposium on Remote Sensing of Environment, April 20-26, Manila, Phillppines, 1978, pp. 1241-1257, ERIM, Ann Arbor.

Kawata, Y., Y. Haba, T. Kusaka and S. Ueno. 1978a. Atmospheric Correction for the Earth Imagery by Air-Borne sensor, in 1978. Proceedings of the International Conference on Cybernetics and Society, Tokyo-Kyoto, JAPAN, Nov. 3-7, 28CH 1306-0 SMC, pp. 667-671.

Kusaka, T., Y. Haba, Y. Kawata, Y. Terashita, and S. Ueno. 1978. Removal of the atmospheric blurring from remotely sensed earth imagery, in Proceedings of the 4th International Joint Conference on pattern Recognition, 1978, Kyoto, Japan, pp. 931-935.

Odell, A. P. and J. A. Weinman. 1975. The effect of atmospheric haze on images of the earth's surface, J. Geophys. Research, Vol. 80, pp. 5035-5040.

Potter, J. F., and M. Shelton. 1974. Effect of atmospheric classification of ERTS 1, in Proceedings of Ninth International Symposium on Remote Sensing of Environment. University of Michigan, Ann Arbor, Michigan. pp. 865-874.

Potter, J. F., and M. A. Mendlowitz. 1975. On the determination of haze levels from LANDSAT data, in Proceedings of the Tenth International Symposium on Remote Sensing of Enviroment. University of Michigan, Ann Arbor, Michigan, pp. 695-703.

Takashima, T. 1975. A new approach of the adding method for the computation of emergent radiation in an inhomogeneous, plane-parallel planetary atmosphere. Astrophys. Space Science, Vol. 36, pp. 319-328.

Turner, R. E., and M. N. Spencer. 1972. Atmospheric model for correction of spacecraft data, in Proceedings of Eighth International Symposium of Remote Sensing of Environment. University of Michigan, Ann Arbor, Michigan. pp. 895-934.

Turner, R. E. 1975. Signature variations due to atmospheric effects, in Proceedings of the Tenth International Symposium on Remote Sensing of Environment. University of Michigan, Ann Arbor, Michigan. pp. 671-682.

Ueno, S. 1976. Invariant imbedding and diffuse reflection of radiation by multidimensional slabs. Memoires of Kanazawa Institute of Technology, Vol. 6, pp. 1-41.

Ueno, S. 1977. Contrast transmittance at the top of an atmosphere bounded by a horizontally non-uniform diffuse refrector, in Proceedings of the Tenth Lunar and Planetary Symposium, Institute of Space and Aeronautical Science, University of Tokyo, pp. 166-170.

Ueno, S., and S. Mukai. 1977. Atmospheric effects on remotely sensed data from space, in Proceedings of IFAC Symposium on Environmental Systems Planning, Design and Control. H. Akashi, Editor, Pergamon Press, Oxford, pp. 423-428.

Ueno, S., Y. Haba, Y. Kawata, T. Kusaka and Y. Terashita. 1978. The atmospheric blurring effct of remotely sensed Earth

imagery, in Remote Sensing of the Atmosphere: Inversion
Methods and Applications. (A. L. Fymat and V. E. Zuev, Eds.),
Elsevier Scientific Publ. Co., Amsterdam, pp. 305-310.

Ueno, S., and Y. Terashita. 1979. The atmospheric scattering
effects on the remotely sensed earth's imagery, in Evaluation
and Control of Environmental System, (Edited by G. Sawaragi),
Kyoto, pp. 75-80.

INVERSE PROBLEMS IN REMOTE SENSING

INTRODUCTION

In recent years researchers tried to estimate unknown parameters in a distributed parameter system from the noisy measurements. Such inverse problems that small perturbations in the
observed functionals may result in large errors in the corresponding solutions are often called ill-posed in the classical sense. By
imposing certain additional restrictions on the admissible solutions, ill-posed problems may possess stable solutions relative to
data perturbations. Such problems are known as conditionally
well-posed. A need for approximate solutions of conditionally
well-posed problems with inaccurate data gave rise to an innovation
with a regularization by which the identification problems are
reduced to the minimization of performance (quadratic) functionals.

483

In a series of papers (cf. Bellman et al. 1965, Kagiwada 1974, Kagiwada et al. 1975), it has been shown how to use powerfully quasilinearization and invariant imbedding for the estimation of the optical properties of the free atmosphere, in the least square sensed, from the noise measurements of the total spectral radiance emerging from the top. In the present section, extending the above algorithm to the case of atmosphere bounded by a diffuse-and-specular reflector, we show how to get least-square estimation of the optical thickness, the underlying surface albedo and others, using the noisy measurements of the total spectral radiance at the top (cf. Ueno 1978a, 1978b).

It should be finally mentioned that in recent years the inverse problems have been discussed controversially from meterological aspects (cf. Fymat 1977, Deepak 1977, Twomey 1977, Fymat and Zuev 1978).

Basic Equations

Suppose that a plane-parallel, inhomogeneous, anisotropically scattering atmosphere of optical thickness x bounded by a diffuse reflector (or a specular reflector) is in a direction Ωo illuminated by parallel rays of net flux πF per unit area normal to itself from above. The equation of transfer governing the radiation field $I(t,\Omega)$ takes the form

$$v \frac{dI}{dt}(t,\Omega) + I(t,\Omega) = \frac{\lambda(t)}{4\pi} \int_{4\pi} p(t;\Omega,\Omega') \, I(t,\Omega')d\Omega'. \qquad (3.1)$$

where $0 \leqq t \leqq x$ and $d\Omega' = dv'd\phi'$. Equation (3.1) should be solved subject to the boundary conditions

$$I(x,-\Omega) = \pi F \delta(\Omega-\Omega o), \qquad (3.2)$$

$$I(0,+\Omega) = \int_{2\pi} v'k(\Omega,\Omega')I(0,-\Omega')d\Omega'/v, \qquad (3.3)$$

where the bidirectional reflectance $k(\Omega,\Omega')$ is given by

$$k(\Omega,\Omega') = A^L v/\pi, \qquad (3.4)$$

$$k(\Omega,\Omega') = \delta(\bar{\Omega}-\Omega')A^S. \qquad (3.5)$$

In Equation (3.4) A^L represents the albedo of the background in accordance with the Lambert's law, whereas in Equation (3.5) A^S denotes the labedo of specular reflector and $\bar{\Omega}$ stands for $(\theta,\phi-\pi)$.

An initial value solution of Equation (3.1) is obtained with the aid of invariant imbedding.

Atmosphere Bounded by Diffuse Reflector

The total spectral radiance in the direction $+\Omega$ emerging from the top is expressed in terms of the upwelling intensity of radiation directly transmitted and the diffusely reflected intensity (Bellman and Ueno 1972: Kagiwada, Kalaba and Ueno, 1975).

$$I(x,+\Omega) = \frac{F}{4v} S(x;\Omega,\Omega o). \tag{3.6}$$

where S-function is the scattering function of Chandrasekhar's planetary problem. In expanding the phase function in Legendre polynomials

$$p(t;\Omega,\Omega') = \sum_{m=0}^{M} (2-\delta om)\{ \sum_{1=m}^{M} C_1^m(t)P_1^m(v)p_1^m(v')\} \cos m(\phi-\phi'), \tag{3.7}$$

where $p_1^m(t)$ is the associated Legendre function of degree 1 and order m

$$C_1^m(t) = C_1(t)\frac{(1-m)!}{(1-m)!} \quad (1=m,\ldots,M, \; 0<m<M) \tag{3.8}$$

$\delta om = 1$ for $m = 0$

$\quad\quad = 0$ otherwise,

then the scattering function is expressed in terms of the Fourier components

$$S(x;\Omega,\Omega o) = \sum_{m=0}^{M} S^{(m)}(x;v,u)\cos m(\phi-\phi o). \tag{3.9}$$

In a manner similar to our preceding case (cf. Bellman and Ueno 1972), it is shown that the Fourier component of the scattering function $S^{(m)}(x;v,u)$ fulfills

$$\frac{\partial S^{(m)}}{\partial \chi} +(\frac{1}{v} + \frac{1}{u})S^{(m)}=\lambda(x)(2-\delta om) \sum_{1=m}^{M} (-1)^{1+m}C_1^m(x)\Psi_1^m(x,v)\Psi_1^m(x,u), \tag{3.10}$$

where

$$\Psi_1^m(x,v) = P_1^m(v) +\frac{(-1)^{1+m}}{2(2-\delta om)} \int_o^1 S^{(m)}(x;v,w)P_1^m(w)\frac{dw}{w}, \tag{3.11}$$

for $m=0,1,2\ldots\ldots,M$. In Equation (3.11), $E_2(x)$ is the second exponential integral defined by

$$E_2(x) = \int_o^1 e^{-x/v} dv. \tag{3.12}$$

Furthermore, the function $S^{(m)}(s;v,w)$ possesses the symmetry property

$$S^{(m)}(x;v,w) = S^{(m)}(x;w,v). \tag{3.13}$$

Equation (3.10) should be solved subject to the initial condition

$$S^{(m)}(o;v,u) = 4A^L vu. \tag{3.14}$$

On using the Gaussian quadratures on the interval $(0,1)$, the integral in Equation (3.11) is replaced by sums, where the Fourier components $S^{(m)}$-function is replaced by a function of the optical thickness x

$$S^{(m)}_{ij}(x) = S^{(m)}(x;v_i,v_j). \tag{3.15}$$

In Equation (3.15) $\{v_i\}$ is the set of N roots of the shifted Legendre polynomial of degree N $p_N^*(v) = P_N(1-2v)$. Then the function $S^{(m)}_{ij}(x)$ fulfills the system of ordinary nonlinear differential equations (3.10)

$$\frac{ds^{(m)}_{ij}(x)}{dx} = -(\frac{1}{v_i} + \frac{1}{v_j}) S^{(m)}_{ij} + \lambda(x)(2-\delta om) \sum_{l=m}^{M} (-1)^{1+m} c_1^m(x) \psi_{1i}^m \psi_{1j}^m, \tag{3.16}$$

for $m=0,1,2...,M$; $i,j,=1,2,...,N$,

$$\psi_{1i}^m = p_1^m(v_i) \frac{(-1)^{1+m}}{2(2-\delta om)} \sum_{j=1}^{M} S^{(m)}_{ij} P_1^m(v_j) \frac{w_j}{v_j}. \tag{3.17}$$

In Equation (3.17) W_j is the Christofell weight corresponding to the value of v_j. Equation (3.16) should be solved subject to the initial condition

$$S^{(m)}_{ij}(0) = 4A^L vu \tag{3.18}$$

Atmosphere Bounded by Specular Reflector

In the case of specular reflection the upwelling intensity of radiation diffusely reflected at the top takes the form

$$I^*(x,\Omega) = \frac{F}{4v} S(x;\Omega,\Omega o). \tag{3.19}$$

In a manner similar to the case of a diffuse reflector, the Fourier component of the scattering function $S^{(m)}(x;v,u)$ is governed by

Equation (3.10), whose ψ_1^m-function fulfills

$$\psi_1^m(x,v) = p_1^m(v) + p_1^m(v)A^S e^{-2x/v} +$$

$$\frac{(-1)^{1+m}}{2(2-\delta om)} \int_0^1 S^{(m)}(x;v,w)p_1^m(w)\frac{dw}{w}, \qquad (3.20)$$

IDENTIFICATION OF THE EARTH-ATMOSPHERE SYSTEM

In a series of papers (cf. Bellman et al 1965), the least squares estimation of such optical properties as the optical thickness, phase function, and others has been done with the aid of quasilinearization and invariant imbedding, assuming an atmosphere bounded by a completely absorbing background. In what follows, we shall consider the estimation of the optical properties such as the optical thickness, and the albedo of the reflecting underlying surfaces.

Estimation of the Optical Thickness

The inverse problem which we wish to solve is to estimate the optical thickness x of atmospheres in the previous section, in the least squares sense, using the noisy total spectral radiance measurements. In other words, it is expected to get the best agreement between the estimated solution using the ordinary differential equations for the scattering function and the observed reflection pattern. Mathematically speaking, we wish to minimize the expression

$$\sum_{i,j,k}[I(x;v_i,\phi_k;v_j,\phi_o) - b_{ijk}]^2, \qquad (3.21)$$

over all choices of the unknown parameter x, provided that the phase function of the atmosphere and the albedo of the diffuse reflector are together known. In Equation (3.21) b_{ijk} represents the intensity of the diffusely reflected light, $b_{ijk} \cong I_{ijk}(\tau)$, for incident directions $(v_i,\phi o), i=1,2,\ldots,N$. The subscript k denotes the k-th component of the azimuth of the view angle. In Equation (3.21) $I_{ijk}(x)$ is given by

$$I(x;v_i,\phi_k;v_j,\phi_o) = \frac{F}{4v_i} \sum_{m=0}^{M} S^{(m)}(x;v_i,v_j)\cos m(\phi_k-\phi_o). \qquad (3.22)$$

On making use of quasilinearization, the problem will be successively solved. Putting the function $S_{ij}^{(m)}$ as S_{mij} and similarly, writing ψ_{ki}^m as Ψ_{mki}, $p_k^m(v_i)$ as P_{mki}, we get a system of linear differential equations for the $(n+1)^{st}$ approximation of S_{mij} and x. Put T as the end point of the range of integration, i.e., $0<x\leq T$. Introducing a new independant variable σ,

$$x = \sigma T \quad (0<\sigma\leq 1.0), \qquad (3.23)$$

so that the integration interval is fixed. Then it takes the form

$$\frac{\partial S_{mij}^{n+1}}{\partial\sigma} = T^n\{f(S_{mij}^n, T^n, \sigma) + \sum_{i,j} [S_{mij}^{n+1} - S_{mij}^n] \frac{\partial f}{\partial S_{mij}^n} +$$

$$(T^{n+1}- T^n) \frac{\partial f}{\partial T^n} \}, \qquad (3.24)$$

$$\frac{\partial T^{n+1}}{\partial\sigma} = 0, \qquad (3.25)$$

where

$$f(S_{mij}^n, T^n,\sigma) = -(\frac{1}{v_i} + \frac{1}{v_j})S_{mij}^n + \lambda(\sigma T^n)(2-\delta_{om}) \sum_{1=m}^{M} x$$

$$(-1)^{1+m}C_1^m(\sigma T^n)\Psi_{mli}\Psi_{mlj}. \qquad (3.26)$$

In Equation (3.26) Ψ_{mli} is given by Equation (3.20) in Gaussian quadrature form, where S_{mij}^n satisfies the principle of reciprocity

$$S_{mij}^n = S_{mji}^n, \qquad (3.27)$$

There are basically $(MN^2 + 1)$ differential equations, which reduce to $\{MN(N+1)/2+1\}$ differential equations with the aid of the reciprocity relation Equation (3.27), whereas the full set of values S_{mij}^n representing MN^2 matrix is always available.

The solution is subject to the initial condition (3.14) and the boundary condition (3.21) with Equation (3.22).

Since S_{mij}^{n+1} is a solution of a system of linear differential equations, we express it in terms of a linear combination of a particular solution q_{mij} and a homogeneous solution h_{mij}

$$S_{mij}^{n+1} = q_{mij}(X) + Th_{mij}(X).\qquad(3.28)$$

The system of differential equations for q_{mij} (σT) is obtained by inserting the appropriate component of q wherever s^{n+1} takes place in Equations (3.24) or (3.25), whose initial conditions are $q(0)=0$. Similarly, the system of equations for the homogeneous solution is provided, dropping all terms not involving the $(n+1)^{st}$ approximation, whose initial vector $h(0)$ has all of the components zero except for the last, which is unity. Then, the initial condition, Equation (3.14), is identically fulfilled. The solutions $q(\sigma T)$ and $h(\sigma T)$ are obtained by numerical integration on the interval $(0,\sigma T)$.

By simple differentiation, the value of T^{n+1} which minimizes Equation (3.21) is given by

$$T^{n+1}= \sum_{i,j,k}[4v_i b_{ijk} - \sum_{m=0}^{M} q_{mij}\cos m\phi_k] \; [\sum_{m=0}^{M} h_{mij}\cos m\phi_k]$$

$$/\sum_{i,j,k}[\sum_{m=0}^{M} h_{mij}\cos m\phi_k]^2,\qquad(3.29)$$

where $x = T^n$, $F=1$, and $\phi_o=0$.

Equation (3.29) is the required equation, which permits us to have the improved version of the optical thickness using the noisy total spectral radiance emitted from the top.

Estimation of the Surface Albedo

Case of a Diffuse Reflector

In a manner similar to the case of the previous section, provided that such optical properties as the optical thickness, the albedo for single scattering, and the phase function are known, we should like to estimate the surface albedo, A^L, in the least squares sense (cf. Ueno 1978). Assuming the same atmosphere as in the above section, and proceeding with the similar approach, we get a system of linear differential equations for the $(n+1)^{st}$ approximation to S_{mij} and A $(=A^L)$

$$\frac{dS_{mij}^{n+1}}{dx} = f(S_{mij}^n, A^n) + \sum_{i,j} [S_{mij}^{n+1} - S_{mij}^n] \frac{\partial f}{\partial S_{mij}^n} +$$

$$(A^{n+1} - A^n) \frac{\partial f}{\partial A^n}, \tag{3.30}$$

$$\frac{dA^{n+1}}{dx} = 0, \tag{3.31}$$

where

$$f(S_{mij}^n, A^n) = -(\frac{1}{v_i} + \frac{1}{v_j}) S_{mij}^n +$$

$$\lambda(x)(2-\delta_{om}) \sum_{l=m}^M (-1)^{1+m} C_1^m(x) \psi_{mli} \psi_{mlj}. \tag{3.32}$$

In Equation (3.32) ψ_{mli} is given by

$$\psi_{mli} = P_{mli} + \frac{(-1)^{1+m}}{2(2-\delta_{om})} \sum_{j=1}^M S_{mij}^n P_{mlj} \frac{w_j}{v_j}, \tag{3.33}$$

where S_{mij}^n satisfies the reciprocity principle, Equation (3.27). The initial condition of Equation (3.24) is Equation (3.14) and the boundary condition is

$$\min_{A^{n+1}} \{\sum_{i,j,k} \sum_{m=0}^M S_{mij}^{n+1} \cos m\phi_k - 4v_i b_{ijk}]^2\}. \tag{3.34}$$

In a manner similar to Equation (3.28), we put

$$S_{mij}^{n+1}(x) = q_{mij}(x) + A h_{mij}(x). \tag{3.35}$$

By a simple differentiation, we get the value of A^{n+1} minimizing Equation (3.34)

$$A^{n+1} = \sum_{i,j,k} [4v_i b_{ijk} - \sum_{m=0}^M q_{mij} \cos m\phi_k][\sum_{m=0}^M h_{mij} \cos m\phi_k]$$

$$/\{\sum_{i,j,k} (\sum_{m=0}^M h_{mij} \cos m\phi_k)^2\}. \tag{3.36}$$

Equation (3.36) is the required equation, permitting the estimation of the background reflectance using the noisy total radiance emanating from the top.

Case of a Specular Reflector

In this subsection, we consider an estimation of an albedo of a specular reflector which bounds an inhomogeneous, anisotropically scattering atmosphere of optical thickness x (cf. Ueno, 1978). In this case the total spectral radiance is expressed in terms of the Fourier components of S-function

$$I^*(x;v_i,\phi_k;v_j,\phi o) = \frac{F}{4v_i} \sum_{m=0}^{M} S^{(m)}(x;v_i,v_j)\cos m(\phi_k-\phi o). \quad (3.37)$$

where the $(n+1)^{st}$ approximation to $S^{(m)}$-function is governed by Equations (3.30) and (3.31), respectively. In Equation (3.28) Ψ_{mli} is given by Equation (3.20) in Gaussian quadrature form. On making use of Equation (3.28), the value A^{n+1} minimizing the boundary condition

$$\min_{A} \sum_{i,j,k} [I^*(x,v_i,\phi_k;v_j,\phi o) - b_{ijk}]^2, \quad (3.38)$$

gives the required recursive solution for the calm sea surface reflectance.

The Numerical Experiments

Specifically, the isotropic case is considered in which λ has the form (Kagiwada et a., 1975),

$$\lambda (z) = 0.5 + az + bz^2, \quad 0 \le z \le x \quad (3.39)$$

where a and b are constants. In the first numerical experiment, 49 accurate measurements are given for the case x=1.0, a=2.0, b=-2.0, and A^L=0.5. The parameters are initially estimated to be x=1.5, a=2.2, b=-1.8, and A^L=0.2. These are then refined. The results of a two-minute computation on an IBM 7044 are given in Table 1.

Next, a series of ten experiments is done, involving 49 noisy measurements per experiment. The relative amount of noise in each measurement is a random number drawn from a Gaussian distribution with zero mean and 0.01 standard deviation. The estimates of x, a, b, and A^L averaged over the ten trials are in error by only 1 to 7 percents.

TABLE 1: Sequence of Parameter Estimates in a Controlled Experiment.

Iteration	x	a	b	A^L
0	1.5	2.2	-1.8	0.2
1	1.28	1.29	-1.05	0.73
2	1.03	1.82	-1.65	0.45
3	0.999	1.995	-1.98	0.501
True values	1.0	2.0	-2.0	0.5

In a manner similar to the previous section, we considered an isotropically scattering atmosphere, whose λ changes according to the expression

$$\lambda \ (z) = 0.5 + 0.1 \tanh \ (x-0.5), \qquad (3.40)$$

and the surface albedo A^L is put to be 1.0. The measurements $\{b_{ij}\}$ are produced computationally by the use of λ given above, and with the aid of differential Equation (3.10), integrating out to a thickness x=1.0. The observations which this produces are given in Table 2.

TABLE 2: The Measurements $\{4v_i b_{ij}\}$.

	v_1	v_2	v_3	v_4	v_5	v_6	v_7
v_1	0.008150	0.014661	0.017347	0.019219	0.020871	0.022221	0.023060
v_2	0.014661	0.047927	0.071550	0.087765	0.100487	0.110122	0.115892
v_3	0.017347	0.071550	0.125648	0.171158	0.210581	0.241708	0.260664
v_4	0.019219	0.087765	0.171158	0.259468	0.349436	0.426665	0.475131
v_5	0.020871	0.100487	0.210581	0.349436	0.506871	0.648194	0.739307
v_6	0.022221	0.110122	0.241708	0.426465	0.648194	0.851887	0.984459
v_7	0.023060	0.115892	0.280664	0.475131	0.739307	0.984459	1.144686

The results of three experiments with different initial guesses of the optical thickness x are shown in Table 3. The initial approximation is generated in each run by integrating the nonlinear Equation (3.10) with the values of x listed as approximation zero. Even with ninety percent errors in the initial approximations, the value of x is determined to 0.04% in four iterations. An integration step size used is 0.01, with the Hamming's predictor corrector method. On making use of the symmetry of S, 58 differential equations have to be integrated for the particular and homogeneous solutions for each approximation.

TABLE 3: Estimation of the surface albedo.

Approximation	Run	Run	Run
0	0.1	0.5	1.5
1	0.4779	0.7813	0.7078
2	0.7632	0.9731	0.9857
3	0.9752	0.9998	1.0000
4	0.9996	1.0000	1.0000
True Value	1.0	1.0	1.0

REFERENCES

Bellman, R., H. Kagiwada, R. Kalaba, and S. Ueno. 1965. ICARUS, 4, 119/126.
Bellman, R., and S. Ueno. 1972. Astrophys. Space Sci., 16, 241/248.
Deepak, A. 1977. Inversion Methods in Atmospheric Remote Sounding, Academic Press, New York.
Fymat, A. L. and V. E. Zuev. 1978. Remote Sensing of the Atmosphere: Inversion Methods and Applications, Elsevier Scientific Publ. Co., Amsterdam.
Fymat, A. L. 1977. An Inversion Methods in Atmospheric Radiation Research: I. Temperature Sounding, Jet Propulsion Laboratory.
Kagiwada, H. 1974. System Identification: Methods and Applications, Addison-Wesley Publ. Co., Reading, Mass.
Kagiwada, H., R. Kalaba, and S. Ueno. 1975. Multiple Scattering Processes: Inverse and Direct. Addison-Wesley Publ. Co., Reading, Mass.

Twomey, S. 1977. Introduction to the Mathematics of Inversion in Remote Sensing and Indirect Measurements, Elsevier Scientific Publ. Co., Amsterdam.

Ueno, S. 1978a. Inference of the atmospheric optical thickness from the total spectral radiance, in 11th Lunar and planetary Symposium (edit. by M. Shimizu), Inst. Space Aeronant. Sci., Tokyo University, 6/12.

Ueno, S. 1978b. Estimation of the surface albedo using the noisy total radiance measurements, in Proc. 4th Int. Joint Conf. on Pattern Recognition, 935.

OILSPILL DETECTION BY THE USE OF LANDSAT MSS DATA

INTRODUCTION

In recent years, oil pollution in oceans has become a world-wide problem. Transport of oil in super tankers and increased offshore oil drilling have also been discussed in relation to the oil pollution on the sea surface because of these threatening sources. Then, it is desirable to develop a monitoring system for the occurrence and magnitude of oil spills. A system of surveillance would require the determination of such parameters as oil thickness, age, the wave-length of observation, illumination angle and view angle, which affect the spectral signature for oil. The transfer problem of solar radiation in a model of the atmosphere -ocean system has been solved in terms of the Monte Carlo (cf. Plass and Kattawar, 1969; Plass et al., 1976) and an iterative method (cf. Raschke, 1972), where the roughness of the sea surface has been taken into account, and a matrix method (cf. Tanaka and Nakajima, 1977), in which the effects of optical parameters of oceanic hydrosols have been examined in detail.

In this section (cf. Mukai and Ueno, 1978), in order to analyze the earth's remotely sensed data, the transfer problem in the atmosphere-ocean system has been solved with the aid of super-position algorithm (cf. Mukai, 1973), which is based on the invariance principles (see next section) and is also derived from the auxiliary equation with the aid of an integration equation approach (cf. Ueno and Mukai, 1972). The merit of this method is that it expresses the laws of diffuse reflection and transmission for the atmosphere-ocean system in terms of scattering and transmission functions for the standard problem (cf. Chandrasekhar, 1950). A numerical computation of simulation model appropriate to this case has been done. A model of two layers separated by a specular reflector is treated, where both layers are homogeneous and plane parallel. Computations are performed at wavelengths of $\lambda = 0.45\mu m$ and $0.7\mu m$ in accordance with the following processes: 1) scattering by molecules and aerosols in the atmosphere, 2) scattering and absorption by water molecules and hydrosols in the ocean, and 3) reflection and refraction at the sea surface. Thus the angular distributions of the intensity of radiation at the top of the atmosphere and just above and below the sea surface are obtained. From the study we obtain useful information for remote sensing purposes, e.g., the optical thickness of the ocean and the pollution of the sea surface by oil slicks.

BASIC EQUATIONS

In this section we consider the diffuse reflection problem of an Atmosphere-Ocean system, whose atmosphere is bounded by an interacting quiet sea surface acting as a specular reflector. For the sake of simplicity, we assume that the Atmosphere-Ocean system consists of two homogeneous layers of finite optical thickness τ_A (Atmosphere) = x and τ_m (Ocean) = y, where multiple scattering of unpolarized photons takes place in both two layers. At a specular reflector an interaction of radiation is governed by Fresnel's formula. Furthermore, it is assumed that the bottom of ocean at z=x+y is bounded by a perfect absorber, and the polar angle of direction denotes the inclination to the outward normal at the top z=0. In other words, in the homogeneous case the direction of the upwelling and downwelling intensities are designated as positive and negative, respectively (Fig. 1).

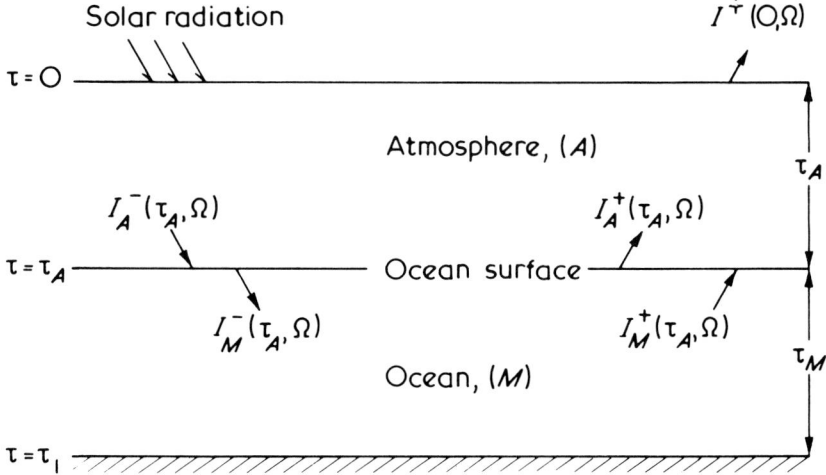

FIGURE 1: Diagram showing symbols used for solution of transfer problem in a model of the atmosphere-ocean system.

The required total spectral radiance of the Atmosphere-Ocean system emerging from the top, $I^*(0,\Omega)$, takes the form in the diffuse radiation field

$$I^+(0,\Omega,\Omega o) = \frac{F}{4v} S(x;\Omega,\Omega o) + I^+(x,\Omega;A)\exp[-x/v]$$

$$+ \frac{1}{4\pi v} \int_{2\pi} T(x;\Omega,\Omega')I^+(x,\Omega';A)d\Omega' \qquad (4.1)$$

$$+ \frac{F}{4v} T(x;\Omega,\Omega_o)\exp[-x/u]r(u),$$

where r(u) is the reflection coefficient of radiation at a calm sea surface. On the right-hand side of Equation (4.1) the first term denotes the intensity of diffusely reflected radiation by a free

atmosphere, the second term is due to the direct transmission of
the upwelling intensity of radiation just above the level z=x, the
third term comes from the diffuse transmission of upwelling intensity of radiation just above the sea surface, and the last term
represents the diffuse transmission of upwards radiation, while the
downwards directly transmitted radiation is reflected specularly by
the sea surface. In Equations (4.1) through (4.5) $S(x;\Omega,\Omega o)$ and
$S(y;\Omega,\Omega o)$ are respectively the scattering and transmission functions of the upper atmospheric layer, and the scattering function
of the lower sea layer. Let the upwelling and downwelling intensities of radiation just above the sea surface be denoted by
$I^+(x,\Omega;A)$ and $I^-(x,\Omega;A)$.

Furthermore, let the upwelling and downwelling intensities of
radiation just below the sea surface be denoted by $I^+(x,\Omega;M)$ and
$I^-(x,\Omega;M)$. Based on the invariance principles, we get the recursive relations

$$I^-(x,\Omega;A) = \frac{F}{4v} T(x;\Omega,\Omega_0) + \frac{F}{4v} \exp[-x/u]r(u)S(x;\Omega,\Omega_0) +$$

$$\frac{1}{4\pi v} \int_{2\pi} S(x;\Omega,\bar{\Omega}') \frac{1-r(w')}{n^2} I^+(x,\Omega';M)d\Omega', \tag{4.2}$$

$$I^+(x,\Omega;A) = \frac{(1-r(\bar{v}))}{n^2} I^+(x,\bar{\Omega};M) + I^-(x,\Omega;A)r(v), \tag{4.3}$$

$$I^-(x,\bar{\Omega};M) = (1-r(v))n^2 I^-(x,\Omega;A) + r(\bar{v}) I^+(x,\bar{\Omega};M), \tag{4.4}$$

$$I^+(x,\bar{\Omega};M) = \frac{1}{4\pi\bar{v}} \int_{2\pi} S(y;\bar{\Omega},\Omega') I^-(x,\Omega';M)d\Omega' +$$

$$\frac{F}{4\bar{v}} \exp[-x/u] (1-r(u)) n^2 S(y;\bar{\Omega},\bar{\Omega}_0), \tag{4.5}$$

where $\bar{\Omega}' = ([1-n^2(1-w'^2)]^{\frac{1}{2}},\phi')$, $\bar{\Omega} = ([n^2-(1-v^2)]^{\frac{1}{2}}/n,\phi)$, $\bar{\Omega}_0 = ([n^2 -(1-u^2)]^{\frac{1}{2}}/n,\phi_0)$.

In Equations (4.2) through (4.5) $r(v)$ is given by Fresnel's formula.

$$r(v) = \frac{1}{2} \left[\frac{\sin^2(\theta_2 - \theta_1)}{\sin^2(\theta_2 + \theta_1)} + \frac{\tan^2(\theta_2 - \theta_1)}{\tan^2(\theta_2 + \theta_1)} \right], \tag{4.6}$$

where $v = \cos\theta_1$, $u = \cos\theta_2$,

and furthermore

$$n\sin\theta_2 = \sin\theta_1, \tag{4.7}$$

where n is the refractive index of the second medium to the first one.

Once when the scattering and transmission functions of the atmospheric layer, the scattering function of sea layer, and the index of refraction have been given, Equations (4.1) through (4.5) permit us to compute numerically the total spectral radiance emerging from the top.

An allowance of sun glint gives the following contribution to the total radiance $I^+(0,\Omega)$:

$$\pi F \delta(v-u)\delta(\phi-\phi_0-\pi)\exp[-2x/v]r(u).$$

Corrections for specular reflection should be made for the measurement of spectral signatures of sea surface.

In a manner similar to the preceding section, the azimuth dependence of the upwelling and downwelling intensities in Equations (4.1) through (4.5) is removed by decomposing into the Fourier components. In our paper (cf. Mukai and Ueno, 1978), starting with Equations (4.2) through (4.5), the recursive relations between the upwelling and downwelling intensities at the sea surface are made tractable and are used for the numerical computation of apparent contrast of radiance.

ATMOSPHERE-OCEAN MODEL

In the atmospheric layer Rayleigh molecular scattering and Mie aerosol scattering are taken into account. Similarly, Rayleigh scattering of water molecules and Mie scattering of hydrosols are considered in the sea layer. Furthermore, absorption of these agents is also allowed for. Because of the strong dependence on the wave-length λ, we used $\lambda = 0.7$ µm.

Atmospheric Model

The model of the Earth's atmosphere is constructed on the basis of tables compiled by Elterman (1968), which give the aerosol and molecular distribution with height. The aerosols are represented by the Haze M model, which corresponds fairly well to the distribution of coastal aerosols (cf. Deirmendjian, 1964), i.e., the size distribution of aerosols is as follows:

$$n(a) = Cae^{-8.9443\sqrt{a}}, \tag{4.8}$$

where a is the radius of a spherical particle and C is a normalization constant. The phase function for this distribution is calculated exactly from the Mie theory. The phase function in the atmosphere is given by:

$$p(\theta) = (1-f_g)P_a(\theta) + f_g P_g(\theta), \tag{4.9}$$

where θ is the scattering angle, P_a and P_g represent phase function by aerosols and molecular gas, respectively, f_g is the ratio of the molecular scattering cross-section to the total scattering cross-section, and P_g is given by the Rayleigh phase function:

$$P_g(\theta) = \frac{3}{4}(1+\cos^2\theta). \tag{4.10}$$

Other parameters in the atmosphere, optical thickness x, albedo for single scattering ω_A, and f_g are listed in Table 1.

TABLE 1: Parameters of atmospheric and oceanic models.

	Atmosphere		Ocean	
λ(µm)	0.45	0.7	0.45	0.7
f_g	0.439	0.148	0.218	0.0494
ω	1.0	1.0	0.514	0.0498
τ	0.508	0.250	1.0	7.56

Oceanic Model

It is clear that there are many different types of particles, both organic and inorganic in ocean water, and the amount of these particles is strongly dependent on location and time. Therefore, more realistic models of ocean water based on measurements at different depths and from different locations are required.

Useful work on the optical properties of hydrosols has been done by Tanaka and Nakajima (1977). Here, however, we are using the clean ocean model (cf. Otterman and Fraser, 1976). The hydrosols are represented by spheres with n=1.15 and k=0.001, where n and k are, respectively, the real and imaginary parts of the refractive index with respect to water, and the size distribution for hydrosols is:

$$n(a) = \text{const.} \qquad a<1\mu m$$
$$= C_a{}^6 e^{-2a} \qquad a>1\mu m. \qquad (4.11)$$

Other parameters are listed in Table 1, where the optical thickness y of the ocean is assumed to be 1.0 at $\lambda=0.45\mu m$. The bottom of the ocean is assumed to be a perfect absorber.

NUMERICAL RESULTS

Total Spectral Radiance

The intensities of radiation at the top of the atmosphere and just above and below the sea surface have been calculated on the basis of the above models.

Figure 2 shows the downward intensity at the bottom of the atmosphere (upper curve) and the upward intensity at the top of the atmosphere (lower curve) for solar zenith angle $\theta_0=16.5°$ and azimuth angle $\phi=0°$. In all graphs in this section, the ordinate represents the intensities in units of incident solar flux F, the abscissa is the emergent polar angle θ in degrees. We can compare the radiation field of a single atmosphere model with that of a combined atmosphere-ocean model by using Figure 2. So far as the downward intensity at the bottom of the atmosphere is concerned, the contribution from the ocean and sea surface is negligible. On

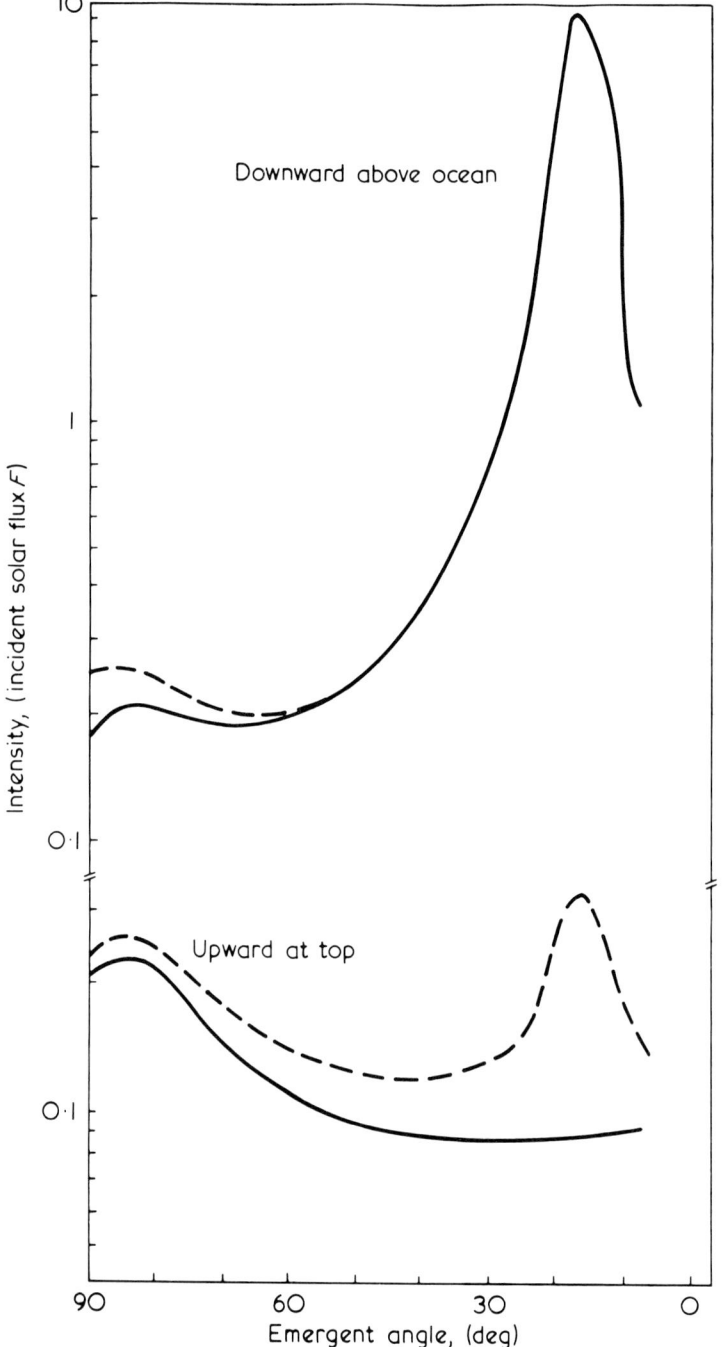

FIGURE 2: (a) Downward intensity at bottom of atmosphere. (b) Upward intensity at top of atmosphere. Solar zenith angle, $\theta_0 = 15.5°$; azimuth angle, $\phi = 0°$, $\lambda 0.45\mu m$. (———), intensity contributed from atmosphere alone; (- - -), intensity contributed from combined atmosphere-ocean system.

the contrary, the upward intensity at the top of the atmosphere is strongly affected by the introduction of a combined model. The maximum values at $\theta=16.5°$ represent the contribution from the direct solar radiation, namely the incident solar radiation directly reaching the bottom of the atmosphere without scattering if it has not been scattered in the atmosphere, and then reflected at the sea surface towards the specular point.

Figure 3 shows the variation of the downward intensity just above the sea surface with the azimuthal angle ϕ for $\theta_0=16.5°$. The curve in the principal plane ($\phi=0°$) has a large peak at $\theta_0=16.5°$ due to direct solar radiation. As the azimuthal angle begins to deviate from the principal plane, the intensity curves become flatter because scattering diffuses the direct solar radiation.

Apparent Contrast of Total Radiance on Oil Polluted Sea Surface

The present calculations on multiple scattering are available for application to the remote sensing data of the earth. First, we will apply these results to estimate the optical thickness of the ocean in terms of the remotely sensed data.

Figure 4a shows the upward intensity at the top of the atmosphere and Figure 4b shows that just above the sea surface for $\theta_0=16.55°$, and $\phi=0°$. The solid and dashed curves represent cases in which the optical thickness of the ocean is y=1.0 and 0.1, respectively. In the intensity above the sea surface there is a large difference between the two cases. When the measurements of upward radiance are carried out above the sea surface, we can roughly estimate the optical thickness of the ocean. It is not so easy to estimate it from space.

Next we compare the oil polluted ocean with natural sea water. A widely spread oil slick will change the refractive index of the sea surface; for example, a value of 1.45 applies for light oil compared to 1.34 for natural water at $\lambda=0.45\mu m$ (Cox and Munk, 1955). It has been shown that the presence of oil slicks on the sea surface changes the refractive index of the sea surface and the roughness of the sea waves (cf. Otterman and Fraser, 1976; Plass and Kattawar, 1969). Here, however, we consider the quiet sea, and

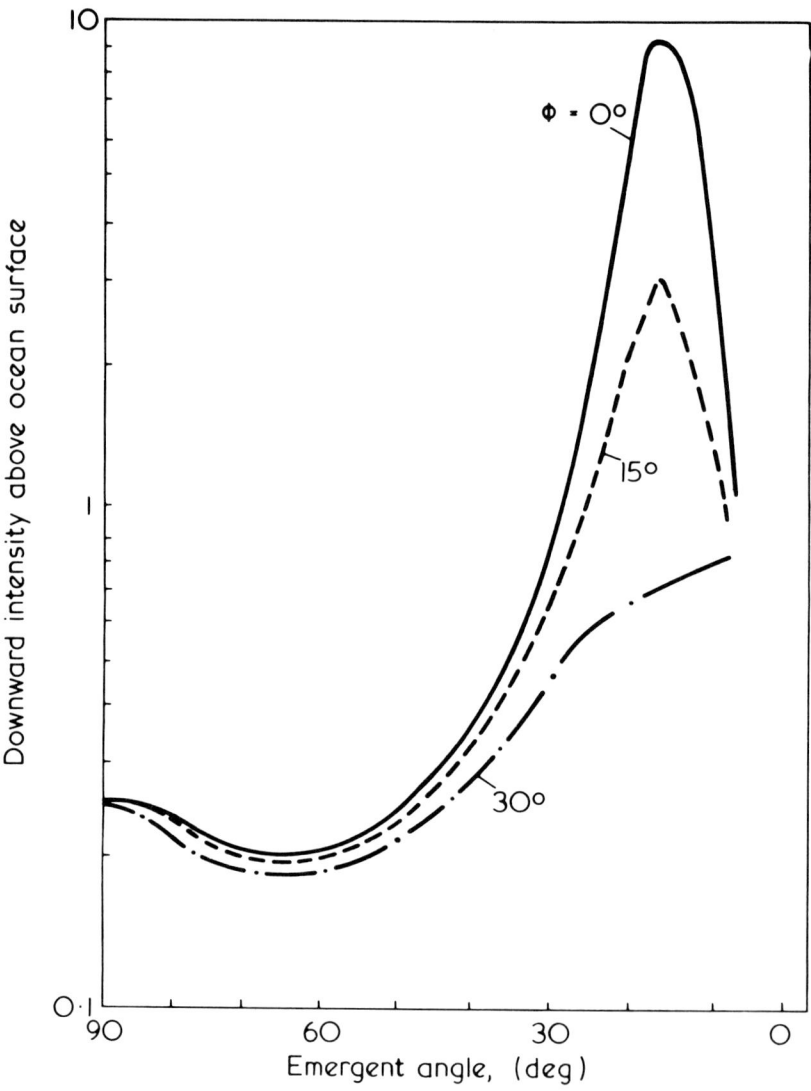

FIGURE 3: Downward intensity just above sea surface for $\theta_0=16.5°$. See Figure 2 for key.

the roughness of the sea surface is not taken into account. The solid and dashed curves in Figure 5 represent the intensities of the reflected sunlight from the atmosphere-ocean system with a natural water surface and with oil slicks, respectively, for (A), $\theta_0=6°$, and (B), $\theta_0=16.5°$ and $\phi=0°$. The reflected sunlight in the

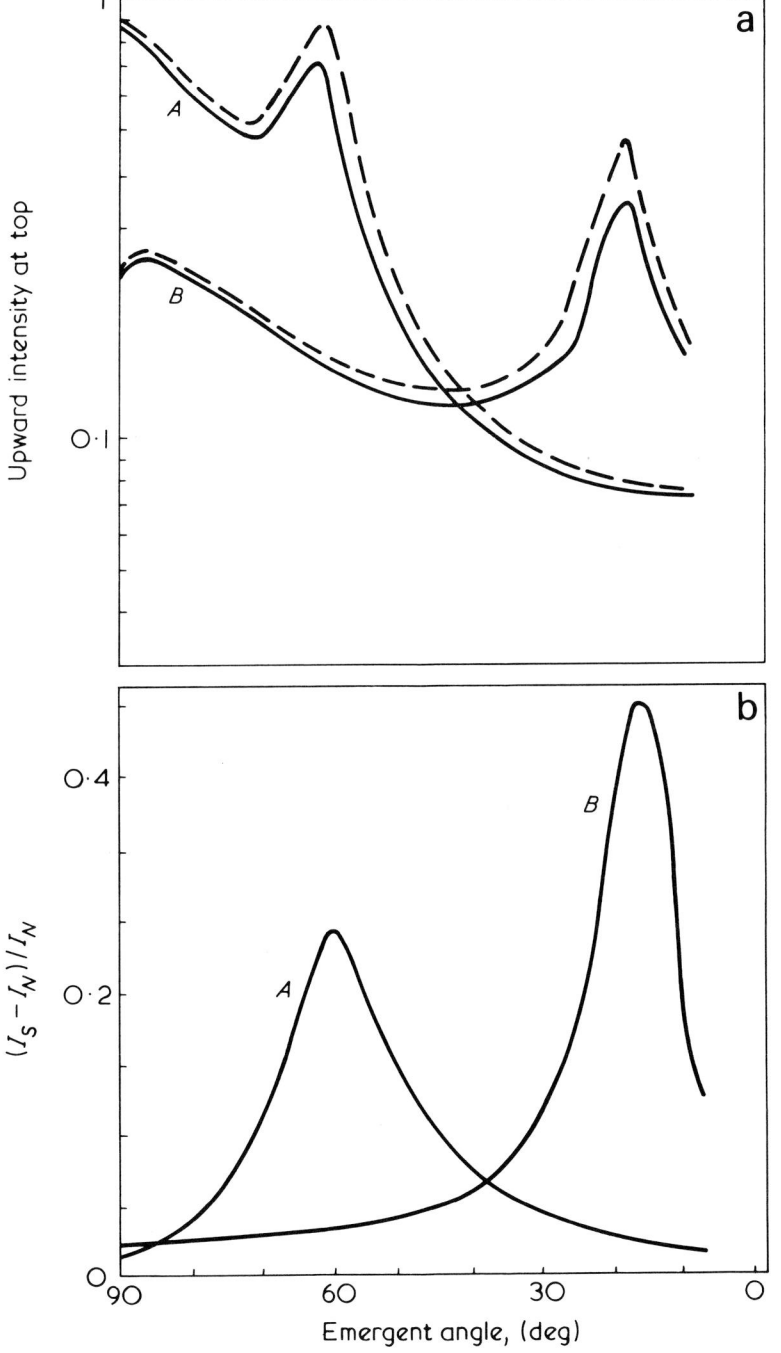

FIGURE 4: Upward intensity at top of atmosphere for A, $\theta_0=60°$; B, $\theta_0=16.5°$ and $\phi=0°$, (a) as in Figure 2. (———), natural water surface, I_N; (---), oil polluted sea surface I_S. (b) Apparent contrast between oil polluted and uncontaminated oceans.

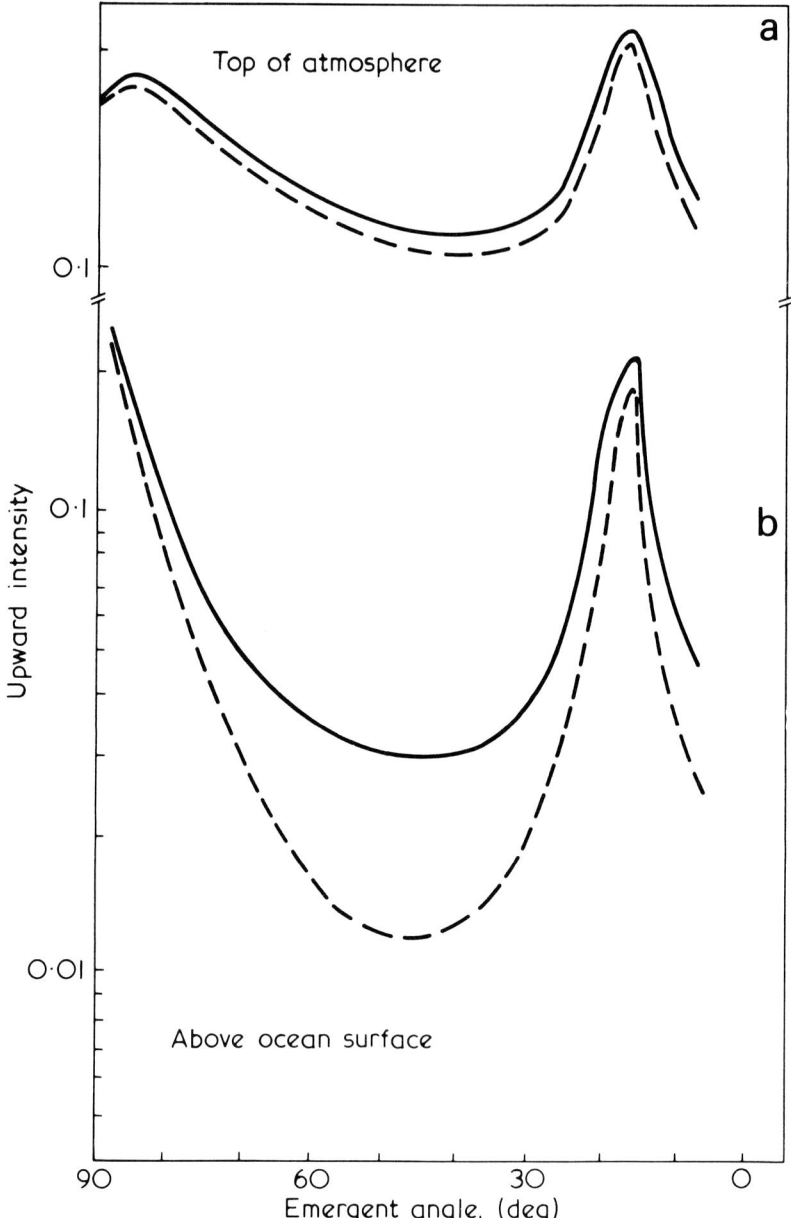

FIGURE 5:   (a) Upward intensity at top of atmosphere.   (b) Just
above sea surface for $\theta_0=16.5°$ and $\phi=0°$.   (————), optical thickness of ocean, $\tau_M=1.0$; (———), optical thickness of ocean, $\tau_M=0.1$.
See Figure 2 for key.

case of the oil polluted ocean is brighter than that of the uncontaminated water surface. Figure 5 shows the visibility of oil slicks to observation by satellite or airplane. This represents a contrast of intensity between the polluted and uncontaminated oceans as follows:

Apparent contrast of oil slicks $= (I_S-I_N)/I_N$ $\qquad$ (4.12)

where I is the upward intensity at the top of the atmosphere, and subscripts S and N represent slick water and natural water, respectively. The apparent contrast is most remarkable at the specular point, i.e., 24.2% for $\theta_0=60°$ and 40.2% for $\theta_0=16.5°$. That is to say, an oil slick lying at the sun's glitter pattern has especially large brightness compared with the uncontaminated water.

The contrast of oil slicks at $\lambda=0.7\mu m$ is shown in Figure 6, where notations (A) and (B) denote $\theta_0=60$ and 16.5°, respectively, as in Figure 5, and the solid and dashed curves represent two kinds of oil slicks, i.e., oil slick A and B (see Table 2). A discrepancy between two kinds of oil slicks is the difference of reflectivity at $\lambda=0.7\mu m$. That is, type B has much more reflectivity than type A has. It is shown that the contrast of oil slicks at $\lambda=0.7\mu m$ is higher than that at $\lambda=0.45\mu m$, i.e. the maximum values are 30.5% for $\theta_0=16.5°$ (oil slick A), and 42.4% for $\theta_0=60°$ and 85.1% for $\theta_0=16.5°$ (oil slick B). This is explained by the difference of the wavelength dependence of the refractive index between oil slicks and natural water.

This difference appears more clearly in the ratio of the upward intensity at the top of the atmosphere at $\lambda=0.7\mu m$ to that at $\lambda=0.45\mu m$ for $\theta_0=60°$ and $\phi=0$. We call this ratio a colour index, and use the values given by Allen (1973) (p.171) for the flux of solar radiation at $\lambda=0.45\mu m$ and $0.7\mu m$. The difference between the colour indices of polluted ocean and uncontaminated ocean is always positive.

From these results we draw the following conclusions:

(i)   We can roughly estimate the optical thickness of the ocean by using the measurements of radiance just above the sea surface.

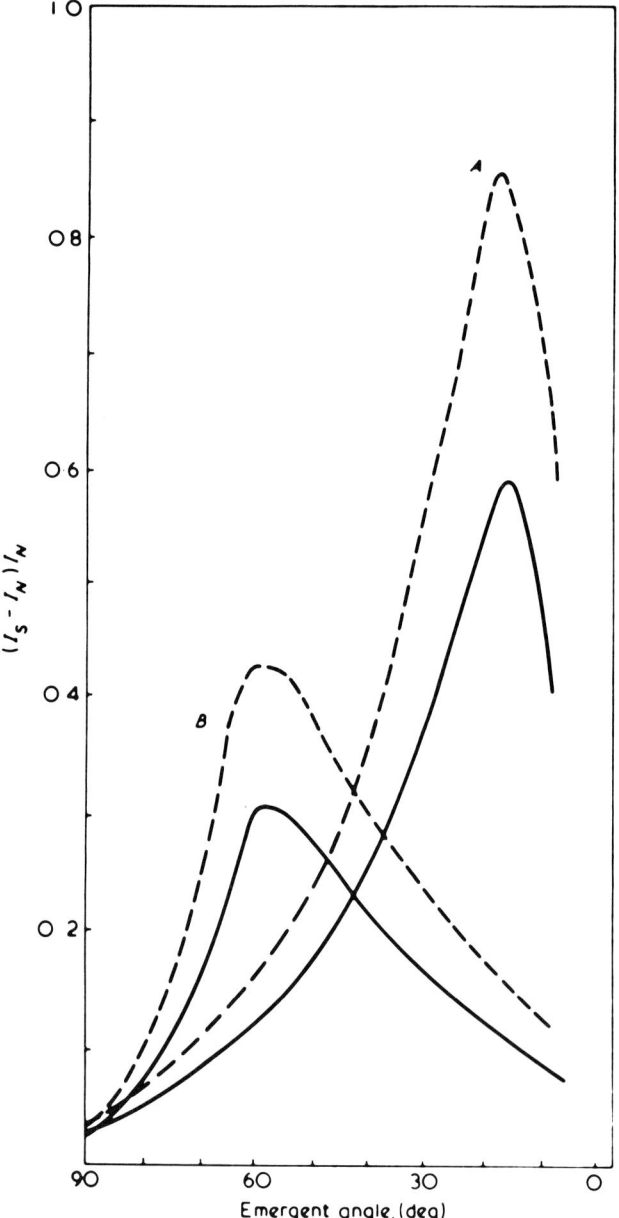

FIGURE 6: Apparent contrast as in Figure 4b at λ=0.7μm for A, θ_0=60° and B, θ_0=16.5°. (———), oil slick A; (---), oil slick B (see Table 2).

TABLE 2: Refractive indices of sea surface.

λ (μm)	0.45	0.7
Natural water	1.34	1.33
Oil slick A	1.45	1.45
Oil Slick B	1.45	1.50

(ii)   The visibility of the oil polluted ocean is higher than that of uncontaminated ocean. Therefore, on the basis of this enhancement of the visibility, or what may be called the apparent contrast of oil slicks, the sea pollution can be measured from space, i.e., by a satellite-borne detector.

(iii)   The colour index of the reflected sunlight is a useful index of sea pollution.

Our present knowledge of the optical constants of a sea pollutants is very sketchy. If these values, i.e. refractive indices, are obtained for a wide wavelength region, we will be able to examine sea pollution not only by oil slicks but also by other kinds of pollutants or organic matter, using the most effective colour index for each pollutant.

REFERENCES

Allen, C. W. 1973 "Astrophysical Quantities" (3rd Ed.), Athlon Press, London.
Casti, L. J., R. Kalaba, and S. Ueno 1969. J. Quant. Spectrosc. Radiat. Transfer 9, 537.
Chandrasekhar, S. 1950. "Radiative Transfer", Oxford University Press, London.
Cox C. and W. Munk 1955. J. Mar. Res., 14, 63.
Deirmendjian, D. 1964. Appl. Opt. 3, 187.
Elterman, L. 1968. AFCRL-68-0153.
Mukai, S. 1973. Publ. Astron. Soc. Japan, 25, 385.
Mukai, S. 1977. Astrophys. Space Sci. 51, 165.
Mukai, S., and S. Uneo 1978. Appl. Math. Modelling, 2, 254.
Otterman, J. and R, S, Fraser, 1976. Remote Sensing of Environment, 5, 247.

Plass, G. N. and G. W. Kattawar 1969. Appl. Opt., 8, 455.
Plass, G., et al. 1976. J. Appl. Opt., 15, 3161.
Raschke, E. 1972. Beit. Phys. Atmosp., 45, 1.
Tanaka, M., and T. Nakajima 1977. J. Quant. Spectrosc. Radiat. Transfer, 18, 99.
Turner, R. E. 1975. 10th Int. Symp. on Remote Sensing of Environment, Envir. Res. Inst. of Michigan, Ann Arbor, Michigan, 2, 671.
Ueno, S., and S. Mukai 1973. J. Quant. Spectrosc. Radiat. Transfer, 13, 1261.

Comparative Study of Remote Sensing Observations and Coastal Processes in Different Geographic Areas

L.J. Rouse, Chairman, A.P. Cracknell,
P.Y. Deschamps, R.E. Dugdale, E.C. Monahan,
J. H. Simpson, C. Tomassini and S. Ueno

INTRODUCTION

Because of the wide range of coastal processes that are amenable to remote sensing observations, the study group chose to examine the specific set of coastal phenomena known as fronts. Fronts can be loosely defined as regions of sustained intensified horizontal gradients in one or more discernible properties of the ocean. The scale of fronts ranges from the edges of major oceanic currents to the edges of the tidal discharge plumes of estuaries the world around.

Two types of fronts were chosen for further consideration:

a) Thermal convection fronts - shelf sea fronts induced by convection mixing driven by surface heat loss.

b) Tidal mixing fronts - shelf sea fronts induced by spatial variations in tidal mixing rates.

KEY ISSUES AND RECOMMENDATIONS

The most commonly observed identifying surface manifestations of these two types of fronts is the horizontal temperature gradient--a "linear" feature. Other differences across the fronts are color changes and variations in surface wave characteristics such as significant wave height, whitecapping coverage, presence or absence of capillary waves, and refraction. In addition, concentrations of phytoplankton have been found to be associated with tidal mixing fronts. Current variations and accumulations of flotsam occur in the vicinity of some fronts. Differences in the air that lies above the front may also occur. Such differences are in water vapor content and liquid water (sea spray) and fog content.

Table 1 relates these characteristics of surface fronts to a suite of remote sensors.

Some of the questions about fronts that may be addressed by remote sensing are:

1)   Do fronts occur in a chosen area?
2)   Are eddies generated along shelf fronts, and what are the characteristics of these eddies?
3)   What is the role of eddies in cross frontal exchange?
4)   Are plankton blooms trapped along fronts?
5)   What are the seasonally and fortnightly geographic shifts of frontal positions?
6)   What is the three-dimensional character of the fronts?

The first five questions can be investigated with present-day systems, allowing for the spatial and temporal resolution of these sensors. The answer to the last question is not possible today from remote sensing, but research efforts in Raman and acoustic techniques indicate that it may be possible in the future to observe the 3-D structure of surface front.

In order to present the general concepts that it was felt should be part of any remote sensing experiment, the study group has formulated an outline for a study that would examine the structure of coastal fronts. The specific conduct of an experiment will

TABLE 1: Sensor-Surface Characteristic Correlation

Characteristic	Thermal Infrared	Visible (Color) Imagery	Microwave Radiometric Images	SAR	OTH Radar	Radar or Laser Altimeter	LIDAR
Horizontal Temp. Change	X		X				X
Horizontal Color Change		X					
Wave Characteristics							
Whitecaps		(X)	X	X			
H 1/3					X		
Capillary		(X)					
Wave Refraction		(X)		X			
Atmospheric Liquid Water	(X)		X				X
Water Vapor	(X)		X				X
Current Variations					X		
Flotsam Lines along Fronts		(X)					

(X) - Secondary or derived parameter

depend upon the area under study and the amount of funding available. In order to adequately address the problem, both remote sensing and conventional methods should be used. In spite of the fact that the remote sensing information may be available infrequently, due to such things as cloud cover or repetition rate, its synoptic two-dimensional perspective overcomes the major deficiency of conventional methods.

The envisioned experiment will consist of measurements made from ships, aircraft, moored buoys, drifting buoys, offshore towers, and, of course, satellites. The two readily available satellite remote sensing measurements that will apply to the experiment are infrared and visible imagery from such systems as TIROS-N, Coastal Zone Color Scanner, and LANDSAT. It is also suggested that advantage be taken of other satellite sensors such as Synthetic Aperture Radars and microwave radiometers when these are available.

An assessment of the seasonal and climatological variability of the front is an invaluable preliminary step to the design of the experiment. Sources of information include previous regional oceanographic studies, fisheries and other marine environmental surveys, and historic satellite data when available. These data give an indication of the spatial and temporal variations of the front and hence are useful in the design of the cruise tracks of the ships involved in the study. Provisions should also be made for the acquisition of "quick look" imagery starting several days before the commencement of conventional measurements so that ship tracks can be modified to respond to changes in the structure of the front. Such imagery is available in near real time as part of standard weather forecast information, but a more desirable alternative is direct contact with appropriate satellite receiving stations so that specially enhanced images of the study area are produced.

Ship tracks should be designed so that, whenever possible, conventional measurements of horizontal profiles of surface properties of the front are acquired in time coincidence with satellite overflights. Ideally the ship, at this time, should be steaming across the front obtaining continuous measurements.

Measurements acquired by ships, with suggested methods in parentheses, should include:

1) Profiles of sea surface temperature (thermal radiometer and towed surface following thermistor)

2) Vertical profiles of temperature and salinity on a grid appropriate to the spatial and temporal scale of the front (CTD casts)

3) Horizontal profiles of chlorophyll and Gelbstoff (continuous flow fluorometer)

4) Vertical profiles of chlorophyll, inorganic suspended matter, and nutrients (bottle casts on the grid)

5) Downwelling and upwelling radiation (sensors matched in precision, reliability, and spectral resolution with satellite sensors)

6) Surface meteorological measurements at each station (air temperature, relative humidity, wind speed and direction).

Instrumented aircraft should be available to fly coincident with ship and for satellite measurements when the weather permits. These aircraft should have instruments such as visual and thermal infrared scanning radiometers that are compatible in accuracy and reliability with the satellite sensors. Other remote sensing devices such as synthetic aperture radar, laser altimeter, microwave radiometers, and atmospheric lidar should be used when they are available. In addition to the remote sensors, conventional measurements of air temperature and humidity and launches of air-dropped expendable bathythermographs should be made.

Offshore towers should be used whenever they are available. At such sites continuous measurements of parameters that affect the front or that are effects of the front can be made. These parameters include wind speed and direction, wind stress, air temperatures, relative humidity, barometric pressure, wave amplitude and direction spectra, and atmospheric structure. The structure of the atmosphere can be observed by using lidar and/or acoustic radar. These offshore towers should also be used as launch sites for radiosondes to gather atmospheric profiles of temperature and relative humidity. These data serve as inputs to programs that calculate the transmission of the atmosphere for radiation in the visible and infrared regions so that proper interpretation of the satellite data can be made.

The final set of measurements that should be acquired is information about the currents in the study area. This is best done by an array of moored current meters positioned so as to sample the major flows that are expected. To supplement this array, radio or satellite tracked drifting buoys should be deployed so as to allow measurement of the Lagrangian flow of the area and to allow any unusual frontal motion to be monitored.

This report has addressed the types of data that should be gathered in an oceanographic experiment that is to take maximum advantage of remote sensing. This experiment is basically conventional with the addition of remote sensing instrument and the logical changes in measurement schedules so that maximum use of the remote sensing information can be made.

Appropriate Statistical Methods Applicable to Remote Sensing and Image Interpretation

D.J. de Priest, Chairman, B. Apolloni,
J.S. Bailey, G.J. Davison, R.J. Dodd,
D. Gaver and Z. Piteri

INTRODUCTION

Several problem areas in marine remote sensing were identified where appropriate statistical methodology should be applicable. The measurement requirement issue listed in the guidance for study groups was not discussed in great detail because this group thought that the measurement requirements depended to a great extent on the capabilities and functions of particular remote sensing systems. However, it was generally agreed that statistical inquiries and explorations are needed to understand the nature of measurement errors associated with remotely sensed data.

Image interpretation concerns the automatic or manual extraction of information from an image. More specifically, it deals with the identification of spectral relationships in the data that are caused by the distribution of objects and their inherent vari-

517

ability. It is often mentioned in the context of environmental considerations such as atmospheric effects.

KEY ISSUES

The key issues identified and characterized by members of Study Group C as areas where improved statistical methods are needed include the following:

1) Better and more mathematically sophisticated methods for image enhancement and image classification.

2) Statistical models for estimation under environmental constraints.

3) Efficient algorithm development for data compression and image classification.

In the area of image enhancement, a promising approach appears to be in the development of multi-dimensional Markov random fields as models of digitized image texture representation. It is also expected that multi-dimensional time series analysis can be utilized to enhance images and assist in the improvement of the resolution of current and future sensor systems. With regards to image classification, there is a strong need for more efficient pattern recognition algorithms, further development and utilization of stochastic tree grammars and better clustering methods and multiple decision analysis.

It has been mentioned several times that for certain remote sensing systems, the radiation is readily affected as a consequence of environmental effects such as water vapor, aerosols, carbon dioxide and ozone. Although atmospheric absorption reduced the intensity of the radiation received by the sensors on the platform or satellite, relative measurements of ocean parameters can be obtained within the limitations of noise and spatial variation. In order to correct for these effects, appropriate numerical methods and statistical models must be developed to approximate the distribution of water vapor, aerosols, carbon dioxide and ozone in the atmosphere relative to time. The difficulty of obtaining good estimates of ocean parameters when there are clouds present has also been mentioned several times. In many operational situations, it may be possible to use contaminated data which are now being discarded to obtain estimates of ocean parameters (seasurface temperature, ocean color, etc.). One technique which should be explored is three-dimensional filtering of multi-spectral imagery to remove cloud cover effects. This involves finding suitable nonlinear transforms that make the cloud information (noise) and ground reflection (signal) additive, estimating the cloud statistics and implementing a three-dimensional Wiener filter. Other approaches which may be helpful include (1) utilization of appropriate parametric and nonparametric techniques and (2) development and utilization of mixture models for prediction and estimation.

Potentially the most frustrating problem confronting users of remote sensing data is the large quantity of data that must be retrieved, stored, processed and assimilated. A high percentage of these data are redundant and therefore utilize precious storage space and computing time. Clearly, research is needed on the computational aspects of image processing with particular attention given to the development of efficient algorithms.

RECOMMENDATIONS

Several recommended actions were proposed for addressing pertinent issues considered by this group. These include:

1) Statisticians and/or mathematicians should become knowledgable and actively involved in all aspects of processing and preprocessing of remotely sensed data.

2) More sophisticated statistical techniques and efficient image classification and pattern recognition algorithms must be developed to handle and lyze large remote sensing data sets.

3) There should be some agency with sufficient funds to support and encourage researchers to explore and develop better statistical methods for current and future complex satellite systems.

4)   A repository or library of standard image classification,
     image enhancement, image interpretation software and
     other standard statistical/mathematical methodologies
     should be created, developed and maintained for current
     and future utilization.

Signal Processing Methods, Sensor Signal Enhancement and Minicomputers

E.J. Wegman, Chairman, P. Baylis, A. Deepak,
C.R. Francis and E.J. Kibblewhite

INTRODUCTION

For purposes of this report, we understand a signal to be the output of a sensor in whatever form it is available, digital, analogue, etc. This study group is concerned with the signal and its subsequent manipulation until it reaches an interim or ultimate consumer in some form. We are not concerned with the reason for a sensor being on board a satellite and are only concerned with sensor design and engineering to the extent that such design affects the processing of the signal. Nor shall we consider, in general, the final stages of computation as this may be done in various ways depending on the scientific purpose to be achieved. We are, in short, concerned with the great middle region.

Signal Processing is the manipulation and transmission of the sensor output until it is presented (displayed) to the user. It is

521

what happens to the signal in the great middle region between
sensor and user. Signal Enhancement is a special type of processing which existing information is manipulated to emphasize certain
features of interest, e.g., to increase S/N ratio or contrast. Of
course, the information contained in a signal is fixed, and all
types of processing will reduce that information content. It is
important to understand therefore that no additional information
may be extracted from the signal by signal exhancement procedures.
The enhanced signal may still contain more information than the raw
signal, however, since most procedures are based on some mathematical model. The assumptions on which the model is based add
information to the enhanced signal. But to the extent that the
assumptions underlying the model may be wrong, the enhanced signal
will also be wrong and the information gain illusory. Signal
enhancement must be done with great care in a remote sensing context. Artifacts introduced by careless or thoughtless use of
signal enhancement may be hard to detect and correct because of the
nature of remote sensing.

Mini- and microcomputers are hardware devices which play an
increasingly important role in the signal processing of remotely
sensed data. They offer the possibility of access to remotely
sensed data by relatively small groups of scientists and other
users who would not otherwise have access to such data. In addition to the conventional electronic devices, we include in our
understanding of minicomputers, electro-acoustic and electro-
optical devices.

KEY ISSUES

Perhaps the most obvious issue involved in remote sensing
involves the large size of the data sets. More data is available
than can be processed and studied by present scientists and sometimes more data than can even be recorded and stored. The most
pressing issue is then:

1)   How to handle large quantities of incoming signal data

for telemetry, storage, archiving, retrieval and processing?

While key issue number 1 is likely to have an engineering solution,
there is a parallel mathematical question

2) How to handle the computational complexities involved in
such large data sets?

Problems abound when scaling up mathematical algorithms to handle
large data sets including for example round-off error. Insuffi-
cient attention has often been paid to these problems so that
uncertainty remains in the validity of inferences drawn from some
remotely sensed data.

APPROACHES AND RECOMMENDATIONS

Several important and potentially fruitful approaches to
ameliorating the problems raised as key issues have been made.
These include:

1) The development of hardware-implemented algorithms for
filtering, inversion, etc. At least part of the difficulty in remote sensing signal processing is that software implementations of signal processing algorithms can
be slow even on the most modern computers, because of the
unavoidable bookkeeping aspects of software. Hardware
implementation would create many-fold improvements in
processing times.

2) The development of modular processing systems. Improvements in hardware and software will make an inadequate
contribution to overall improvement if the human labor
involved in the development of hardware and software must
be repeated for each satellite system at each earth-based
processing establishment. Design of modular, off-the-
shelf systems would greatly expand the number of investigators involved.

3) The development of better and cheaper display devices,
both video type and hardcopy. While great advances have

been made in microprocessor technology, exciting advances
have not been made in the development of peripheral
devices. These support devices are often the limiting
factor in otherwise state-of-the-art systems.

4) The development of relatively inexpensive man-computer
interactive systems for use by small groups, schools and
individuals. Even where the equipment for processing
remote sensing signals is inexpensive and available, it
is often technologically too complex and sophisticated
for potential users. Properly designed interactive
systems could, for example, make available, cheaply,
images of a city to its council for the purposes of
monitoring growth and the progress of construction projects or images of a farm to a farmer for the purposes of
monitoring his crops.

5) The development of charge-coupled devices and surface
acoustic wave devices for remote sensing processing.
Such devices would allow the improvement of sensors whose
signals would be easier and cheaper to process. Sensor
design improvement can play an important role in reducing
the complexity of signal processing.

6) A careful study of the feasibility and desirability of
on-board processing techniques. On board signal pro-
cessors may be limited by power and cooling requirements,
but such preprocessing could dramatically reduce the
telemetry and on-ground processing requirements. Also,
preprocessed data could simplify design of the interactive systems mentioned in 4 above and facilitate the
use of remote sensing data by small groups. On board
processing implies the loss of some information and the
possible loss of benefits of using the data for unantici-
pated purposes.

7) The development of novel signal processing algorithms
(particularly for imaging systems). Conventional techniques are, for the most part, well developed. Further
improvements in many of these algorithms could result in

only marginal improvements. Radically different approaches should be taken not only for processing and transmission, but also for retrieval of archived data. Examples of such approaches include the use of two-dimensional splines as images or semi-imaging approaches based on time series methods.

8) Further research on the computational aspects of signal processing using efficient algorithms. While many algorithms and techniques are optimal for accomplishing the desired mathematical goal, they are not always optimized from a numerical or computational point of view. Careful and systematic work must be done on these computational aspects.

9) The development of multidimensional signal processing algorithms. Often data from multiple sensors is treated as so many streams of univariate data and analyzed accordingly. This ignores the richness of the multi-variate structure and the possible synergistic effect of a simultaneous, multivariate signal processing approach.

These nine approaches to addressing some of the key issues offer a wide variety of potential work for the scientist, engineer, and mathematician. We commend these approaches to the attention of the appropriate groups for their thoughtful consideration and to the attention of scientific managers who should be alert to opportunities to fund work compatible with these approaches.

Remote Sensor Instruments and Requirements in the Next Decade

D.B. Trizna, Chairman, E. Khedouri,
P. Mather, J. Seeley and D.P. Thomas

INTRODUCTION

The responsibility of Study Group E was to organize material
relating to available and proposed sensors useful for remote sensing of coastal waters and the open seas for reference by PRIMARS
participants for future experiment planning purposes.

KEY ISSUES AND RECOMMENDATIONS

Table 1 represents the output of the committee as organized at
the PRIMARS meeting. We have organized the sensors according to
those which will be available upon a satellite platform sometime in
the next decade, the data of which should be available for public
consumption. Obviously, most of these can be made operational from
aboard an aircraft platform for specific applications, but depend

527

then upon the individual investigator for use, either by way of actual implementation or contractual service.

We have indicated the precision or accuracy, as appropriate, for each sensor, and these may vary depending upon the satellite in question. The pixel size determines the spatial resolution over which the instrument averages. Additional comments are provided regarding satellites.

High Frequency (HF) radars have been included in the list and bear some explanation for those interested in measuring wind waves and surface currents. Those radars listed as land based are large experimental facilities which may be available for cooperative programs of research. The ONR Chapel Bell array has a 0.5 degree beamwidth at 30 MHz; the SRI WARF array has a 0.25 degree beamwidth, and can be operated in the Gulf of Mexico and off the Eastern U.S. coast in addition to its Pacific Ocean coverage. The Chapel Bell array is operated in conjunction with the MADRE transmitter. These systems and the SEA ECHO radar on San Clemente Island, off the coast of San Diego, CA, are operated by the U.S. Naval Research Laboratory. The SEA ECHO radar is the only high angular resolution radar operating in the surface wave mode which is suitable for basic studies in ocean waves. Stanford University has a broader beam portable radar, and also operates a synthetic aperture receiver in conjunction with any LORAN C station, as does Birmingham University. The NOAA portable surface wave current measuring radar, under the development of Donald Barrick, is a single frequency system as is soon to be available commercially. Interest in any of the above systems should be directed toward D. B. Trizna of the U.S. Naval Research Laboratory for further information.

In addition to the table assembled by Study Group E, we have included information from a report by V. E. Noble of NRL which is complementary to Table 1. Table 2 is a list of projected earth remote sensing satellites for times early into the next decade. Table 3 indicates the sensors planned for each of these satellites. Tables 4, 5, and 6 show microwave and visible/infrared sensors for these satellites and their characteristics and purpose.

Finally, Table 7 presents projected satellite ocean boundary measurement capabilities for the near term. With the demise of SEASAT, the projected U.S. National Ocean Sensing Satellite System (NOSS) will be very similar in sensor characteristics to SEASAT, but without a synthetic aperture radar abroad according to current plans.

TABLE 1:

SENSOR	PLATFORM TYPE	PRECISION/ACCURACY	PIXEL	COMMENTS
μ Wave Altimeter	Sat/A.C.	Δh = 10 cm	$(5 \text{ km})^2$	(SEASAT) Sea Surface Topography/Geoid
Synthetic Aperture Radar	Sat/A.C.		$(25 \text{ m})^2$	Macroscopic features/Wave Spectrum (SEASAT). 100 km Swath Width (ERS[3])
Scatterometer	Sat/A.C.	± 20° Wind Angle ± 20° Wind Angle	$(50 \text{ km})^2$	(SEASAT) Wind Speed/Direction 1500 km Swath
Multi Beam Altim.	Sat	Δh = 10 cm	$(5 \text{ km})^2$	(NOSS) Sea Surface Topography/Geoid vertical Beam + Forward/Backward/ Right/Left beams for Surface Slope determination.
SMMR	SAT	0.5°C ± 2 m/s	$(80 \text{ km})^2$	(Nimbus 7, ERS[3]) Sea Surface Temp/ Wind Speed Salinity (1-2 GHz)
Radiometer (IR)	Sat/A.C.	0.1° (A.C.) 0.25° (TIROS N) 1.0 (GEOS)	$(.5 \text{ km})^2$ $(98 \text{ km})^2$	2/day (Polar Orb) Sea/Coastal Temp. ½ hrly (Geostationary) % Cloud Cover
		MMS MSS-Bendix A	$(80 \text{ m})^2$ $(20 \text{ m})^2$	Landsat: Chlorophyll, Algae, Sediment French SPOT: Geldstoffer (yellow matter) Land Sea Boundaries
HF Radars	Land Based			ONR Chapel Bell (Atlantic - 1100m)
Skywave		2 kt Winds/Waves 5 cm/s Currents	$(3 \text{ km})^2$	NRL MADRE (Atlantic), NRL, SEA ECHO (Pacific, SRI WARF (Pacific-2440m) 750-3000 km Coverage
Surface Wave			(1 km)	NRL SEA ECHO-150 km Range Coverage
S. W. Synth. Aper.				Stanford Un., Birmingham Un. (Syn. Aper.)

TABLE 1 (Continued):

SENSOR	PLATFORM TYPE	PRECISION/ACCURACY	PIXEL	COMMENTS
S. W. Portable				Currents - NOAA Portable (Surface Wave)
LIDAR Sat(?)/A.C.				Ocean Waves, Bathymetry, Resonance Florescence
Double SAR	A.C.			Surface Imaging-2 antennae on A.C. (Fore/aft) to cancel turbulence blurring of Atmosphere. (Suggested by Dag Gjessing)
2(N) frequency radar	Sat (?) A.C./ Landbased/Tower			Cross-Cor. 2 radar returns at X/C/L band. Separated in Frequency by 1-20MHz Sense Waves/Currents 15m-300m Wavelength
Satellite	Sat Interogation of Surface Instr.		Point	Buoys or remote Towers as Surface Platforms Measure Wave Height, Currents, Temperature, Turbidity, Salinity, Conductivity, and Subsurface Parameters
Instrumented Towers	Tower		Point	Winds and Turbulence, Wave Height, Currents, via Surface Temperature μ-wave radars wind guages, wave
Underwater	Hydrophones			Subsurface thermal Structure using Seismic Type Measurements for R.S. of Warm Volumes of Water (90% of Earth's Storage Capability) - useful for location of marine life and climatology.

TABLE 2:   Earth Remote Sensing Satellites (Approximate Launch Bracket 1978 - 1980)

Satellite/Program Name	Approximate Launch Date	Program Type
TIROS-N	13 Oct 78	Meteorological-prototype for spacecraft in NOMSS series
SEASAT-A	26 June 78	Sea/Ocean Dynamics - first of planned operational satellite system
LANDSAT-C (ERTS-C)	5 Mar 78	Land/Earth Resources - third in operational satellite system
NIMBUS-G	24 Oct 78	Meteorological - R & D satellite, basis for operational system
NOAA-A (TIROS-N follow-on)	1st Qtr 79	Environmental - Third generation operational satellite system
NOAA-B (TIROS-N follow-on)	3rd Qtr 79	Environmental - Third generation operational satellite system
GOES-D (SMS-F)	Aug 80	Meteorological/Environmental - geosynchronous, NASA developed NOAA operated
DMSP-5D (4)	1st Qtr 79	DoD Meteorological satellite
DMSP-5D (5)	2nd Qtr 80	DoD meteorological satellite
DMSP-5D (6)	1st Qtr 81	DoD meteorological satellite
DMSP-5D (7)	3rd Qtr 82	DoD meteorological satellite

ACRONYMS: TIROS   -   Television Infrared Observations Satellite (NASA)
          SEASAT   -   Sea/Ocean Dynamics Satellite (NASA)
          LANDSAT/ERTS -   Land Satellite/Earth Resources and Technology Satellite (NASA)
          NIMBUS   -   Meteorological Satellite (NASA)
          GOES/SMS   -   Geosynchronous Operational Environmental Satellite/Synchronous Meteorological Satellite (NASA/NOAA)
          DMSP   -   Defense Meteorological Satellite Program (DoD)
          NOMSS   -   National Operational Meteorological System

TABLE 3: Primary Sensors of Earth Remote Sensing Satellites (Approximate Launch Bracket 1978 - 1980)

TIROS-N (4 primary sensors)
    1. AVHRR   - Advanced Very High Resolution Radiometer
    2. TOVS   - TIROS Operational Vertical Sounder
      a. BSU - Basic Sounding Unit
      b. SSU - Stratospheric Sounding Unit
      c. MSU - Microwave Sounding Unit

SEASAT-A (5 primary sensors)
    1. SAR   - Synthetic Aperture Radar
    2. SASS   - Radar Scatterometer
    3. ALT   - Radar Altimeter
    4. SMMR   - Scanning Multispectral Microwave Radiometer
    5. VIRR   - Visible/Infrared Radiometer

LANDSAT-C (ERTS-C) (2 primary sensors)
    1. MSS   - Multispectral Scanner
    2. RBV   - Return Beam Vidicon

NIMBUS-G (9 primary sensors)
    1. CZCS   - Coastal Zone Color Scanner
    2. SMMR   - Scanning Multispectral Microwave Radiometer
    3. THIR   - Temperature Humidity IR Radiometer
    4. ERB   - Earth Radiation Budget
    5. LIMS   - Limb IR Monitoring of Stratosphere
    6. SBUV   - Solar and Backscatter Ultraviolet Spectrometer
    7. SAMS   - Stratospheric and Mesospheric Sounder
    8. SAMM II - Stratospheric Aerosol Measurement
    9. TOMS   - Total Ozone Mapping System

NOAA-A, -B (2 primary sensors)
    1. AVHRR   - Advanced Very High Resolution Radiometer
    2. TOVS   - TIROS Operational Vertical Sounder
                 (Compound Sensor)

GOES-D (SMS-F) (1 primary sensor for earth remote sensing)
    1. VAS   - VISSR for Atmospheric Sounding
                 (VISSR = Visible IR Spin Scan Radiometer)

DMSP-5D (4) (2 primary sensors)
    1. OLS-1 - Optical Line Scanner
    2. SSH   - Special Sensor H - Temperature/H_2O
                 Vapor/Ozone Radiometer

DMSP-5D (5) (3 primary sensors)
    1. OLS-1 - Optical Line Scanner
    2. SSH-2 - Special Sensor H
    3. SSM   - Special Sensor M - (Passive Microwave
                 Temperature Sounder)

DMSP-5D (6) (2 primary sensors)
    1. OLS-1
    2. SSH-2

DMSP-5D (7) (3 primary sensors)
    1. OLS-2
    2. SSH-2
    3. SSM

TABLE 4: Performance Characteristics - Microwave Sensors Earth
Remote Sensing Satellites (Launch Bracket 1978 - 1980)

Sensor	Platform or Program	Approximate Launch Date	Frequency(ies) (GHz)	Resolution*	IFOV[†]	Type[‡]
SAR	SEASAT-A	6/78	1.275	25m	0.03 mrad	A/I/S
ALT	SEASAT-A	6/78	13.5	10cm	-	A/NI/NS
SASS	SEASAT-A	6/78	14.595	50km	25°	A/NI/NS
SMMR	SEASAT-A	6/78	37,21,18,10.7,6.6	25-125 km	31°	P/I/S
SMMR	NIMBUS-G	10/78	37,21,18,10.7,6.6	30-150 km	31°	P/I/S
SSM	DMSP-5D (4)	1st Qtr 79	50-60 range 7 chan.	-	-	P/NI/S
	(5)	2nd Qtr 80	50-60 range 7 chan.			
	(6)	1st Qtr 81	50-60 range 7 chan.			
	(7)	3rd Qtr 82	50-60 range 7 chan.			
TOVS/MSU	TIROS-N	10/78	50.3,53.74,54.96,57.95	-	-	P/NI/S

*Resolution is surface resolution except for altimeter, where altitude resolution is cited

[†]IFOV - instantaneous field of view (1° = 17.453 mrad)

[‡]Type: A = Active,   P = Passive
         I = Imaging,   NI = Nonimaging
         S = Scanning, NS = Nonscanning

TABLE 5:   Performance Characteristics - Visible/Infrared Sensors
Earth Remote Sensing Satellites (Launch Bracket ~ 1978 - 1980)

Sensor	Platform or Program	Approximate Launch Date	Wavelength(s) (μm)	Surface Resolution	Type[*]
VIRR	SEASAT-A	6/26/78	0.52-0.73	4 km	I/S
MSS	LANDSAT-C	3/5/78	0.5-0.6, 0.6-0.7 0.7-0.8, 0.8-1.1	80 m	I/S
			10.4-12.6	240 m	I/S
RBV	LANDSAT-C	3/5/78	0.55-0.72 (3 cameras)	40 m	I/NS
CZCS	NIMBUS-G	10/24/78	0.433-0.453, 0.51-0.53 0.7-0.8, 10.5-12.5	830 m	I/S
THIR	NIMBUS-G	10/24/78	6.5-7.0	22 km	I/S
			10.3-12.5	8 km	I/S
SR	ITOS-1 (NOAA-F)[†] ITOS-J (NOAA-G)[†]	Deleted (see NOAA-A)	0.52-0.73 10.5-12.5	4 km 8 km	I/S I/S
VHRR	ITOS-1 (NOAA-F)[†] ITOS-J (NOAA-G)[†]	Deleted (see NOAA-A)	0.6-0.7 10.5-12.5	1 km 4 km	I/S I/S
VAS	GOES-D (SMS-F)	8/80	0.55-0.73 12.7-14.7	~1 km 8 km	I/S I/S
AVHRR	TIROS-N NOAA-A NOAA-B	10/13/78 1st Qtr 79 3rd Qtr 79	0.55-0.90 0.725-1.0 10.5-11.5 3.44-3.93	1 km 1 km 1 km 4 km	I/S I/S I/S I/S
OLS	DMSP-5D (4 salellites #4 #7)	79-82	0.4-1.1 8.13	0.6 km (day) 2.7 km (night) 0.6 km	I/S I/S

[*]Type   I = Imaging,   NI = Nonscanning
         S = Scanning,   NS = Nonscanning

[†]ITOS   Improved TIROS Operational Satellite

TABLE 6:   Performance Characteristics - Visible/Infrared Sensors
Earth Remote Sensing Satellites (Launch Bracket ~ 1978 - 1980)

Sensor	Platform or Program	Approximate Launch Date	Wavelength(s) (µm)	Surface Resolution	Type[*]
ERB	NIMBUS-G	10/24/78	0.2-50 (22 chan.)	Earth radiation budget	NI/S
LIMS	NIMBUS-G	10/24/78	6.22(NO_2), 6.75 (H_2O), 9.54 (O_3), 11.3 (HNO_3), 14.93 and 15.21 (CO_2)	Atmospheric profiles	NI/S
SBUV	NIMBUS-G	10/24/78	0.25-0.34 (12 chan.) 0.16-0.40	O_3 profiles O_3 profiles	NI/NS NI/S
TOMS	NIMBUS-G	10/24/78	0.3125-0.38 (6 chan.)	Total O_3 content	NI/S
SAMS	NIMBUS-G	10/24/78	4.3 and 15 (CO_2), 5.3 (NO), 7.7 (CH4), 4.5 (CO) 2.7 and 25-100 (H_2O), 7.7 (N_2O)	Stratosphere, mesosphere sounder	NI/S
SAM II	NIMBUS-G	10/24/78	1.0	Stratosphere, tropospheric aerosols	NI/S
TOVS/BSU	NIMBUS-G	10/24/78	3.6-29.4 (14 chan.)	Total O_3, H_2O vapor and temperature profiles	NI/S
TOVS/SSU	NIMBUS-G	10/24/78	14.97 (3 chan.)	Stratospheric temperature profiles	NI/S
VIPR	ITOS-1	Deleted	8.0-18.7 (8 chan.)	Temperature profiles 0-30 km	NI/S
TOVS	TIROS-N NOAA-A NOAA-B	10/78 1st Qtr 79 3rd Qtr 79	3.8-30 (17 chan.)	Temperature O_3 profiles	NI/S
SSH	DMSP-5D (4 satellites #4-#7)	79-82	several IR chan.	Temperature, H_2O profiles, total ozone	NI/S

[*]   I = Imaging,   NI = Nonimaging
      S = Scanning,   NS = Nonscanning

TABLE 7: Projected Satellite Ocean Boundary Measurement
Capabilities: 1978-1980

Measurement	Precision	Accuracy	Range	Resolution	Swath	Remarks
SST						
High-resolution IR	0.5°C	3-6°C	-10° to 40°C	< 1 km[*]	~1800 km	Regional coverage, TIROS-N, NOAA, DMSP
Global IR	0.5°C	3-6°C	-10° to 40°C	5 km[*]	~1800 km	TIROS-N, NOAA, DMSP
Corrected IR	0.5°C	1°C	- 5° to 35°C	<40 km	~1800 km	Expected capability, derived data product TIROS-N, DMSP
Geosynch, IR	0.5°C	3-6°C	-10° to 40°C	10 km[*]	50° cone	30-min repeat cover- age, 55° lat., GOES
SMMR (μ wave)	1-1.5°C	1.5°C	- 5° to 35°C	125-150 km	600 km	SEASAT, NIMBUS-G
SMMR-B (μ wave)	1°C	1°C	- 5° to 35°C	70 km	600 km	Propoed for SEASAT-B
ROMS (μ wave)	0.5°C	1°C	- 5° to 35°C	10 km	1000 km	Research objective - 1982
SURFACE WINDS						
SMMR-wind speed	2m/sec	10%	4-30m/sec	50 km	600 km	SEASAT, NIMBUS-G
SASS wind speed	2m/sec	10%	4-25m/sec	50 km	1000 km	SEASAT
wind direction	20%	Ambiguity	0-360°	50 km	1000 km	
SURFACE WAVES						
Sig. wave ht. GEOS-3	1m		0-15 m	10 km	Nadir	
Sig. wave ht SEASAT	0.5 m		0-20 m	7 km	Nadir	
SAR, direct wave spectra						
h	TBD					Research objective
λ	25 m		50-700 m		25 km	SEASAT is proof-of-
θ	10°				frame	concept experiment
OCEAN SURFACE TOPOGRAPHY						
GEOS-3	0.5 m	†		10 km	Nadir-only	
SEASAT	0.1 m	†		7 km	Nadir-only	

[*]Nadir

†Accuracy is a function of orbit determination and error contribution due to
time-varying environmental factors.

INDEX